Fire Strategies
From Principles to Practice

By

Duncan Winsbury

MAPLE
PUBLISHERS

Fire Strategies – From Principles to Practice

Author: Duncan Winsbury

ISBN 978-1-80723-006-7 (Paperback)
978-1-80723-007-4 (Hardback)
978-1-80723-008-1 (E-Book)

Book Layout and Cover Design by:
White Magic Studios
www.whitemagicstudios.co.uk

Published by:
Maple Publishers
Fairbourne Drive, Atterbury,
Milton Keynes,
MK10 9RG, UK
www.maplepublishers.com

A CIP catalogue record for this title is available from the British Library.

This book is dedicated to my Grandchildren
Caleb, Ellie and Rowan

Acknowledgements

I would like to thank the following, but for whose sharp eyes and assistance, producing this book would have been even more difficult and time consuming:

Manson Ting: The best fire engineer I know for reviewing documents; his sharp eye and attention to detail are outstanding.

Colleen, *my wife, for supporting me through the long hours involved in the production of this book and through life as well.*

Foreword

Fire has shaped human history. It has given us warmth, light, and progress; but it has also brought tragedy, destruction, and loss. From ancient cities consumed by flames to modern disasters such as Grenfell Tower, one lesson remains constant: when fire safety is neglected or misunderstood, the consequences are devastating.

This book has been written to help ensure that those lessons are not forgotten. My aim is to provide clarity about what a Fire Strategy is, why it matters, and how it can be applied effectively throughout the lifecycle of a building. A Fire Strategy is not simply a technical report or a requirement to satisfy regulators; it is the foundation of fire safety assurance. It links design, engineering, management, and accountability, ensuring that buildings remain safe not just at the point of construction, but throughout their occupation and adaptation over time.

In writing this book, I have drawn on experience, research, and the many lessons hard-won by our industry. The text is intended for a wide audience — from architects and engineers to building managers, regulators, insurers, and students. My goal has been to make the subject accessible, practical, and relevant, regardless of the reader's background.

I believe that fire safety must never be seen as an obstacle or an afterthought. Instead, it should be understood as an enabler of safe, resilient, and sustainable design. If this book helps even one professional to approach fire strategies with greater clarity, confidence, and commitment, then it will have served its purpose.

I invite you to read on with an open mind and a shared determination to make buildings safer for all who live and work in them.

Introduction

Fire strategies are fundamental to modern building safety. They are not optional extras or bureaucratic paperwork, but structured documents that define how fire safety is achieved, maintained, and demonstrated across a building's entire lifecycle.

This book has been written to provide a clear, practical, and comprehensive understanding of fire strategies: what they are, why they matter, and how they can be effectively developed for different building types. It is both a reference guide and a practical tool for professionals at every stage of a building's design, construction, and management.

Purpose of the Book

The aim of this book is threefold:

1. To explain the **foundations** of fire strategies, setting out the legal, historical, and technical context.
2. To examine the **core elements** that make up a robust fire strategy, from compartmentation and means of escape to detection, suppression, and firefighting facilities.
3. To explore how strategies are applied in **different building types**, recognising that the challenges of a high-rise residential block, a hospital, or a heritage site are very different.

Who This Book is For

This book is designed for a broad audience, including:

- **Architects and Designers**, who must integrate fire safety into their creative vision.
- **Engineers and Contractors**, who deliver the systems and details that make strategies work in practice.
- **Building Managers and Responsible Persons**, who are accountable for maintaining safety throughout a building's operation.

- **Regulators, Fire Authorities, and Insurers**, who verify compliance and resilience.
- **Students and Early-Career Professionals**, who need a solid grounding in the principles and practice of fire strategies.

Structure of the Book

The book is divided into three parts:

- **Part 1: Foundations** – introduces the concept of fire strategies, their purpose, their historical development, and their place in the building lifecycle.
- **Part 2: Core Elements of a Fire Strategy** – examines the technical and management measures that underpin safety, including evacuation, compartmentation, smoke control, fire service access, and human behaviour.
- **Part 3: Strategies for Different Types of Buildings** – explores how fire strategies are tailored to specific occupancies, such as high-rise residential, healthcare, education, transport hubs, commercial premises, and heritage buildings.

How to Use This Book

You can read this book in two ways:

- **Sequentially**, building up a complete picture of fire strategies from first principles to advanced applications.
- **As a reference**, consulting individual chapters when facing specific challenges, such as retrospective strategies, atriums, or high-risk occupancies.

Final Thoughts

A fire strategy is more than a technical report — it is the **blueprint of safety**. It connects law, engineering, management, and accountability in a single framework that protects lives, property, and communities. My hope is that this book will equip you with the knowledge and confidence to treat fire strategies as living documents, central to the resilience of every building we design, construct, and occupy.

CONTENTS

PART 1
Foundations

Chapter 1

Introduction to Fire Strategies

1.1 What is a Fire Strategy?

A fire strategy is a structured, comprehensive document that defines how fire safety is to be achieved, maintained, and demonstrated throughout the lifecycle of a building. It provides a logical framework that brings together:

- Building design and layout.
- Fire protection measures (passive and active).
- Management procedures for occupants and staff.
- Compliance with fire safety legislation, standards, and guidance.

The fire strategy explains how fire safety objectives are embedded into the building's design, construction, and operation. It ensures that both prescriptive requirements (such as those found in *Approved Document B* of the Building Regulations) and performance-based solutions (such as the fire engineering methodologies in *BS 7974*) are met.

Fire Strategy vs Fire Risk Assessment

It is important to distinguish a fire strategy from a fire risk assessment (FRA):

- An FRA focuses on identifying hazards, assessing risks, and recommending mitigation during the operational phase of a building. It is dynamic, reactive, and site-specific, dealing with how risks are managed in day-to-day use.
- A fire strategy, by contrast, is a proactive design document. It describes the *intent and rationale* of the fire safety provisions built into the design. It demonstrates why and how the fire precautions are adequate to meet legal obligations and life safety objectives.

While a fire risk assessment might evolve frequently (especially if there are layout, occupancy, or management changes), the fire strategy forms the foundation upon which those risk assessments are based.

Retrospective Fire Strategies

Although fire strategies are ideally developed during the design stage of a project, there are many cases where a retrospective fire strategy is required. This occurs when:

- **Older buildings** are involved that were constructed before modern regulations or without a formal strategy.
- **Building use has changed**, such as an office converted into flats or a warehouse adapted for leisure use.
- **Major refurbishments** have altered layouts, materials, or fire systems.
- **Regulatory or insurance requirements** demand a formal strategy document, even for existing premises.

Retrospective fire strategies are particularly valuable for:

- **Hospitals, care homes, or schools** built decades ago, where compartmentation or evacuation provisions may not reflect current best practice.
- **Heritage and listed buildings**, where fire protection must be balanced against conservation constraints.
- **Residential high-rises** post-Grenfell, where façade systems, stay-put policies, and firefighting facilities need clear documentation.

In these cases, the retrospective fire strategy:

- Reviews existing construction and systems against current codes and standards.
- Identifies where the building does not meet modern prescriptive guidance, and sets out compensatory measures (e.g., enhanced management, additional detection, or suppression systems).
- Provides a defensible framework for regulators, insurers, and responsible persons.
- Acts as the baseline for future fire risk assessments.

Retrospective fire strategies often require site surveys, destructive inspections (to confirm compartmentation or service penetrations), and coordination with fire authorities. They are typically more evidence-heavy than design-stage strategies because they must prove adequacy without the benefit of original design assumptions.

Who Uses a Fire Strategy?

A well-prepared fire strategy is not just a compliance document — it is a living reference used by a wide range of stakeholders:

- **Designers and Architects** – to embed fire safety in the building's form and function.
- **Engineers and Contractors** – to specify and install passive and active systems correctly.
- **Building Managers and Responsible Persons** – to maintain safety throughout occupation.
- **Regulators and Building Control Authorities** – to verify compliance with legislation.
- **Fire and Rescue Services** – to understand how a building is designed to perform in fire conditions.
- **Insurers** – to assess risk and set premiums based on the robustness of fire precautions.

A Living Document

A fire strategy should not be filed away once approved. Instead, it must:

- Be accessible to those responsible for fire safety.
- Be reviewed and updated following material alterations, refurbishments, or changes of use.
- Provide the **"Golden Thread"** of information required under the *Building Safety Act 2022* for higher-risk buildings.

When properly maintained, the fire strategy becomes the central reference point for all fire safety decisions across the building's lifespan.

1.2 Why are Fire Strategies Important?

The importance of fire strategies can be understood across several interconnected dimensions: legal compliance, life safety, resilience, economic security, and public confidence. A well-prepared fire strategy ensures not only that buildings meet minimum regulatory requirements, but also that they function safely, sustainably, and reliably throughout their life.

Legal Compliance

In the UK, the Building Regulations 2010 (as amended) mandate that all buildings must be designed and constructed with adequate fire precautions. Compliance is demonstrated through:

- **Approved Document B (ADB)** – prescriptive guidance on means of escape, structural fire protection, and fire service facilities.
- **British Standards** (BS 9991, BS 9999, BS 7974) – alternative or performance-based approaches.
- **The Regulatory Reform (Fire Safety) Order 2005** – requires ongoing fire risk assessments during occupation.

A fire strategy provides the **documented evidence of compliance**. It is scrutinised by:

- **Building Control bodies** (Local Authority or Approved Inspectors) at design and construction stage.
- **Fire and Rescue Services** during consultation (especially for complex or high-risk occupancies).
- **The Health and Safety Executive (HSE)** and the new **Building Safety Regulator** for higher-risk buildings under the Building Safety Act 2022.

Without a fire strategy, compliance is difficult to prove. In disputes or enforcement cases, it provides the legal defence that fire precautions were properly considered and documented.

Life Safety

The most fundamental purpose of a fire strategy is to safeguard lives. This extends beyond simple evacuation routes to a holistic framework that ensures:

- **Safe evacuation** for all occupants, including disabled persons, children, the elderly, or those requiring assistance.
- **Protection of escape routes** through compartmentation, smoke control, and emergency lighting.
- **Reliable early warning** via detection and alarm systems.
- **Time for rescue and intervention** by the fire service.

Real-world experience shows that poorly conceived fire strategies have devastating consequences. Events such as **Grenfell Tower (2017)**, the **Station Nightclub Fire (USA, 2003)**, and various hospital and care home fires illustrate that:

Simply providing exits is not enough.

Occupant behaviour, communication, and management must be considered.

Vulnerable populations require tailored evacuation strategies.

A properly prepared fire strategy provides confidence that if a fire occurs, people can survive.

Resilience and Property Protection

While life safety is paramount, resilience is increasingly recognised as a crucial element of fire strategy. Buildings are not only places of occupation — they are centres of economic activity, cultural heritage, and community value.

A fire strategy supports resilience by:

- **Protecting critical infrastructure** – e.g., server rooms, plant rooms, data centres.
- **Maintaining business continuity** – ensuring organisations can recover quickly after a fire.
- **Reducing insurance liabilities** – insurers often require evidence of robust fire strategies.
- **Preserving heritage assets** – listed buildings, museums, and archives demand strategies that balance conservation with protection.

Example: A warehouse storing critical medical supplies may not only threaten lives if destroyed but also disrupt healthcare services nationally. A resilient fire strategy ensures continuity of essential functions.

Reputation and Accountability

In the post-Grenfell regulatory landscape, the public and regulators expect accountability and transparency. A fire strategy:

- Demonstrates that safety has been proactively designed into the building.
- Provides assurance to residents, staff, and visitors.
- Protects organisations, owners, and directors from reputational damage or litigation in the event of fire.

Buildings without robust fire strategies risk not only fines and enforcement, but also the erosion of public trust.

A Tool for All Stakeholders

Fire strategies are important because they serve different but overlapping needs:

- **Designers and Architects** – to ensure fire safety is integrated, not retrofitted.
- **Contractors** – to build correctly with certified products.
- **Building Managers** – to run safe buildings day-to-day.
- **Residents and Occupants** – to live and work with confidence.
- **Regulators and Fire Authorities** – to check compliance and plan interventions.
- **Insurers and Investors** – to assess financial risk and resilience.

This multi-stakeholder relevance makes the fire strategy a core document for every building, not just a technical formality.

Summary

The importance of fire strategies lies in their ability to:

- **Prove compliance** with UK Building Regulations and fire safety law.
- **Save lives** through reliable evacuation and protection systems.
- **Ensure resilience** by protecting property, continuity, and heritage.
- **Protect reputation** by providing transparency and accountability.
- **Support stakeholders** across the building lifecycle.

In short, a fire strategy is not simply a compliance exercise — it is the foundation of fire safety assurance, linking legal duties, life safety, and long-term resilience.

1.3 Fire Strategies vs Fire Risk Assessments

It is essential to distinguish between **fire strategies** and **fire risk assessments (FRAs)**. While they are related and complementary, they serve different purposes, are created at different stages, and are written by different professionals. Confusing the two can lead to gaps in safety or non-compliance.

Fire Strategy

A **fire strategy** is a structured document prepared to guide fire safety throughout the lifecycle of a building.

Key Features

- **Timing** – Developed at the design stage, ideally during concept or planning, but should be updated at handover, during occupation, and after refurbishments or change of use.
- **Purpose** – Provides a blueprint for fire safety. Demonstrates compliance with Building Regulations, Approved Document B, British Standards (BS 9991, BS 9999, BS 7974), and other statutory requirements.
- **Content** – Defines both passive measures (compartmentation, fire resistance, escape routes) and active systems (alarms, sprinklers, smoke control). Sets out the evacuation methodology, fire service access, and management assumptions.
- **Use** – Submitted to Building Control bodies, Approved Inspectors, Fire and Rescue Services, and forms part of the "Golden Thread" of information under the Building Safety Act 2022.

Who Writes It?

- Fire strategies should be prepared by a competent person with detailed knowledge of fire engineering, building regulations, and performance-based design.
- For complex buildings (e.g., high-rise, healthcare, mixed-use, atrium buildings), a fire engineer should always be consulted.
- Architects may prepare a basic fire strategy for simple, low-risk buildings (e.g., small offices), but these should still be validated by a fire engineer if deviations from guidelines are proposed.

When to Consult a Fire Engineer?

- At the concept design stage – before layouts are fixed, to avoid costly redesigns.
- When performance-based solutions (BS 7974) are needed to justify deviations from prescriptive codes.
- For higher-risk or complex occupancies – high-rise, hospitals, shopping centres, transport hubs, heritage buildings.
- When retrospective strategies are required for older buildings, particularly post-Grenfell, to evidence safety and compliance.

Fire Risk Assessment (FRA)

A fire risk assessment is a systematic review of fire hazards and safety measures in an occupied building.

Key Features

- **Timing** – Conducted once the building is in use and occupied. Must be regularly updated (commonly annually) and whenever significant changes occur (e.g., new layouts, different occupancy, major refurbishments).
- **Purpose** – Identifies hazards, evaluates risks, and ensures that fire safety measures and management arrangements are effective in practice.
- **Content** – Covers ignition sources, combustible materials, means of escape, signage, emergency lighting, alarm audibility, staff training, and maintenance records.
- **Legal Requirement** – Required under the Regulatory Reform (Fire Safety) Order 2005 (RRO). The "Responsible Person" has a duty to carry it out or commission a competent assessor.

Who Writes It?

- FRAs are usually conducted by fire risk assessors, who may be consultants, competent in-house staff, or external specialists.
- For higher-risk premises (e.g., care homes, high-rise residential), assessors should be third-party accredited (e.g., under BAFE SP205, IFE registers or IFSM Tiered register).

How They Work Together

Fire strategies and FRAs are not interchangeable — they are two sides of the same coin:

- The Fire Strategy is the blueprint:
    - It sets out the design intent, how the building meets fire safety objectives, and what systems are required.
- The FRA is the audit and review:
    - It checks, in real-world conditions, whether the fire strategy is still valid, whether systems are being maintained, and whether occupants can still escape safely.

In practice:

- A new hospital will have a fire strategy developed during design, showing compartmentation, PHE (progressive horizontal evacuation), and suppression. Once in use, the FRA ensures that staff training, evacuation drills, and maintenance of fire doors align with the strategy.
- A converted office-to-residential block may require a retrospective fire strategy to document how compartmentation and evacuation work in the new use. The FRA then ensures those measures remain effective in operation.

Summary Table: Fire Strategy vs Fire Risk Assessment

Feature	Fire Strategy	Fire Risk Assessment (FRA)
Stage	Design stage (updated throughout lifecycle)	Occupation stage (ongoing)
Focus	Compliance, design intent, systems, evacuation methodology	Hazards, risks, management, maintenance
Prepared by	Fire engineer / competent person	Fire risk assessor / competent person
Audience	Building Control, Fire Authorities, designers, managers	Building managers, Responsible Person, regulators
Update Frequency	On material changes (design or use)	At least annually, or after significant changes
Purpose	Blueprint for fire safety	Audit of how fire safety is working in practice

1.4 Stakeholders and Benefits

A fire strategy is not just a technical report for engineers; it is a communication and coordination tool that ensures every stakeholder involved in a building's lifecycle understands their role in achieving fire safety. It translates complex technical requirements into a practical framework that links design, construction, regulation, and management.

Stakeholders in Fire Strategies

1. **Developers and Owners**
    - Fire strategies provide assurance that the building is designed to meet legal obligations and insurance requirements.
    - They demonstrate that safety has been considered from the outset, reducing the risk of costly redesigns, enforcement notices, or reputational harm.
    - For owners of existing buildings, retrospective fire strategies provide a defensible framework for compliance with current legislation and a foundation for fire risk assessments.
2. **Architects and Designers**
    - Fire strategies explain how fire safety integrates with form, function, and aesthetics.
    - They allow architects to understand constraints on layouts, compartmentation, and escape routes early in the design process.
    - Importantly, they also show where performance-based solutions may allow greater design flexibility without compromising safety.
3. **Structural, Mechanical, and Electrical Engineers**
    - Fire strategies clarify the technical requirements for structural fire resistance, smoke control, pressurisation systems, detection, alarms, and suppression.
    - They ensure that engineers coordinate their designs with fire safety needs, reducing clashes between disciplines (e.g., ductwork penetrating fire compartments).
4. **Contractors and Specialist Installers**
    - Fire strategies provide clear specifications for the installation of passive and active systems.
    - They reduce ambiguity during construction, helping to ensure that fire doors, cavity barriers, dampers, sprinklers, and alarms are installed correctly and certified.
    - They also provide the baseline for commissioning tests and certification handover.

5. **Building Control and Approved Inspectors**
    - Fire strategies form the primary submission document to demonstrate compliance with Building Regulations.
    - They enable regulators to assess whether the design meets prescriptive or performance-based standards, reducing disputes at approval stage.
6. **Fire and Rescue Services**
    - Fire strategies provide critical operational information on:
        - Firefighting shafts and lifts.
        - Dry and wet risers.
        - Smoke control systems.
        - Expected building performance under fire conditions.
    - This allows firefighters to plan interventions and assess risks before incidents occur.
7. **Insurers and Investors**
    - A fire strategy demonstrates risk management and resilience, often reducing premiums and improving investment confidence.
    - For high-value assets, insurers may insist on a formal fire strategy before cover is agreed.
8. **Building Managers and Responsible Persons**
    - Fire strategies provide a management framework once the building is occupied.
    - They set out responsibilities for maintenance, testing, and evacuation procedures.
    - They form the baseline for fire risk assessments and staff training.
9. **Occupants and End Users**
    - While occupants may never read the fire strategy in full, they benefit from clear evacuation procedures, signage, and safe conditions that stem from its design.
    - In residential buildings, strategies provide assurance that "stay put" or evacuation policies are based on robust design, not assumption.

Benefits of a Robust Fire Strategy

The benefits of a properly prepared fire strategy extend across the building lifecycle:

- **Design and Planning Stage**
  - Reduces disputes between architects, engineers, and regulators by providing a clear, agreed framework.
  - Identifies potential issues early, avoiding costly redesigns.
  - Supports planning applications and fire authority consultations.
- **Construction Stage**
  - Provides contractors with unambiguous specifications.
  - Reduces errors, rework, and disputes between trades.
  - Simplifies commissioning by setting out system performance requirements.
- **Approval and Handover**
  - Streamlines Building Control approval by demonstrating compliance.
  - Provides fire and rescue services with the information they need for pre-planning.
  - Forms the foundation of the building's Golden Thread of safety information under the Building Safety Act 2022.
- **Occupation and Management**
  - Ensures building managers and responsible persons understand their duties.
  - Provides clarity on evacuation strategies (e.g., simultaneous, phased, progressive horizontal, stay-put).
  - Forms the baseline against which fire risk assessments are conducted.
  - Helps insurers assess liability and resilience, often reducing premiums.
- **Long-Term Benefits**
  - Promotes safer buildings with reduced risk of catastrophic fire.
  - Improves business continuity by protecting critical systems and assets.

- Enhances reputation and accountability for owners and operators.
- Provides regulators and occupants with confidence and transparency.

Summary

A fire strategy is more than a technical document — it is a shared language of safety. By aligning developers, architects, engineers, contractors, regulators, fire services, managers, insurers, and occupants, it reduces conflict, accelerates approval, clarifies responsibilities, and delivers safer, more resilient buildings.

In short:

- Developers see compliance and reduced risk.
- Designers gain flexibility and clarity.
- Contractors get clear instructions.
- Regulators and fire services see proof of safety.
- Managers gain a framework for ongoing safety.
- Occupants gain confidence and protection.

The benefits are both immediate (fewer delays, smoother approvals, safer construction) and long-term (safer occupation, reduced insurance costs, reputational protection).

1.5 Historical Development of Fire Strategies

The concept of embedding fire precautions into the design and regulation of buildings is not new. Human civilisation has repeatedly learned the hard way that poorly planned cities, flammable building materials, and inadequate firefighting access can lead to catastrophic fires. Over centuries, each major incident has shaped the principles that underpin modern fire strategies.

Early Origins: Ancient and Medieval Approaches

- **Ancient Rome** – Following repeated urban fires, including the Great Fire of 64 AD, Rome introduced measures to limit fire spread. Regulations addressed street widths, separation between buildings, and the use of less combustible materials (e.g., masonry rather than timber in some districts). The Roman Vigiles (public fire brigades) were also established, combining firefighting with law enforcement.

- **Medieval Cities** – Dense timber construction and narrow streets made medieval towns across Europe particularly vulnerable. Cities such as London, Paris, and Edinburgh suffered repeated devastating fires. By the late medieval period, rudimentary building bylaws were introduced, including requirements for stone party walls, tiled roofs instead of thatch, and restrictions on wooden projections over streets.

These early attempts at fire safety were often reactive, emerging only after disaster struck.

The Great Fire of London (1666) and the First Building Acts

The **Great Fire of London** destroyed more than 13,000 houses and much of the medieval city. The response was the **London Building Act of 1667**, one of the earliest examples of codified fire safety regulation. It required:

- Buildings to be constructed in brick or stone.
- Restrictions on wooden eaves, overhangs, and combustible façades.
- Wider streets to act as firebreaks and improve firefighting access.

This was effectively the beginning of formalised building control. Similar acts followed in other growing cities, embedding fire safety into urban planning.

Industrialisation and 19th Century Fires

The 18th and 19th centuries saw rapid industrialisation, with new building types such as mills, warehouses, and rail stations. Fires in these occupancies were often catastrophic due to high fire loads and unprotected structural ironwork.

Notable lessons included:

- The need for **compartmentation** in mills and factories.
- Early adoption of **sprinkler systems** in 19th-century textile mills in northern England.
- The importance of **fireproof construction** (cast iron and brick), though tragically many mills collapsed when iron columns failed in fire.

These events pushed governments to expand fire safety bylaws, but regulation remained largely prescriptive and reactive.

Codification of Building Regulations (20th Century)

By the early 20th century, fire precautions were consolidated into formal Building Regulations. In the UK, progressive versions of the London Building

Acts evolved into a nationwide framework, eventually forming the Building Regulations 1965, the first truly national set.

Codification introduced consistency but remained highly prescriptive:

- Specific limits on **travel distances** to exits.
- Minimum **fire resistance periods** for walls and floors.
- Requirements for **firefighting access.**

This was the first step toward the modern fire strategy, as it required explicit demonstration of compliance. However, it offered little flexibility for complex or innovative architecture.

Rise of Fire Engineering (Late 20th Century)

By the mid-to-late 20th century, modern architecture began to outgrow prescriptive codes. Increasing use of:

- **High-rise towers** with complex evacuation needs.
- **Large atriums** with interconnected volumes.
- **Transport hubs** and underground stations.
- **Heritage buildings** requiring conservation-sensitive solutions.

These building types could not be addressed by prescriptive regulations alone. This gave rise to fire engineering as a profession, with methods based on **BS 7974** in the UK and NFPA/IBC frameworks internationally. Fire strategies began to document:

- Performance-based justifications using fire and smoke modelling.
- Evacuation analysis comparing ASET (available safe egress time) with RSET (required safe egress time).
- Integration of active systems (sprinklers, smoke control) with passive measures.

At this point, the fire strategy document became a recognised requirement in complex projects.

Learning from Major Fires

Several catastrophic fires reinforced the need for systematic documentation of fire safety measures:

- **Summerland Leisure Centre, Isle of Man (1973)** – 50 deaths in a fire exacerbated by combustible acrylic cladding and poor evacuation planning. Highlighted the dangers of innovative materials without fire testing.

- **King's Cross Underground Fire (1987)** – 31 deaths from a fire in a wooden escalator, worsened by the "trench effect" in escalator shafts. This event drove improvements in underground train station fire safety and smoke control and introduced the requirement for comprehensive train station fire strategies.
- **Grenfell Tower (2017)** – 72 deaths following façade fire spread in a high-rise residential block. This tragedy fundamentally changed UK fire safety regulation, banning combustible cladding on high-risk buildings, creating the Building Safety Regulator, and reinforcing the importance of fire strategies as living documents forming part of the Golden Thread.

Each incident expanded the scope and expectations of what a fire strategy must address:

- Not just evacuation and structure, but also materials performance, smoke control, management assumptions, and retrofit safety.

The Modern Fire Strategy

Today, the fire strategy is recognised as a core document in the building lifecycle:

- Required for design approval, forming the basis of consultation with building control and fire services.
- Used during construction to guide contractors and ensure correct installation.
- Handed over at completion as part of the Golden Thread of safety information.
- Referenced by building managers and Responsible Persons during operation.
- Updated during refurbishment or change of use.
- Created retrospectively for older or heritage buildings to align them with modern safety expectations.

Summary

The historical development of fire strategies reflects a journey from reactionary measures after catastrophic fires (Rome, medieval London,

Summerland, Grenfell) to a structured, professional discipline that integrates prescriptive codes, fire engineering, and modern safety management.

Whereas early regulations focused on materials and spacing, the modern fire strategy is a holistic, multidisciplinary document:

- Rooted in legislation.
- Guided by engineering analysis.
- Adapted through lessons from disasters.
- Maintained as a living record throughout a building's life.

1.6 Fire Strategies in the Building Lifecycle

A fire strategy is not a static report that ends once Building Control approval is granted. Instead, it is a dynamic reference document whose relevance extends across the entire lifecycle of a building — from concept design through construction, operation, and eventual refurbishment or repurposing.

At each stage, the fire strategy provides a different function for a different set of stakeholders. Its continuity ensures that design assumptions are not lost, safety systems are correctly implemented, and future users understand how to manage risks.

Design Stage – *Blueprint for Safety*

At concept and design stage, the fire strategy is the foundation document for fire safety.

- **Purpose**
    - Sets out the fire safety objectives (e.g., life safety, property protection, business continuity).
    - Defines design assumptions, such as occupancy levels, evacuation methods, fire service response times, and fire loads.
    - Establishes whether the design will follow prescriptive guidance (e.g., Approved Document B) or adopt performance-based fire engineering solutions (e.g., BS 7974).
- **Stakeholders**
    - **Architects** use it to shape layouts and escape routes.
    - **Structural and Mechanical & Electrical (M&E) engineers** rely on it for compartmentation, fire resistance, smoke control, and suppression requirements.

- **Clients/Developers** gain assurance that compliance and insurance requirements are embedded early.
- **Building Control and Fire Services** review it to confirm compliance.

- **Critical Point**
    - Early consultation with a fire engineer can save significant cost by avoiding late design changes and enabling innovative layouts.

Construction Stage – *From Paper to Reality*

During construction, the fire strategy becomes the instruction manual for contractors and specialists.

- **Purpose**
    - Guides correct installation of fire-resisting materials, fire doors, cavity barriers, dampers, and penetration seals.
    - Specifies correct positioning and standards for active systems such as alarms, sprinklers, risers, and smoke control fans.
    - Acts as a benchmark for quality control and inspection.
- **Stakeholders**
    - **Main contractors** use it to ensure trade coordination (e.g., M&E penetrations do not compromise fire stopping).
    - **Specialist installers** (fire door, alarm, sprinkler contractors) work to the specifications it contains.
    - **Clerk of Works / Fire Engineer** may conduct site inspections to verify that the design intent is being delivered.
- **Critical Point**
    - Deviations from the strategy (e.g., substituting a cheaper material) must be reviewed by the fire engineer to avoid non-compliance.

Handover Stage – *The Golden Thread Begins*

At completion, the fire strategy becomes part of the handover documentation to the client or building manager.

- **Purpose**
    - Provides a clear record of all fire safety systems installed and how they are intended to function.

 - Documents the management procedures required to keep the building safe (e.g., evacuation plans, testing regimes).
 - Forms part of the Golden Thread of safety information required by the Building Safety Act 2022.

- **Stakeholders**
 - **Building managers / facilities teams** inherit the document as their baseline reference.
 - **Insurers** may review it to assess risk.
 - **Fire services** may use it for familiarisation and operational planning.
- **Critical Point**
 - A strategy that is not properly handed over (or is left on a shelf unread) represents a broken link between design and occupation.

Operation Stage – *Living Reference Document*

Once occupied, the fire strategy becomes the baseline reference for fire safety management.

- **Purpose**
 - Provides the foundation against which fire risk assessments (FRAs) are conducted.
 - Supports maintenance schedules for alarms, sprinklers, emergency lighting, and smoke control.
 - Clarifies the building's evacuation strategy (e.g., simultaneous, phased, stay-put).
 - Informs staff training, drills, and management procedures.
- **Stakeholders**
 - **Responsible Person (under RRO 2005)** ensures that management aligns with the strategy.
 - **Fire risk assessors** use it to benchmark operational compliance.
 - **Occupants and residents** benefit from the safety framework it defines.

- **Critical Point**
    - If the fire strategy is ignored during occupation, operational practices may drift away from design assumptions, undermining safety.

Refurbishment or Change of Use – *Adapt and Revalidate*

As buildings evolve, fire strategies must be reviewed and updated.

- **Purpose**
    - Reflects new layouts, occupancies, or hazards introduced by refurbishment.
    - Validates that existing systems are still adequate or identifies where upgrades are required.
    - Provides evidence of compliance for regulatory submissions during change of use.
- **Examples**
    - An office block converted into flats requires a new strategy to confirm compartmentation and evacuation are appropriate for residential use.
    - A heritage building refurbished as a hotel must have a strategy balancing conservation with modern fire safety.
    - A warehouse adapted into a leisure centre requires entirely different assumptions for evacuation and fire load.
- **Critical Point**
    - Failing to update the fire strategy after major change can leave owners and managers legally exposed and occupants at risk.

The Continuous Golden Thread

Across all stages, the fire strategy forms part of the Golden Thread of safety information:

- Created during design.
- Verified during construction.
- Handed over at completion.
- Maintained during occupation.
- Updated during refurbishment.

This ensures transparency, accountability, and traceability throughout the building's life.

Summary

A fire strategy is a living document that must evolve with the building. Its value lies not just in design approval but in its continuity across all lifecycle stages:

- **Design** – blueprint for compliance and safety.
- **Construction** – technical guide for implementation.
- **Handover** – record for managers and the Golden Thread.
- **Operation** – baseline for management and FRAs.
- **Refurbishment** – updated to reflect change.

When properly managed, the fire strategy ensures that safety is not a one-off achievement but a sustained outcome over decades of use.

1.7 Common Challenges in Developing Fire Strategies

Despite their critical role, fire strategies often suffer from recurring challenges that reduce their effectiveness. These pitfalls can undermine both compliance and safety, leaving buildings vulnerable and duty holders exposed to legal and reputational risk. Understanding these common failings is essential to producing strategies that are robust, practical, and enforceable.

1. **Insufficient Detail –** ***Generic Documents vs Site-Specific Strategies***

One of the most frequent failings is the production of generic templates rather than genuinely site-specific strategies.

- Some strategies contain only copied text from regulations without analysis of how it applies to the specific building.
- Key details such as occupant load calculations, evacuation assumptions, compartment layouts, and smoke control designs are often omitted or glossed over.
- This results in a "tick-box" document that satisfies a submission requirement but provides little real value to designers, contractors, or managers.

Consequence: Lack of site-specific detail can lead to unsafe design assumptions, delayed approvals, or conflicts during construction.

2. **Lack of Integration with Design – *Fire Safety in Isolation***

A common challenge is the preparation of fire strategies in isolation from the wider design process.

- Sometimes strategies are written late, once layouts are already fixed, which limits options and forces compromises.
- Fire safety is occasionally seen as an "add-on" rather than an integral design driver, creating conflicts between aesthetics, structure, and safety.
- Poor coordination between architects, engineers, and fire consultants can result in clashes, such as services cutting through fire-resisting walls or insufficient space for firefighting shafts.

Consequence: Fire safety becomes a barrier rather than an enabler, leading to costly redesigns, disputes, or unsafe compromises.

3. **Poor Version Control and Documentation – *Losing the Golden Thread***

With multiple revisions during design and construction, it is easy for fire strategies to suffer from poor version control.

- Revisions may not be clearly dated or numbered.
- Contractors may be working from out-of-date versions, creating inconsistencies on site.
- Approved versions may differ from those handed over at completion, breaking the Golden Thread of information.

Consequence: Confusion during approvals, loss of confidence from regulators, and gaps in the record that undermine accountability under the Building Safety Act 2022.

4. **Disconnect with Building Management – *Strategy vs Reality***

Fire strategies too often remain as static documents that are never updated once a building is in use.

- When layouts change (e.g., offices subdivided into flats, shops reconfigured), the fire strategy is not revised to reflect new risks.
- Management arrangements such as evacuation procedures are often different from the design assumptions written in the strategy.
- Retrospective strategies for older buildings are sometimes neglected, leaving duty holders without a defensible framework.

Consequence: The fire strategy becomes irrelevant in practice, exposing the building to risk and leaving managers non-compliant.

5. **Over-Reliance on Prescriptive Guidance – *One Size Doesn't Fit All***

Another challenge is the rigid application of prescriptive codes without considering performance-based alternatives.

- Complex buildings such as atriums, transport hubs, and mixed-use towers often cannot comply with prescriptive guidance alone.
- In such cases, fire engineers may need to use **BS 7974 methodologies** (smoke modelling, evacuation analysis, structural fire engineering).
- Without this, designers may be forced into unnecessary compromises on usability or aesthetics.

Consequence: Either non-compliance (if prescriptive rules are ignored) or over-design (if prescriptive rules are applied too conservatively).

6. **Limited Stakeholder Engagement – *Lack of Buy-In***

Fire strategies are most effective when all stakeholders — designers, owners, fire services, insurers, and building managers — are engaged.

- Too often, strategies are prepared without meaningful input from the fire and rescue service or without consulting the eventual Responsible Person.
- This creates disconnects between design intent and operational management.

Consequence: Occupants and managers may not understand the evacuation policy and fire services may be unprepared for the building's firefighting features.

7. **Inadequate Consideration of Human Behaviour – *People Don't Always Act as Planned***

Many strategies focus heavily on systems and structure but underestimate human factors.

- Occupants often head for the exit they entered by, even if it is not the closest.
- Vulnerable occupants (disabled, elderly, children) may require different provisions.
- Communication failures during an incident (e.g., alarms not heard, unclear voice messages) can drastically affect evacuation.

Consequence: Overly optimistic evacuation assumptions can collapse in real emergencies, leading to unacceptable risks.

8. **Resource and Competency Gaps**

Finally, many challenges stem from the competence of those preparing strategies.

- In simple buildings, basic strategies may be produced by architects or consultants using Approved Document B.
- However, in complex projects, failing to involve a qualified fire engineer can lead to unsafe or non-compliant solutions.
- Competency frameworks (e.g., IFE, IFSM, BAFE) are increasingly recognised as essential.

Consequence: Poorly prepared strategies lack credibility, leading to delays, enforcement action, or — in worst cases — unsafe occupation.

Addressing the Challenges

A key theme throughout this book is how to overcome these pitfalls by ensuring that fire strategies are:

- **Detailed and site-specific**, not generic.
- **Integrated into design**, not written in isolation.
- **Properly managed with version control**, forming part of the Golden Thread.
- **Kept up to date**, reflecting operational and occupancy changes.
- **Based on both prescriptive and performance-based methods**, as appropriate.
- **Developed by competent professionals**, with early fire engineer involvement.
- **Grounded in real human behaviour**, not just theoretical models.

Common challenges in fire strategy development include lack of detail, poor integration, weak version control, disconnect with management, and inadequate engagement with stakeholders or human behaviour.

By recognising these pitfalls and addressing them proactively, fire strategies can evolve from paper exercises into living, practical tools that truly protect people, property, and reputation across the building lifecycle.

1.8 Summary

Fire strategies are foundational to modern fire safety practice. They are not optional extras or technical appendices but core documents that define how a building addresses the risk of fire throughout its entire lifecycle.

A well-prepared fire strategy:

- **Bridges the gap** between law, engineering, and management — translating abstract legal requirements into practical, enforceable design and operational measures.
- **Provides clarity** by setting out, in a structured manner, how passive protection, active systems, evacuation methods, and management responsibilities combine to achieve safety.
- **Safeguards lives** by ensuring that evacuation, compartmentation, and fire service access are robust and tailored to the building's unique risks and occupants.
- **Protects assets** and resilience by embedding measures that minimise disruption, protect critical infrastructure, and reduce insurance exposure.
- **Assures stakeholders** — from developers and designers to regulators, fire services, insurers, managers, and occupants — that fire safety has been properly considered, documented, and integrated.

Importantly, fire strategies are living documents. They must evolve in line with the building:

- At **design stage**, they guide architects and engineers.
- During **construction**, they direct contractors and ensure correct installation.
- At **handover**, they provide managers with a baseline record and part of the "Golden Thread" of building safety information.
- During **operation**, they form the reference point for fire risk assessments and maintenance.
- At **refurbishment or change of use**, they must be reviewed and updated to remain relevant.

Without an up-to-date and coherent fire strategy, buildings risk falling into the common pitfalls outlined earlier: generic documents, lack of integration, poor version control, or disconnection from real-world management. The consequences of such failings are not merely bureaucratic delays, but potentially catastrophic losses of life, property, and public trust.

Looking Ahead

In the following chapters, we will explore:

- The **regulatory framework** that underpins fire strategies in the UK and internationally.
- The **structure and contents** of a robust fire strategy document, including mandatory elements and best practice additions.
- The **technical principles** — compartmentation, means of escape, detection, suppression, smoke control, and firefighting facilities — that form the backbone of fire engineering solutions.
- How strategies are **applied across different building types**, from high-rise residential towers and healthcare facilities to underground structures and historic buildings.

By the end of this book, it will be clear that a fire strategy is far more than a compliance document: it is a **blueprint for safety, resilience, and accountability**, central to protecting people, property, and society.

Key Takeaways – Chapter 1

- A **fire strategy** is a structured, living document that sets out how fire safety is achieved across a building's **design, construction, operation, and refurbishment**.
- Fire strategies are distinct from **fire risk assessments**: strategies provide the **blueprint**, while FRAs check that the **blueprint is working in practice**.
- They are vital for **legal compliance, life safety, and resilience**, providing assurance to developers, regulators, insurers, managers, and occupants.
- Fire strategies have evolved historically from **reactionary regulations after disasters** (e.g., Great Fire of London, King's Cross, Grenfell) to today's professional discipline, embedded in the **Golden Thread of building safety information**.
- Common pitfalls include **generic documents, poor integration, weak version control, and lack of updates** — this book will show how to avoid these and ensure strategies remain practical and enforceable.

Chapter 2

Legislation, Standards And Guidance

2.1 Introduction

Fire strategies are not created in a vacuum; they must be grounded in the **legal, statutory, and technical frameworks** that govern fire safety. A strategy's credibility and enforceability depend on its ability to demonstrate compliance with the relevant regulations, guidance documents, and recognised standards.

In the **United Kingdom**, the framework for fire safety is shaped primarily by the **Building Regulations 2010 (as amended)**, specifically **Part B – Fire Safety**. Compliance with Part B is often demonstrated through **Approved Document B (ADB)**, which provides prescriptive guidance on issues such as means of escape, internal fire spread, external fire spread, access and facilities for firefighting. Alongside ADB, a suite of **British Standards** (e.g., BS 9991, BS 9999, BS 7974, BS 5839, BS 9251) and **industry codes of practice** provide detailed benchmarks and methodologies.

Beyond the UK, the picture varies but follows similar principles. **Globally**, fire safety regulation is underpinned by a mix of:

- **Prescriptive codes** – setting fixed requirements (e.g., maximum travel distances, minimum fire resistance periods, sprinkler coverage).
- **Performance-based methodologies** – allowing flexibility through engineering analysis, modelling, and risk-based solutions where prescriptive rules are insufficient.

For example:

- In the **United States**, compliance is demonstrated through the **International Building Code (IBC)** and **National Fire Protection Association (NFPA) standards**, particularly NFPA 101 (*Life Safety Code*). These set prescriptive baselines but allow performance-based fire engineering solutions where justified.

- In **Europe**, harmonised **Eurocodes** and guidance documents supplement national regulations, with countries such as Sweden and the Netherlands adopting performance-based approaches more widely.
- In **Asia-Pacific regions** like Australia, Singapore, and Hong Kong, **performance-based codes** are increasingly embedded, requiring fire strategies to justify compliance through risk assessment and engineering studies rather than prescription alone.

The result is that fire strategies must always be tailored to the **regulatory context** in which they are prepared. A strategy that satisfies UK Building Control might look very different from one prepared under NFPA rules or Eurocodes, even if the technical solutions are similar.

The growing global trend is toward **hybrid approaches**, where prescriptive requirements establish a baseline, but performance-based fire engineering methods are used to demonstrate safety in **complex, unusual, or innovative buildings**. Fire strategies therefore serve as the bridge between rigid codes and flexible engineering, ensuring that safety is achieved without stifling architectural ambition or operational needs.

In this chapter, we will explore:

- The legal basis for fire strategies in the UK and internationally.
- The role of **Approved Documents, British Standards, and Eurocodes**.
- The growing importance of **performance-based fire engineering**.
- How **international codes (e.g., IBC, NFPA, European standards)** influence best practice.

This framework forms the **bedrock of every fire strategy**, ensuring it is not only technically robust but also **legally defensible, enforceable, and globally relevant**.

2.2 UK Building Regulations Framework

The legal foundation for fire safety in buildings in England and Wales is provided by the **Building Act 1984**, which grants the enabling power for the creation and enforcement of Building Regulations. The regulations themselves are periodically updated, with the current framework set out in the **Building Regulations 2010 (as amended)**.

Within this framework, fire safety is primarily addressed in **Part B – Fire Safety** of Schedule 1. The functional requirements of Part B are expressed in broad, performance-oriented terms, leaving scope for compliance to be demonstrated through prescriptive guidance, alternative standards, or performance-based fire engineering.

Core Functional Requirements of Part B

A building must be designed and constructed so that, in the event of fire:

1. **Means of Escape**
    - Occupants can escape safely without external assistance.
    - This includes the provision of adequate exit routes, suitable travel distances, protected escape staircases, and, where relevant, refuges for disabled persons.
2. **Structural Stability**
    - The **loadbearing capacity of the structure** is maintained for a **reasonable period**, preventing premature collapse.
    - Fire resistance periods are specified depending on building height, use, and occupancy risk.
3. **Limitation of Fire Spread**
    - Fire and smoke spread within the building is restricted through **compartmentation, cavity barriers, and fire-stopping**.
    - External fire spread is limited by controls on external wall performance, roof coverings, and separation distances between buildings.
4. **Access and Facilities for the Fire and Rescue Service**
    - Adequate **access, water supplies, firefighting shafts, lifts, and smoke control facilities** must be provided.
    - This ensures that firefighters can intervene effectively without undue risk.

Together, these functional requirements form the legal backbone of any fire strategy prepared in England and Wales.

Approved Document B (ADB)

To assist compliance, the government publishes **Approved Document B (Fire Safety)**, split into two volumes:

- **Volume 1** – Dwellings.
- **Volume 2** – Buildings other than dwellings.

ADB provides **prescriptive guidance** on how to achieve compliance with the functional requirements of Part B. This includes numerical limits (e.g., maximum travel distances), tabulated fire resistance ratings, and technical guidance on matters such as smoke control, cavity barriers, and firefighting shafts.

Importantly:

- ADB is **not mandatory** – it represents one way to demonstrate compliance.
- Alternative approaches may be adopted using **British Standards (e.g., BS 9991 for residential buildings, BS 9999 for non-residential)** or **fire engineering analysis under BS 7974**.

Other Relevant UK Legislations

- **Regulatory Reform (Fire Safety) Order 2005** – Governs fire safety management during occupation. While distinct from Building Regulations, fire strategies must align with the ongoing requirements of the Fire Safety Order.
- **Fire Safety Act 2021** and **Fire Safety (England) Regulations 2022** – Strengthened requirements for high-rise residential buildings post-Grenfell, particularly around external walls, flat entrance doors, and resident information.
- **Building Safety Act 2022** – Established the **Building Safety Regulator** and the requirement for a **Golden Thread of safety information**, in which the fire strategy is a central document.

Devolved Nations

It is important to note that building regulations are devolved within the UK:

- **Scotland** – Governed by the **Building (Scotland) Regulations 2004**, supported by the **Technical Handbooks**.
- **Wales** – Has its own Approved Document B, diverging from England in some areas.

- **Northern Ireland** – Operates under the **Building Regulations (Northern Ireland) 2012**, supported by **Technical Booklet E (Fire Safety)**.

While the principles are similar, details differ; fire strategies must therefore be prepared with the correct jurisdiction in mind.

Role of the Fire Strategy in Demonstrating Compliance

A fire strategy must explicitly reference the functional requirements of Part B and demonstrate how they are satisfied. This may include:

- Confirming that means of escape routes meet ADB requirements or providing engineering justification for extended travel distances.
- Demonstrating structural fire resistance using Eurocodes, BS 476, or other recognised testing standards.
- Documenting compartment layouts, cavity barriers, and external wall performance.
- Setting out the provision of firefighting shafts, risers, access routes, and water supplies.

Where deviations from ADB are proposed, the fire strategy must provide a **clear performance-based justification**, supported by calculations, modelling, or risk assessments.

Summary

The UK Building Regulations Framework, underpinned by the Building Act 1984 and the Building Regulations 2010, provides the legal foundation for fire strategies. Part B of the Regulations sets out the functional requirements, while Approved Document B and British Standards provide routes to compliance.

A robust fire strategy must explicitly align with these requirements, bridging the gap between law, design, and operation, and ensuring that fire safety is not only achieved but can be demonstrated to regulators, fire services, and building managers.

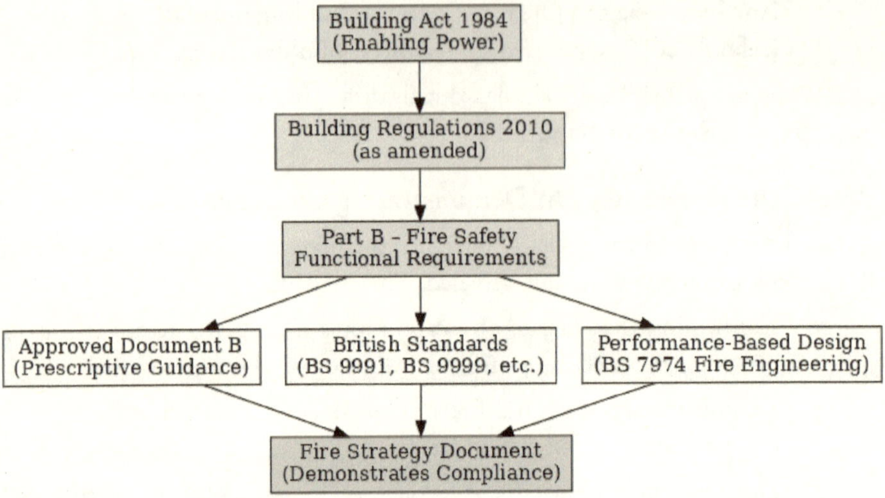

2.3 Approved Document B (ADB)

Approved Document B (ADB) is the UK government's official statutory guidance for demonstrating compliance with **Part B (Fire Safety)** of the Building Regulations. It provides prescriptive provisions on how buildings should be designed and constructed to meet the functional requirements of the law.

It is important to note that **ADB is not in itself law** — compliance with the Building Regulations is the legal requirement. ADB provides one commonly accepted route to compliance. Designers are free to adopt **alternative solutions**, provided they can demonstrate that these are at least equivalent in safety terms.

Structure of ADB

Since 2006, ADB has been split into two volumes to make it easier to apply to different types of buildings:

- **Volume 1 – Dwellings**

Covers **houses, flats, and maisonettes**. It addresses issues specific to residential buildings such as escape windows, protected stairways, smoke alarms, and guidance for loft conversions.

- **Volume 2 – Buildings other than Dwellings**

Applies to all other building uses, including offices, schools, hospitals, shopping centres, and industrial buildings. It provides more complex provisions for multi-occupancy and large-scale buildings, covering extended escape routes, firefighting shafts, atria, and mechanical smoke control.

Each volume is subdivided into sections, with guidance on:

1. **Means of warning and escape** – provision of alarms, travel distances, escape routes, refuges, protected stairways, phased evacuation strategies.
2. **Internal fire spread (linings and structure)** – fire performance of internal linings (surface spread of flame, reaction to fire classifications) and structural elements (minimum periods of fire resistance, compartmentation).
3. **External fire spread** – performance of roofs and walls against fire exposure, restrictions on external cladding, boundary conditions between buildings.
4. **Access and facilities for the fire service** – requirements for fire service vehicle access, hydrants, firefighting shafts, lifts, and smoke control.

Prescriptive Nature of ADB

ADB sets out **clear, prescriptive provisions**:

- Maximum **travel distances** to final exits.
- Minimum **fire resistance ratings** for walls, floors, and doors.
- Minimum sizes and locations of **escape windows**.
- Limits on **unprotected areas** of external walls depending on boundary distance.

For simple and conventional buildings, this prescriptive approach is highly effective: it provides **certainty for designers**, clarity for **building control authorities**, and a straightforward compliance route.

Flexibility through Alternative Approaches

However, ADB recognises that not all buildings fit standard layouts or assumptions. Therefore, it explicitly allows for the use of **alternative fire engineering solutions**, provided equivalence is demonstrated.

This often involves:

- Referencing **British Standards** (BS 9991 for dwellings, BS 9999 for non-residential).
- Using **performance-based design** under **BS 7974**, supported by fire and smoke modelling, evacuation analysis, and risk assessments.

- Combining prescriptive and performance-based approaches in a **hybrid strategy** (e.g., using ADB guidance for compartmentation but BS 7974 analysis for an atrium smoke control system).

In practice, most fire strategies adopt a **mixed approach**, clearly identifying which provisions follow ADB prescriptive guidance and which rely on engineered solutions.

Role of ADB in Fire Strategies

A fire strategy will often need to:

- Explicitly **reference relevant ADB provisions** for key design aspects.
- Demonstrate **compliance with prescriptive requirements** where they are followed.
- Provide **clear justification** where the strategy departs from ADB, showing that safety is achieved through equivalence.

For example:

- ADB may set a **maximum travel distance** of 18 m in an unsprinklered office. If the design requires 22 m, the fire strategy must show — via smoke modelling and evacuation analysis — that occupants can still evacuate safely.
- ADB may require **60 minutes' fire resistance** in a compartment wall. If the wall achieves 90 minutes to Eurocode standards, the fire strategy will explain this as an enhancement beyond ADB.

Limitations and Criticisms of ADB

While ADB remains the cornerstone of UK fire safety design, it has limitations:

- It has been criticised as overly complex and difficult to interpret, particularly after multiple revisions.
- Some provisions are outdated compared to modern construction techniques and materials.
- It is primarily intended for common building types; for complex or innovative buildings, it often cannot provide adequate solutions.
- The **Grenfell Tower Inquiry** highlighted that reliance on prescriptive compliance alone may not be sufficient, especially in high-risk buildings.

These limitations reinforce the importance of fire strategies that **go beyond prescriptive compliance**, embedding performance-based fire engineering and robust management practices.

Summary

Approved Document B provides the most widely used prescriptive guidance for fire safety in England, covering dwellings (Volume 1) and other buildings (Volume 2). It addresses warning and escape, fire spread, and firefighting facilities.

Although prescriptive, it allows flexibility: alternative standards and fire engineering analysis can be used where justified. A well-prepared fire strategy will therefore highlight where ADB has been followed, and where performance-based methods are used to achieve equivalent or enhanced safety.

2.4 Relevant British Standards

While **Approved Document B (ADB)** provides prescriptive statutory guidance, many fire strategies in the UK also reference **British Standards (BS)** to demonstrate compliance, provide flexibility, and adopt recognised best practice. British Standards can supplement, refine, or replace ADB provisions, particularly where designs require greater adaptability or performance-based justification.

BS 9999:2017 – Code of Practice for Fire Safety in the Design, Management and Use of Buildings

BS 9999 provides a **comprehensive, risk-based framework** for fire safety in non-residential buildings.

- **Key Features**:
    - Moves away from rigid prescriptive rules and instead applies a **risk profile** to the building, based on occupancy type, fire load, and management level.
    - Introduces the concept of **risk profiles (A–D)** and **Fire Growth Rate (1–4)**, which adjust parameters such as travel distances, exit widths, and fire protection requirements.
    - Emphasises the role of **management and maintenance** in ensuring ongoing fire safety.
- **Applications**:
    - Used for offices, retail, schools, industrial, leisure, and most non-domestic building types.

  - Particularly valuable when designs deviate slightly from ADB, as it provides structured flexibility.
- **Relevance to Fire Strategies**:
  - Provides an **alternative benchmark** to ADB for design justifications.
  - Demonstrates that risk-based solutions are embedded in design and management.
  - Encourages integration of **design, management, and operational planning** into the strategy.

BS 9991:2024 – Fire Safety in the Design, Management and Use of Residential Buildings

BS 9991 provides guidance specific to **residential occupancies**, recognising the distinct fire safety challenges in housing.

- **Key Features**:
  - Applies to blocks of flats, maisonettes, sheltered housing, student accommodation, and residential care homes.
  - Focuses heavily on **means of escape strategies**, including stay-put, phased evacuation, and full evacuation policies.
  - Provides detailed guidance on **detection and alarm systems**, **sprinklers**, and **smoke control** in multi-residential environments.
  - Recognises the needs of **vulnerable occupants**, such as those in care homes or supported housing.
- **Applications**:
  - Increasingly important post-Grenfell, where residential fire safety is under greater scrutiny.
  - Often referenced in fire strategies for high-rise housing, student halls, or care facilities to justify enhanced measures (e.g., sprinklers, evacuation lifts).
- **Relevance to Fire Strategies**:
  - Provides a more nuanced and occupant-focused approach than ADB for residential use.
  - Forms a critical part of the evidence base for evacuation strategies in fire strategy documents.

BS 7974:2019 – Code of Practice for Application of Fire Safety Engineering Principles to the Design of Buildings

BS 7974 is the core British Standard for performance-based fire engineering. It provides a formalised methodology for developing engineered solutions where prescriptive guidance is insufficient.

- **Key Features**:
    - Sets out a **framework for fire engineering analysis**, including:
        - Fire growth and development modelling.
        - Smoke and heat movement analysis.
        - Structural response to fire.
        - Evacuation modelling and human behaviour.
        - Fire service intervention analysis.
    - Uses the principle of comparing **ASET (Available Safe Egress Time)** with **RSET (Required Safe Egress Time)** to ensure evacuation can be achieved safely.
    - Encourages holistic solutions that integrate multiple systems (active and passive).
- **Applications**:
    - Essential for **complex buildings** such as atria, underground stations, airports, shopping centres, and mixed-use high-rises.
    - Used to justify deviations from prescriptive codes (e.g., extended travel distances, innovative layouts, or use of new materials).
- **Relevance to Fire Strategies**:
    - Provides the technical backbone for engineered solutions.
    - Fire strategies frequently reference BS 7974 when documenting CFD modelling, evacuation analysis, or alternative design solutions.
    - Demonstrates to regulators that solutions are not arbitrary but follow a recognised, peer-reviewed methodology.

Other Relevant British and European Standards

In addition to the three cornerstone standards above, fire strategies often reference a wide range of supporting British and European Standards, including:

- **BS EN 13501** – *Fire classification of construction products and building elements.*

    - Provides the **Euroclass reaction-to-fire classifications** (A1, A2, B, etc.), used widely for cladding, linings, and insulation.
- **BS EN 12101** – *Smoke and heat control systems.*
    - Covers design and testing of smoke vents, smoke curtains, pressure differential systems, and mechanical extract systems.
- **BS 5839** – *Fire detection and fire alarm systems for buildings.*
    - Provides requirements for system design, installation, commissioning, and maintenance.
- **BS 5266** – *Emergency lighting.*
    - Sets out design and testing standards for emergency lighting systems to support evacuation.
- **BS EN 12845** – *Fixed firefighting systems – Automatic sprinkler systems – Design, installation and maintenance.*
- **BS 5306** – *Fire extinguishing installations and equipment on premises.*

Together, these standards provide the **detailed technical requirements** that underpin the measures set out in fire strategies.

Integration into Fire Strategies

When preparing a fire strategy, referencing British Standards provides:

- **Credibility** – they are nationally and internationally recognised.
- **Consistency** – ensures that systems are designed and maintained to an agreed benchmark.
- **Flexibility** – BS 9999, BS 9991, and BS 7974 offer structured alternatives to rigid ADB compliance.

A typical fire strategy will therefore:

- State explicitly which **standards have been applied**.
- Reference **BS 9999** or **BS 9991** where ADB is not sufficient.
- Use **BS 7974** as the justification for performance-based or non-standard solutions.
- Supplement with specific standards (e.g., BS EN 12101, BS 5839) for system design details.

Summary

British Standards play a **central role** in UK fire strategies.

- **BS 9999** – risk-based flexibility for non-residential buildings.
- **BS 9991** – residential-specific guidance, crucial post-Grenfell.
- **BS 7974** – the foundation for performance-based fire engineering.
- **Other BS/EN standards** – provide detailed technical benchmarks for systems, materials, and maintenance.

By referencing these standards, fire strategies gain legitimacy, flexibility, and robustness, ensuring that designs are not only legally compliant but also technically sound and resilient.

2.5 The Regulatory Reform (Fire Safety) Order 2005 (RRO)

While the **Building Regulations** govern the *design and construction* of new and altered buildings, the **Regulatory Reform (Fire Safety) Order 2005 (RRO)** sets out the legal framework for the *management of fire safety once a building is occupied.*

The RRO, often referred to simply as the "Fire Safety Order," consolidated and replaced over 70 pieces of previous fire safety legislation, including the Fire Precautions Act 1971. It was intended to simplify the law, reduce bureaucracy, and place responsibility firmly on those who control or manage premises.

Scope of the RRO

- Applies to **nearly all non-domestic premises** in England and Wales, including offices, shops, factories, warehouses, schools, hospitals, care homes, hotels, and common areas of flats.
- Does **not generally apply to private single dwellings** (houses occupied by a single family), but does apply to the common parts of multi-occupied residential buildings.
- Covers both **commercial** and **public sector** premises.

This makes the RRO one of the most wide-reaching pieces of fire safety legislation in the UK.

The Responsible Person

The RRO places the duty of compliance on the **"Responsible Person"**, who will usually be:

- The **employer** (in a workplace).

- The **owner** or **landlord** of the premises.
- The **managing agent** or **occupier** with control over fire safety arrangements.
- In multi-occupied premises, there may be **more than one Responsible Person**, each required to cooperate.

The Responsible Person must ensure that **fire safety is managed proactively and continuously**.

Fire Risk Assessment (FRA)

A central requirement of the RRO is that the Responsible Person must carry out a **suitable and sufficient Fire Risk Assessment (FRA)**.

- The FRA must identify:
    - Fire hazards (ignition sources, fuel, oxygen).
    - People at risk (employees, visitors, residents, vulnerable persons).
    - Existing fire safety measures (detection, escape routes, compartmentation, extinguishers).
    - Management procedures (training, drills, maintenance).
- The FRA must assess whether measures are adequate and, if not, recommend improvements.
- FRAs must be:
    - **Documented** if five or more people are employed.
    - **Regularly reviewed** (commonly annually).
    - **Updated** when significant changes occur (e.g., refurbishment, change of occupancy).

Other Key Duties under the RRO

In addition to carrying out a Fire Risk Assessment, the Responsible Person must ensure:

- **Maintenance** of fire precautions: alarms, emergency lighting, extinguishers, sprinklers, and passive fire protection must be maintained in working order.
- **Evacuation Planning**: clear evacuation strategies must be established, including arrangements for vulnerable occupants (e.g., disabled people).
- **Staff Training**: all staff must receive fire safety training, including induction and periodic refreshers.

- **Fire Drills**: should be carried out regularly, with records kept.
- **Record-Keeping**: maintenance logs, training records, FRA updates, and drill reports must be available for inspection.
- **Information to Residents/Occupants**: in multi-residential buildings, residents must be given clear instructions on evacuation policies.

Enforcement and Penalties

- The RRO is enforced by **local Fire and Rescue Authorities**. Inspectors may visit premises, review documentation, and assess compliance.
- They may issue:
    - **Alterations Notices** – requiring notification of future changes.
    - **Enforcement Notices** – requiring deficiencies to be remedied.
    - **Prohibition Notices** – immediately restricting or prohibiting use of unsafe premises.
- Non-compliance is a **criminal offence**. Penalties range from fines to imprisonment, particularly where negligence leads to injury or death.

High-profile prosecutions have highlighted the seriousness of compliance, with landlords, employers, and managing agents facing heavy fines or custodial sentences for failing to meet RRO duties.

Interaction between Fire Strategies and the RRO

A fire strategy prepared at the **design stage** plays a crucial role in supporting later compliance with the RRO.

- **Baseline for Risk Assessments**
    - The fire strategy provides the starting point for future FRAs, documenting evacuation assumptions, compartmentation, fire service access, and system requirements.
    - A well-documented strategy ensures assessors understand the original design intent.
- **Management Framework**
    - Many aspects of the fire strategy (e.g., phased evacuation, firefighting lift use, maintenance intervals) become management duties under the RRO.
    - Building managers can use the strategy as a manual to align operational practices with design assumptions.

- **Golden Thread Continuity**
    - The RRO requires ongoing safety management, while the **Building Safety Act 2022** requires the preservation of a **Golden Thread** of safety information.
    - Together, they ensure that the fire strategy remains a living, practical document throughout occupation.
- **Retrospective Fire Strategies**
    - Where older buildings were constructed without a formal strategy, retrospective fire strategies are often commissioned to provide the baseline required for RRO compliance.

Summary

The Regulatory Reform (Fire Safety) Order 2005 is the cornerstone of **fire safety management during occupation**. It applies to almost all non-domestic premises and places responsibility on the **Responsible Person** to carry out Fire Risk Assessments, maintain precautions, plan evacuations, train staff, and keep records.

A fire strategy prepared during design is a vital support tool for RRO compliance, acting as the baseline for FRAs and ensuring that the original fire safety intent is carried through into daily management. Together, the Building Regulations and the RRO provide a **two-pillar system**:

- Building Regulations – ensure fire safety is *designed and built in.*
- Fire Safety Order – ensures fire safety is *maintained and managed.*

2.6 Local Authority and Fire Service Guidance

While the **Building Regulations and statutory guidance** form the national framework for fire safety, many fire strategies also need to take account of localised requirements and the expectations of other stakeholders. A strategy that only references national codes may fall short if it fails to satisfy the practical needs of those responsible for firefighting, approving developments, or underwriting risk.

Local Authority Planning Conditions

Local authorities often impose planning conditions that directly affect fire safety and must be reflected in the fire strategy:

- **Hydrants** – Developers may be required to install additional hydrants to ensure adequate water supply for firefighting. The exact number and locations are often agreed with the local Fire and Rescue Service.
- **Access roads** – Planning consent may be conditional on providing suitable fire appliance access roads, turning circles, or clearances, particularly on large housing estates or commercial developments.
- **Building height restrictions and façades** – Some planning authorities impose additional conditions on external wall finishes or façade treatments, especially in sensitive areas or following local fire incidents.

Failure to address these conditions at strategy stage can lead to planning refusal or costly re-submissions.

Fire and Rescue Service Requirements

The **local Fire and Rescue Service (FRS)** is a key stakeholder in fire strategy development. Their role is not just to enforce legislation, but to ensure buildings are designed with **operational firefighting needs** in mind.

Key operational considerations include:

- **Appliance access** – Roads must be wide and strong enough for fire engines, with turning facilities and parking restrictions to prevent obstruction.
- **Water supplies** – Hydrants must be appropriately located, with adequate flow and pressure. In rural areas, static water supplies (ponds, tanks) may be required.
- **Firefighting facilities** – Firefighting shafts, staircases, lifts, and smoke control systems must be designed in consultation with the FRS, ensuring they are both functional and usable in practice.
- **Pre-incident planning information** – Many fire strategies include operational information for the FRS, such as building layouts, compartment plans, and system summaries, to aid incident response.

Early engagement with the Fire and Rescue Service is particularly critical for **complex, high-risk, or unusual buildings** such as hospitals, transport hubs, and high-rise residential towers.

Insurer Requirements

Insurers often go beyond life safety, focusing on **property protection and business continuity**. Fire strategies may therefore need to demonstrate:

- **Enhanced suppression systems** – e.g., sprinklers, water mist, or gas suppression in server rooms.
- **Compartmentation for asset protection** – beyond minimum code requirements, to protect critical areas.
- **Redundancy and resilience measures** – duplicate power supplies, smoke extraction, or emergency communications.

In large industrial, commercial, or heritage properties, insurers sometimes commission **independent fire risk assessments or engineering reviews**. If insurer requirements are not considered at design stage, premiums may be significantly higher, or cover may even be refused.

Other Stakeholders

- **Heritage bodies** (e.g., Historic England) may impose conditions on fire protection measures for listed buildings.
- **Housing associations and local councils** often have their own standards for residential safety, particularly around evacuation policies and fire detection.
- **Health and safety regulators** may add requirements for specific sectors, such as healthcare, prisons, or educational institutions.

Importance of Early Engagement

Engaging with stakeholders early in the design process provides several benefits:

- **Reduces disputes** – if the Fire Service, planners, and insurers are consulted at concept stage, disagreements are less likely at approval or construction stage.
- **Saves cost and time** – redesigns to accommodate access roads, hydrants, or additional systems are far more expensive once construction has begun.
- **Builds confidence** – regulators, fire services, insurers, and clients all gain assurance that the design has considered their operational needs.

A fire strategy that actively incorporates local and stakeholder requirements is far more likely to be **accepted quickly by approving authorities** and to function as intended in practice.

Summary

Beyond national regulations, fire strategies must reflect:

- **Local authority planning conditions** – such as hydrant provision and appliance access.
- **Fire and Rescue Service requirements** – ensuring operational firefighting is safe and effective.
- **Insurer requirements** – protecting property, continuity, and financial risk.
- **Other stakeholders** – including heritage bodies, housing providers, and regulators.

By engaging stakeholders early, fire strategies become not only compliance documents but **practical, accepted, and enforceable tools** for all parties involved in building safety.

2.7 International Perspectives

Although this book focuses on the **UK context**, fire strategies are part of a wider **global conversation** on fire safety regulation and engineering practice. As construction projects become increasingly international, and as multinational firms design and operate buildings across jurisdictions, it is important for UK practitioners to understand the major international codes and standards that influence strategy development.

International frameworks vary widely in emphasis, but they all aim to balance **life safety, property protection, and resilience**. Some lean heavily on **prescriptive codes** (fixed rules and specifications), while others adopt **performance-based approaches**, allowing engineers to demonstrate safety through modelling and analysis.

NFPA Codes (USA)

The **National Fire Protection Association (NFPA)** produces a wide suite of fire codes and standards used extensively in the **United States** and adopted in many other regions, particularly the Middle East, South America, and parts of Asia.

- **NFPA 101 – Life Safety Code**
    - Focuses on occupant life safety from fire, smoke, and toxic gases.
    - Covers both new and existing buildings, making it adaptable to retrofits.

    - Provides prescriptive requirements for means of escape, fire protection systems, and occupancy classifications.
- **NFPA 5000 – Building Construction and Safety Code**
    - A comprehensive building code, similar in scope to the International Building Code (IBC).
    - Addresses structural fire resistance, egress, systems, and construction types.

Other NFPA Standards

- **NFPA 13** – Sprinkler system design and installation.
- **NFPA 72** – Fire detection and alarm systems.
- **NFPA 92** – Smoke control systems.

Relevance to UK Practitioners: NFPA codes are often the required basis for fire strategies in **international projects**, particularly in oil and gas, high-rise, and infrastructure developments in the Middle East. A UK engineer working abroad must often reconcile NFPA prescriptions with BS/EN or BS 7974 methodologies.

International Building Code (IBC, USA)

The **IBC** is part of the **International Code Council (ICC)** suite and is widely adopted in the United States, but also influential globally.

- Provides prescriptive requirements for building design, fire resistance, egress, and fire protection systems.
- Adopted or adapted in parts of **Asia, the Middle East, and the Caribbean.**
- Often used as a contractual standard on large international projects, particularly where US-based design teams are involved.

Relevance to UK Practitioners: UK firms working in the Middle East or Asia often encounter IBC requirements and must be prepared to benchmark UK fire strategies against IBC compliance.

European Standards (ENs) and Eurocodes

Across the **European Union**, fire safety is influenced by a growing set of harmonised standards:

- **EN Standards** – cover reaction to fire (EN 13501), smoke and heat control systems (EN 12101), and fire testing methodologies.

- **Eurocodes** – structural design codes (e.g., EN 1991-1-2 for actions on structures exposed to fire, EN 1993-1-2 for steel structures in fire).

Although each EU member state still has its own building regulations, EN standards increasingly form the **technical backbone**. Post-Brexit, the UK continues to reference EN standards, and many UK products are dual certified under BS and EN classifications.

Relevance to UK Practitioners: Understanding EN classifications is essential when working with European supply chains, materials, or cross-border projects.

ISO Standards (International Organization for Standardization)

ISO standards provide global reference points for fire safety and engineering methods. While not directly enforced by law in most jurisdictions, they offer internationally recognised methodologies and testing standards.

Key examples include:

- **ISO 23932** – Fire Safety Engineering. Provides principles for applying performance-based design globally.
- **ISO 834** – Standard fire resistance test (basis for many national time-temperature curves).
- **ISO 13571** – Life-threatening components of fire – guidelines for estimating toxic hazard.

Relevance to UK Practitioners: ISO standards are often used in **internationally funded projects** (e.g., World Bank, UN) or where multinational companies seek harmonised standards across regions.

Global Trends in Fire Safety Regulation

- **Hybrid Systems**: Most countries now adopt a **hybrid model** – prescriptive baselines supported by performance-based options for complex or innovative buildings.
- **Performance-Based Fire Engineering**: Countries such as **Australia, New Zealand, Sweden, and Hong Kong** have embraced performance-based design as a standard pathway, often requiring engineers to demonstrate equivalence via ASET/RSET analysis, CFD modelling, and risk assessment.
- **Post-Disaster Reforms**: Just as Grenfell reshaped UK policy, events such as the **Daegu Subway Fire (South Korea, 2003)** and the **Station Nightclub Fire (USA, 2003)** have triggered regulatory changes abroad.

- **International Benchmarking**: Multinational developers increasingly require that fire strategies demonstrate compliance not only with local law, but also with international benchmarks such as NFPA or ISO, to satisfy investors and insurers.

Summary

While the UK framework centres on **Building Regulations, Approved Document B, and British Standards**, fire strategies are increasingly shaped by **international codes and standards**:

- **NFPA (USA)** – prescriptive, widely adopted worldwide, especially NFPA 101 and NFPA 5000.
- **IBC (USA)** – influential in the US and parts of Asia/Middle East.
- **EN and Eurocodes (EU)** – harmonised European standards underpinning testing and structural design.
- **ISO standards** – provide global methodologies for fire safety engineering.

For UK practitioners, familiarity with these systems enables them to:

- Work effectively on **international projects**.
- Benchmark UK practice against global best practice.
- Translate between prescriptive codes and performance-based approaches across different jurisdictions.

2.8 Best Practice in Referencing Legislation and Standards

One of the most common weaknesses in fire strategy documents is the lack of clarity in referencing regulations and standards. Vague or outdated references create uncertainty, delay approvals, and undermine confidence in the strategy. A robust fire strategy must demonstrate not only that it complies with the law but also how it does so, supported by precise and up-to-date references.

Why Clear Referencing Matters

- **For Building Control and Fire Services**: precise citations allow regulators to quickly verify compliance.
- **For Contractors**: accurate references provide unambiguous technical specifications.
- **For Building Managers**: clear citations link operational practices to recognised standards.

- **For Legal Defensibility**: in the event of disputes, enforcement, or litigation, documented references demonstrate due diligence.

Principles of Best Practice Referencing

1. **Be Explicit**
    - Quote the **exact clause, section, or table** rather than general statements.
    - Example: *"Travel distances comply with Approved Document B Volume 2, Table 2.1 (Maximum Travel Distances for Assembly Buildings)."*
    - Poor practice: *"Travel distances are acceptable under ADB."*
2. **Be Up to Date**
    - Always use the **current version** of standards and regulations.
    - Outdated references (e.g., BS 5588 series, withdrawn in 2006) undermine credibility.
    - A strategy should state the **publication year** of the standard used (e.g., BS 9999:2017).
3. **Distinguish between Mandatory and Advisory**
    - Make clear the difference between:
        - **Mandatory requirements** – Building Regulations (Part B functional requirements).
        - **Statutory guidance** – Approved Document B (a common route to compliance, but not compulsory).
        - **Best practice guidance** – BS 9999, BS 9991.
        - **Engineering methodologies** – BS 7974 performance-based design.
    - This distinction helps stakeholders understand which elements are legally binding and which are professional judgments.
4. **Show Equivalence for Performance-Based Solutions**
    - When deviating from prescriptive codes, the strategy must **demonstrate equivalence**.
    - Example:
        - *Prescriptive*: ADB sets maximum unsprinklered office travel distance at 18 m in one direction only

        - *Performance-based*: CFD and evacuation modelling under BS 7974 demonstrate safe evacuation with 22 m in one direction only.
    - Clearly document the **assumptions, methodology, and results** to show compliance with the functional requirement.

5. **Maintain Traceability**
    - Each referenced document should appear in a **reference list or bibliography** at the end of the strategy.
    - Use consistent citation style (e.g., "BS 9999:2017, Section 27.3).
    - Where documents are updated mid-project, ensure version control is maintained and revisions are documented.

Common Pitfalls

- **Generic referencing**: "designed to BS standards" without specifying which standard, edition, or clause.
- **Outdated citations**: referencing BS 5588 instead of BS 9999, or older editions of BS 5839 or BS 5266.
- **Failure to distinguish mandatory vs advisory**: treating ADB guidance as if it were a legal requirement.
- **Unsupported deviations**: claiming equivalence without showing calculations or modelling results.

Examples

- **Good Practice**:
    - *"The compartment wall between flats is designed to achieve 60 minutes' fire resistance, in accordance with Approved Document B, Table B4 (Minimum Periods of Fire Resistance). Construction tested to BS EN 1364-1:2015."*
- **Poor Practice**:
    - *"Walls are fire resistant as required by regulations."*
    - Lacks detail, no standard cited and gives no evidence of compliance.

How This Supports the Fire Strategy

- Clear referencing provides **transparency** and builds trust with approving authorities.
- Reduces **disputes and delays** during consultation and approval.
- Provides a **traceable record** for future fire risk assessments, ensuring that managers know what standards were assumed.
- Creates a legally **defensible document**, demonstrating that the design team acted with competence and diligence.

When preparing a fire strategy, references to legislation and standards must be:

- **Explicit** – quote the relevant clause or section.
- **Current** – use up-to-date standards, avoiding withdrawn documents.
- **Clear** – distinguish between legal requirements and best practice.
- **Justified** – demonstrate equivalence where deviating from prescriptive routes.
- **Traceable** – maintain version control and a complete reference list.

This clarity not only ensures smoother approvals but also provides confidence, transparency, and accountability throughout the building lifecycle.

2.9 Summary

Fire strategies are grounded in a robust framework of legislation, statutory guidance, and technical standards. In the UK, this begins with the **Building Act 1984** and is implemented through the **Building Regulations 2010 (as amended)**, with **Part B (Fire Safety)** providing the functional requirements.

Compliance with these requirements is most commonly demonstrated through **Approved Document B (ADB)**, which offers prescriptive guidance for both dwellings (Volume 1) and all other building types (Volume 2). For more flexible or complex projects, **British Standards** play a central role:

- **BS 9999** – risk-based fire safety in non-residential buildings.
- **BS 9991** – residential-specific guidance for flats, care homes, and other housing.
- **BS 7974** – the cornerstone of performance-based fire engineering, providing the framework for advanced analysis of fire growth, smoke movement, evacuation, and firefighting intervention.

Beyond Building Regulations, the **Regulatory Reform (Fire Safety) Order 2005 (RRO)** ensures that once a building is occupied, fire safety is actively managed. The RRO places duties on the **Responsible Person** to conduct fire risk assessments, maintain precautions, train staff, and plan evacuations. A well-prepared fire strategy at the design stage therefore becomes the baseline document that supports ongoing compliance during occupation.

Fire strategies also need to reflect local and stakeholder requirements. Planning conditions may require hydrants or access roads; Fire and Rescue Services expect clear operational facilities and pre-incident information; insurers often demand enhanced protection and resilience measures; and heritage or housing authorities may impose additional safeguards.

Although this book primarily addresses the UK system, international practice increasingly influences fire strategy development. **NFPA codes (USA)**, the **International Building Code (IBC)**, **European EN standards and Eurocodes**, and **ISO fire safety standards** are frequently encountered on global projects. For UK practitioners, awareness of these systems allows effective collaboration across borders and benchmarking of domestic practice against international norms.

A fire strategy must therefore do more than satisfy national Building Regulations. It must demonstrate an understanding of:

- **UK statutory guidance and British Standards**.
- **Operational and management frameworks** such as the RRO.
- **Local stakeholder expectations** including fire services, planners, and insurers.
- **International influences** that shape best practice.

By integrating these elements into a coherent, well-referenced document, a fire strategy provides regulators, designers, managers, and occupants with a **clear, defensible, and practical roadmap for fire safety across the building lifecycle**.

Looking Ahead

Having established the regulatory and standards framework, the next chapter will turn to the **principles of fire safety engineering**. Here we will explore the scientific and engineering foundations — from fire dynamics and smoke movement to evacuation analysis and structural behaviour — that underpin modern fire strategies and enable them to go beyond prescription towards performance-based, evidence-driven solutions.

Key Takeaways – Chapter 2

- Fire strategies in the UK are underpinned by the **Building Regulations 2010**, with **Part B (Fire Safety)** providing the functional requirements for means of escape, structural stability, fire spread, and firefighting access.
- **Approved Document B (ADB)** offers prescriptive routes to compliance, while **British Standards** (BS 9999, BS 9991, BS 7974) provide risk-based or performance-based alternatives.
- The **Regulatory Reform (Fire Safety) Order 2005** governs fire safety during occupation, requiring ongoing fire risk assessments, maintenance, and staff training — with the fire strategy acting as the baseline.
- **Local stakeholders** (planning authorities, Fire and Rescue Services, insurers, heritage bodies) often impose additional requirements that must be addressed in the fire strategy.
- International frameworks such as **NFPA, IBC, EN/Eurocodes, and ISO standards** influence best practice globally, and familiarity with them allows UK practitioners to work effectively across borders.

Chapter 3
Principles of Fire Safety Engineering

3.1 Introduction

Fire safety engineering provides the **analytical framework** behind modern fire strategies. While prescriptive guidance such as Approved Document B (ADB) offers a straightforward route to compliance for many buildings, it can be too rigid, conservative, or insufficient for more complex designs.

Contemporary architecture increasingly includes features that challenge prescriptive limits:

- **High-rise towers** with phased evacuation strategies.
- **Large atria and interconnected spaces**, where smoke control becomes critical.
- **Transport hubs, stadiums, and hospitals**, with large or vulnerable populations.
- **Heritage buildings**, where invasive prescriptive solutions may not be possible.
- **Mixed-use developments**, where multiple occupancies share risks.

In these cases, fire safety engineering provides the tools to justify **alternative approaches** that still meet the overarching objectives of:

- **Life safety** – ensuring occupants can escape or be protected until rescue.
- **Property protection** – safeguarding assets, infrastructure, and cultural heritage.
- **Business continuity and resilience** – minimising disruption to operations.
- **Firefighter safety and intervention** – ensuring safe conditions for emergency responders.

The Role of Fire Safety Engineering in Fire Strategies

Fire safety engineering enables practitioners to:

- Analyse **fire dynamics** (how fires start, grow, and spread).
- Model **smoke movement** and assess its impact on visibility, toxicity, and evacuation.
- Evaluate **human behaviour** and evacuation times (comparing ASET – Available Safe Egress Time – with RSET – Required Safe Egress Time).
- Assess **structural performance in fire**, ensuring stability under realistic heating scenarios.
- Examine **fire service intervention strategies**, considering access, water supplies, and firefighting facilities.

By combining these analytical tools, fire engineers can create fire strategies that are both flexible and robust: flexible enough to support innovative design, yet robust enough to satisfy regulators, fire services, and insurers.

Fire Engineering vs Prescriptive Guidance

- **Prescriptive approach**: provides certainty and simplicity but can lead to over-design, reduced flexibility, or impractical requirements for unusual buildings.
- **Fire engineering approach**: provides adaptability, allowing designers to prove safety through evidence and analysis rather than by following one-size-fits-all rules.
- **Hybrid approach**: in practice, most fire strategies combine prescriptive provisions (e.g., compartmentation, escape staircases) with performance-based justifications (e.g., extended travel distances supported by evacuation modelling).

Setting the Stage for This Chapter

This chapter introduces the **core principles of fire safety engineering**, showing how they are applied in practice and how they shape the structure of a fire strategy. Key themes include:

- Understanding **fire dynamics and development**.
- Analysing **smoke movement and control strategies**.
- Modelling **human behaviour and evacuation**.
- Assessing **structural fire performance**.
- Considering **fire service operations and intervention**.

Through these principles, fire strategies can be transformed from **checklist documents** into **evidence-based safety frameworks** that withstand scrutiny and adapt to the needs of complex buildings.

3.2 Fundamentals of Fire Behaviour

To apply fire safety engineering, one must first understand the **fundamental dynamics of fire** — how it develops, spreads, and impacts the built environment. Fire is not a single event but a process that passes through distinct **phases**, each with different hazards and implications for life safety, property protection, and firefighting.

Ignition and Growth

A fire begins when a **source of ignition** (such as an electrical fault, naked flame, or hot surface) comes into contact with a **combustible material** and sufficient oxygen is present.

- **Material Properties**
  - Different materials ignite and burn at different rates depending on their **ignition temperature, thermal inertia, and surface finish**.
  - Plastics, foams, and textiles may ignite and spread flame rapidly, while dense, solid timber requires more energy to ignite.
- **Ventilation**
  - In the early stages, fire is usually **fuel-controlled**: the rate of growth is determined by the type and amount of available fuel.
  - Adequate oxygen supply accelerates growth; limited oxygen can result in smouldering combustion and heavy smoke production.

Geometry and Enclosure

  - The shape and size of the compartment strongly influence fire growth. Fires in open-plan layouts develop differently from those in small, sealed rooms.
  - Heat feedback from surrounding surfaces can accelerate growth — leading to **flashover** (a rapid transition where the entire compartment becomes engulfed).

Engineering Relevance: Fire engineers use material data, ventilation conditions, and compartment geometry to model the **heat release rate (HRR)**, which is the fundamental input for smoke, heat, and evacuation modelling.

Fully Developed Fire

Once a compartment fire involves **all available combustible materials**, it enters the fully developed phase.

- **Ventilation-Controlled Fire**
    - At this point, fire growth is no longer limited by fuel but by the availability of oxygen.
    - Heat release rates can exceed **10–20 MW** in large compartments.
    - Flames may extend out of windows and doors, threatening adjacent buildings through **external fire spread**.
- **Thermal Environment**
    - Temperatures in a fully developed fire can exceed **1000°C**, rapidly compromising unprotected structural steel.
    - Radiation and convection drive fire spread to adjacent compartments if barriers fail.

Engineering Relevance: Fire strategies must ensure that structures can withstand these conditions for the **required fire resistance period**, allowing evacuation and firefighting intervention before collapse.

Decay Phase

As fuel is consumed or oxygen becomes depleted, the fire enters a **decay phase**.

- **Reduced Flames, Persistent Hazards**
    - Flame intensity drops, but the environment remains highly hazardous due to residual heat and smoke.
    - Toxic gases such as **carbon monoxide (CO), hydrogen cyanide (HCN), and nitrogen oxides** may accumulate.
    - Flashover can still occur if fresh oxygen enters a smouldering compartment — leading to **backdraft**.

Engineering Relevance: The decay phase is particularly dangerous for firefighters, who may misinterpret a smouldering fire as less hazardous. Strategies must account for this in operational planning and firefighting facilities.

Smoke Movement

Smoke, not flames, is often the greatest threat to life in fire. Most fire deaths result from **smoke inhalation**, not burns.

- **Buoyancy and Stratification**
    - Hot smoke rises, forming a ceiling layer that spreads horizontally across the compartment.
    - If not vented, this layer deepens, reducing visibility and exposing occupants to heat and toxic gases.
- **Spread Pathways**
    - Smoke migrates through **doors, stairwells, lift shafts, ceiling voids, service ducts, and ventilation systems**.
    - In tall atria and shafts, rising smoke plumes entrain cooler air, which can reduce buoyancy and cause stalling before reaching vents.
- **Toxicity**
    - Modern building contents release **highly toxic smoke** when burning. Polyurethane foams, PVC, and synthetic fabrics generate lethal concentrations of CO and HCN rapidly.

Engineering Relevance: Smoke modelling (using tools such as Computational Fluid Dynamics (CFD)) is essential in fire strategies to design **smoke control systems**, size of smoke reservoirs, and ensure **tenable conditions for evacuation**.

Integrating Fire Behaviour into Fire Strategies

Understanding these fire behaviour fundamentals allows engineers to:

- **Model fire growth and spread** under realistic conditions.
- **Predict smoke movement** to design effective ventilation and smoke control systems.
- **Assess tenability conditions** for evacuation and firefighting (temperature, visibility, toxicity).
- **Set structural fire resistance periods** based on realistic heating profiles.
- **Plan firefighting facilities** to account for backdraft, flashover, and smoke logging.

Without this knowledge, fire strategies risk being overly simplistic, relying solely on prescriptive tables rather than scientific analysis.

Summary

The fundamentals of fire behaviour — ignition, growth, fully developed fire, decay, and smoke movement — underpin every fire safety engineering analysis.

By understanding how fires develop and interact with the built environment, fire strategies can move beyond generic compliance documents and become scientifically grounded frameworks that protect life, property, and resilience.

3.3 Heat Release Rate (HRR)

At the core of fire dynamics is the concept of **Heat Release Rate (HRR)** — the rate at which a fire generates energy, typically expressed in **kilowatts (kW)** or **megawatts (MW)**. HRR is the **single most important parameter in fire safety engineering**, as it dictates:

- How quickly conditions in a compartment deteriorate.
- The scale of smoke production and toxic gas release.
- The thermal load placed on structures and fire protection systems.
- The time available for safe evacuation and firefighting.

Definition

HRR is defined as:

$$\dot{Q} = \Delta H_c . \dot{m}_f$$

Where:

- $\dot{Q}$ = Heat release rate (kW).
- ΔH_c = Heat of combustion of the fuel (kJ/kg).
- $\dot{m}_f$ = Mass loss rate of fuel (kg/s).

This formula shows that HRR depends on both the type of fuel and the rate at which it burns.

Phases of HRR Development in a Compartment Fire

1. **Growth Phase**
   - HRR increases as fuel ignites, and flames spread.
   - Initially fuel-controlled, depending on the material properties (e.g., wood vs plastics).
   - Typical peak HRR for single upholstered chair: **200–300 kW**.
2. **Flashover**
   - When radiant heat feedback ignites all combustible surfaces almost simultaneously.

- HRR spikes dramatically, often exceeding **2–5 MW** in a typical room.
- Flashover is a critical design scenario in fire engineering.

3. **Fully Developed Fire**
    - HRR is governed by **ventilation** (size of openings, oxygen availability).
    - Large compartments can reach **10–20 MW or more**, producing untenable conditions in seconds.
4. **Decay**
    - HRR falls as fuel is consumed, or oxygen is depleted.
    - Risk of backdraft remains if oxygen re-enters.

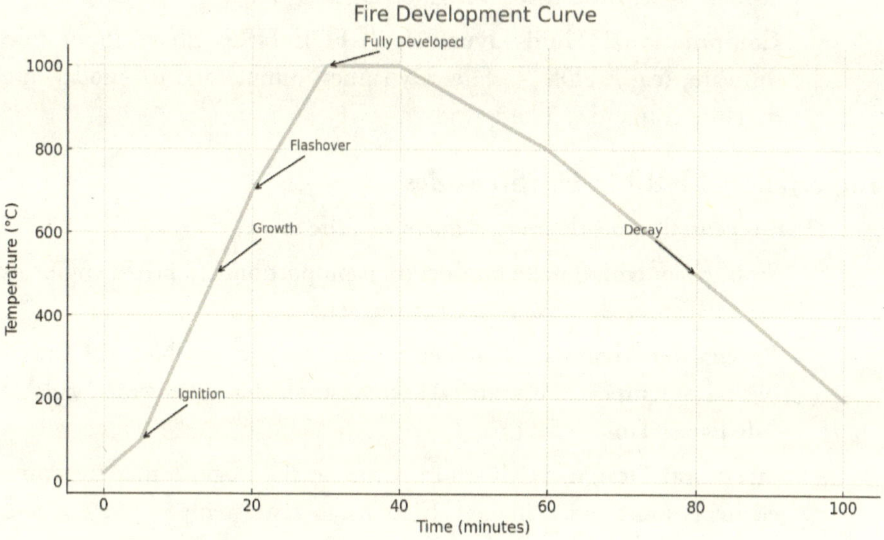

- **X-axis (Time)**: Represents the duration since ignition, usually in minutes.
- **Y-axis (Temperature)**: Represents the temperature within the fire compartment in degrees Celsius (°C).

Each segment of the curve corresponds to a distinct stage:

- **Ignition**: A slow rise in temperature as the fire begins.
- **Growth**: Rapid increase in temperature as the fire spreads and consumes more fuel.

- **Flashover**: A sudden spike when the environment becomes fully involved — nearly all materials ignite simultaneously.
- **Fully Developed**: The peak of the fire, characterized by maximum heat release and structural damage.
- **Decay**: A decline in temperature as fuel and oxygen are exhausted.

Measurement and Modelling of HRR

- **Cone Calorimeter (ISO 5660)**: laboratory test that measures HRR per unit area of a material under a controlled heat flux.
- **Full-Scale Fire Tests**: HRR measured in calorimeter rooms, car parks, aircraft cabins, etc., to provide benchmark data.
- **Empirical Correlations**: e.g., t^2 fire growth models (slow, medium, fast, ultra-fast fires) used in engineering analysis.
- **Computational Fluid Dynamics (CFD)**: HRR curves input into software (e.g., FDS – Fire Dynamics Simulator) to model fire development and smoke spread.

Importance of HRR in Fire Strategies

HRR is central to nearly every element of a fire strategy:

- **Smoke Control**: Smoke production is proportional to HRR; engineers use HRR to size smoke reservoirs and vents.
- **Evacuation Analysis**: HRR determines the rate at which conditions (visibility, temperature, toxicity) become untenable, setting the Available Safe Egress Time (ASET).
- **Structural Design**: HRR informs the heating regime that structural elements must resist; higher HRRs mean faster temperature rise and greater collapse risk.
- **Suppression Systems**: Sprinkler systems are designed based on expected HRR and design fire scenarios (e.g., 2 MW for office, 8 MW for car parks).
- **Fire Service Intervention**: HRR helps predict whether manual firefighting is feasible or whether conditions will be unsurvivable without suppression.

Design Fires

In performance-based design, engineers define **"design fires"** to represent credible worst-case scenarios for a building. These are characterised by:

- **Peak HRR** (e.g., 5 MW office workstation, 8 MW car fire, 20 MW fully developed retail compartment).
- **Growth rate** (slow, medium, fast, ultra-fast).
- **Ventilation conditions** (oxygen-rich vs starved).
- **Fuel load** (calorific content per m^2).

Design fires are critical inputs to evacuation models, smoke control design, and structural fire analysis.

Summary

- Heat Release Rate (HRR) is the most critical variable in fire dynamics, controlling smoke, toxicity, structural load, and tenability.
- HRR evolves through **growth, flashover, fully developed fire, and decay phases**.
- It is measured through laboratory tests, full-scale tests, and modelled in engineering tools.
- HRR underpins every aspect of fire strategies, from **smoke control and evacuation** to **structural fire resistance and firefighting intervention**.

3.4 Prescriptive vs Performance-Based Design

In fire safety engineering, there are two principal approaches to demonstrating compliance and developing a fire strategy: the **prescriptive (code-based) approach** and the **performance-based (fire engineering) approach**. Both have value, but their suitability depends on the building type, complexity, and the aspirations of the design team.

Prescriptive (Code-Based) Approach

The prescriptive approach relies on following the solutions set out in statutory guidance such as **Approved Document B (ADB)** or other established standards.

- **Key Characteristics**
    - Provides **standard solutions** to common design issues.
    - Based on tabulated values (e.g., travel distances, fire resistance periods, exit widths).

  - Requires minimal analysis — compliance is achieved by following rules.

- **Advantages**
  - **Clarity and simplicity** – easy for designers, contractors, and regulators to apply.
  - **Certainty of approval** – regulators are familiar with ADB and similar documents.
  - **Cost-effective** – avoids need for complex modelling or specialist consultancy.
- **Limitations**
  - **Conservatism** – can lead to over-design or excessive cost.
  - **Inflexibility** – may restrict innovative layouts or materials.
  - **Inapplicability to complex buildings** – tall buildings, atria, transport hubs, and heritage properties may not fit the assumptions behind prescriptive rules.
- **Best Applied to**
  - **Low-rise dwellings, offices, schools, warehouses, retail units** – where standard layouts are used and innovation is limited.

Performance-Based (Fire Engineering) Approach

The performance-based approach uses **scientific and engineering methods** to demonstrate that safety objectives are met. Instead of following rules rigidly, it applies **modelling, risk assessment, and scenario testing** to prove that a design satisfies the **functional requirements of Part B**.

- **Key Characteristics**
  - Governed in the UK by **BS 7974** and its supporting documents.
  - Involves defining **design fires**, modelling smoke movement, assessing evacuation times, and evaluating structural fire resistance.
  - Compares **ASET (Available Safe Egress Time)** with **RSET (Required Safe Egress Time)** to demonstrate life safety.
- **Advantages**
  - **Flexibility** – allows innovative architectural features such as large atria, open-plan layouts, and mixed-use buildings.

    - **Optimisation** – can reduce unnecessary costs by tailoring fire protection to actual risks.
    - **Evidence-based** – provides robust technical justification that withstands scrutiny.
- **Limitations**
    - **Complexity** – requires specialist fire engineers and advanced modelling tools.
    - **Time and cost** – more resource-intensive than prescriptive design.
    - **Regulatory acceptance** – some authorities may require extensive justification before approving engineered solutions.
- **Best Applied to**
    - **Complex or innovative buildings** – high-rises, hospitals, airports, underground stations, heritage projects, and developments with unusual geometries.

Hybrid Approaches in Practice

In reality, most fire strategies adopt a **hybrid approach**, combining prescriptive and performance-based methods.

- **Example 1: Office Building with an Atrium**
    - Escape stairs designed prescriptively to ADB.
    - Atrium smoke control system designed using performance-based CFD modelling.
- **Example 2: Residential Tower**
    - Compartmentation and travel distances follow BS 9991.
    - Evacuation strategy justified using evacuation modelling to support phased evacuation or stay-put.

This hybrid model allows the certainty of prescriptive measures to coexist with the flexibility of engineered solutions, producing a strategy that is both robust and innovative.

The Principle of Equivalence

When adopting performance-based solutions, the engineer must demonstrate **equivalence** — that the proposed design achieves a level of safety at least equal to (and often better than) the outcome intended by the prescriptive guidance.

For example:

- ADB may limit unsprinklered office travel distance to **18 m** in one direction. If the design requires 22 m, performance-based evacuation modelling must demonstrate that occupants can still evacuate before conditions become untenable.
- ADB may require **60 minutes' fire resistance** for a wall. If the wall is tested to Eurocode standards and achieves **90 minutes**, equivalence is clearly demonstrated.

This principle ensures that even when departing from prescriptive rules, the fire strategy still aligns with the legal **functional requirements**.

Summary

- **Prescriptive design**: rule-based, simple, cost-effective, but inflexible.
- **Performance-based design**: analytical, flexible, optimised, but requires expertise and regulatory negotiation.
- **Hybrid approaches** are the norm, using prescriptive rules for standard elements and performance-based methods for complex or innovative features.
- The key principle is **equivalence**: engineered solutions must deliver safety outcomes at least equal to prescriptive requirements.

3.5 Engineering Tools and Methods

Fire safety engineering relies on a suite of **analytical tools and modelling techniques** that allow engineers to predict how fires will develop, how smoke will spread, how structures will perform, and how occupants will respond. These tools provide the **evidence base** for performance-based fire strategies and are essential when moving beyond prescriptive code compliance.

Fire Modelling

Fire modelling provides quantitative predictions of **temperature, smoke movement, toxicity, and visibility** in buildings under defined fire scenarios.

- **Zone Models**
    - Simplified models that divide a compartment into zones (typically an upper hot layer and a lower cool layer).

  - *Example*: **CFAST (Consolidated Model of Fire Growth and Smoke Transport) or B-Risk.**
  - Useful for quick assessments, design of smoke reservoirs, or scoping studies.
  - Less accurate for complex geometries, large atria, or interconnected spaces.
- **Computational Fluid Dynamics (CFD) Models**
  - Solve the full equations of fluid flow, heat transfer, and combustion.
  - *Example*: Pyrosim, Smartfire, widely used in the UK and internationally.
  - Provide detailed visualisation of smoke spread, temperature gradients, and toxicity levels.
  - Require careful input assumptions and significant computational resources.
  - Often combined with **tenability criteria** (visibility > 10 m, temperature < 60°C at 2 m height, CO < 500 ppm) to assess evacuation safety.

Application in Fire Strategies: Fire modelling is used to justify **extended travel distances**, design **smoke control systems** in atria and car parks, and assess **firefighter conditions**.

Evacuation Modelling

Understanding how occupants move during an emergency is vital for ensuring safe egress.

- **Evacuation Modelling Software**
  - **Building Exodus, Pathfinder, STEPS, MassMotion** are commonly used tools.
  - Simulate occupant movement, speed, queuing, congestion, and route choices.
  - Allow for consideration of **disabled or vulnerable occupants**.
  - Can model phased evacuations, stay-put policies, or simultaneous evacuation.

- **Behavioural Aspects**
    - Models can incorporate human behaviour factors such as pre-movement time, familiarity with the building, and response to alarms.
    - These behavioural assumptions often dominate evacuation outcomes and must be justified with reference to research (e.g., SFPE Handbook).

Application in Fire Strategies: Evacuation modelling supports **performance-based justifications** for travel distances, exit widths, phased evacuation in high-rises, or complex scenarios like stadiums and airports.

Structural Fire Engineering

Structural fire engineering assesses how **loadbearing elements** respond under fire exposure.

- **Analytical Methods**
    - Hand calculations using simplified heating curves (e.g., ISO 834, Eurocode parametric fires).
    - Estimate temperature rise in structural steel or timber elements.
- **Numerical Methods**
    - Finite element modelling (e.g., **ANSYS, SAFIR**) simulates complex interactions between heat transfer, material degradation, and structural loading.
    - Can predict progressive collapse, thermal bowing, or connection failure.
- **Performance-Based Design**
    - Used to justify reduced fire resistance periods if realistic analysis shows safety is maintained.
    - Increasingly applied to demonstrate **robustness and resilience** in large-span structures, tall buildings, and tunnels.

Application in Fire Strategies: Provides evidence that structural fire resistance is sufficient to maintain stability for evacuation and firefighting.

Risk Assessment Techniques

In some cases, fire safety cannot be fully addressed through deterministic modelling alone.

- **Quantitative Risk Analysis (QRA)**
    - Assigns probabilities to fire scenarios, combining frequency and consequence to produce **risk profiles**.
    - Useful for high-hazard industries (e.g., petrochemical plants, nuclear facilities).
- **Probabilistic Methods**
    - Consider uncertainties in fire growth rates, occupant behaviour, and system reliability.
    - Monte Carlo simulations and event tree analysis can provide **probabilistic safety margins**.
- **Cost-Benefit Analysis**
    - Used to weigh investment in additional fire protection (e.g., sprinklers, smoke control) against reduced risk and potential losses.

Application in Fire Strategies: Risk assessment techniques are often applied to **unusual hazards, heritage buildings, or industrial facilities**, where prescriptive or standard performance methods may not capture the true risks.

Integration of Tools in Fire Strategies

In practice, these tools are rarely used in isolation. A robust fire strategy integrates:

- **Fire modelling** to define smoke conditions.
- **Evacuation modelling** to test egress safety.
- **Structural analysis** to ensure the building remains stable.
- **Risk analysis** to support decisions on unusual scenarios.

Together, they provide a **holistic picture** of fire safety, bridging the gap between regulation, design, and operation.

Summary

Engineering tools and methods transform fire strategies from prescriptive checklists into **evidence-based safety frameworks**.

- **Fire modelling** predicts smoke and heat spread.
- **Evacuation modelling** evaluates occupant safety.
- **Structural fire engineering** demonstrates stability.

- **Risk assessment** supports decision-making in complex scenarios.

By combining these tools, fire engineers can create strategies that are flexible, scientifically robust, and tailored to the unique challenges of each building.

3.6 Fire Engineering Scenarios

To test the robustness of a fire strategy, engineers develop a set of **design fire scenarios** that represent credible worst cases. These scenarios are not random; they are carefully selected to reflect the most significant risks a building may face, ensuring that **life safety, property protection, and firefighter access** are maintained under a range of conditions.

The use of scenarios is central to **performance-based fire engineering**, and is explicitly set out in **BS 7974**. By modelling multiple plausible fires — rather than a single "standard" event — engineers demonstrate that the fire strategy is **comprehensive, resilient, and defensible**.

Fast-Growing Fires

- **Description**: Fires that accelerate rapidly due to highly combustible materials (e.g., upholstered furniture, plastics, foams).
- **Purpose of Scenario**:
    - Tests whether **early detection and alarm systems** activate quickly enough.
    - Assesses whether occupants can begin evacuation before conditions become untenable.
    - Validates sprinkler or suppression system effectiveness in fast ignition environments.
- **Example**: Fire in a hotel bedroom with modern furnishings, reaching flashover in under 5 minutes.

Slow-Growing Fires

- **Description**: Fires that smoulder or develop slowly due to limited ventilation, low-energy ignition sources, or fuel type.
- **Purpose of Scenario**:
    - Tests effectiveness of **detection systems** in identifying low-output fires.
    - Assesses management and staff intervention where evacuation may be delayed.

    - Highlights risks of toxic smoke accumulation before flames are visible.
- **Example**: Fire from faulty electrical wiring in a plant room or storage cupboard.

Ventilation-Controlled Fires

- **Description**: Fires in which growth is limited by lack of oxygen rather than lack of fuel.
- **Purpose of Scenario**:
    - Tests **smoke control systems**, as ventilation-controlled fires produce large volumes of smoke.
    - Examines conditions for **firefighter access**, since these fires can transition suddenly to flashover or backdraft when fresh oxygen is introduced.
- **Example**: Basement fire with limited openings, or a sealed office where windows fail under heat.

Localized Fires

- **Description**: Fires confined to a small area or specific room, often associated with high-risk functions.
- **Purpose of Scenario**:
    - Ensures **compartmentation** prevents fire spread.
    - Tests whether suppression or detection systems in critical areas (e.g., kitchens, server rooms, plant rooms) are effective.
    - Demonstrates that localized fires do not compromise building-wide evacuation.

Example: Grease fire in a commercial kitchen; electrical fire in a server room.

Multiple Fire Scenarios

- **Description**: For very large or complex developments, it is often necessary to model multiple simultaneous or sequential fires.
- **Purpose of Scenario**:
    - Ensures **resilience** in buildings where a single fire event may not represent the worst case.

  - Tests robustness of systems (e.g., whether multiple fire alarms in different zones create confusion or delay).
  - Provides assurance for **large campuses, airports, or shopping centres**, where multiple ignition sources exist.
- **Example**: Fire starting in one retail unit of a shopping centre while a second fire occurs in a service corridor.

Analysis of Scenarios

Each scenario is analysed to verify that **safety objectives are consistently achieved**. This typically involves:

Evacuation analysis: Comparing **Available Safe Egress Time (ASET)** with **Required Safe Egress Time (RSET)**.

Smoke modelling: Ensuring tenable conditions (visibility, temperature, toxicity) are maintained long enough for safe evacuation.

Structural fire analysis: Verifying that compartmentation and load-bearing elements withstand the thermal load.

Firefighting access analysis: Assessing whether the Fire and Rescue Service can safely intervene.

The combination of scenarios ensures that a fire strategy is not dependent on a single assumption and provides confidence to regulators, insurers, and building managers.

Summary

Fire engineering scenarios represent the **"test cases"** of a fire strategy. By modelling **fast, slow, ventilation-controlled, localized, and multiple fire events**, engineers demonstrate that fire safety measures are resilient across a wide range of conditions.

This multi-scenario approach ensures that strategies are robust, defendable, and aligned with real-world risks — providing assurance that safety objectives will be met, regardless of how a fire begins or develops.

3.7 Human Behaviour in Fire

Fire strategies must account not only for the **physical dynamics of fire and smoke**, but also for the **psychological and behavioural responses of occupants**. In many incidents, outcomes are determined less by the speed of fire growth than by how people recognise, interpret, and respond to the threat.

Engineering models often assume **ideal behaviour** — that people evacuate immediately and head to the nearest safe exit. In reality, behaviour is far more complex, influenced by perception, training, communication, and social context. A credible fire strategy must therefore embed **realistic behavioural assumptions** rather than over-simplified models

Recognition and Response Time

- **Pre-movement Time**: After an alarm is raised, people often delay evacuation. They may investigate, collect belongings, or wait for confirmation.
- Research shows that **pre-movement times often exceed travel times**, making them a critical determinant of safety.
- Factors influencing recognition and response:
    - Alarm type (audible tones vs voice alarms).
    - Occupant familiarity with the building.
    - Visibility of smoke or fire.
    - Previous false alarms, which may reduce urgency.

Fire Strategy Implication: Assumptions about immediate evacuation are unrealistic. Fire strategies must allow for **delays in occupant response** when comparing ASET vs RSET.

Mobility Limitations

- Occupants are not homogeneous: buildings may contain **children, elderly people, disabled persons, or those with temporary impairments**.
- Evacuation of wheelchair users, visually impaired occupants, or patients in hospitals requires **dedicated provisions** such as evacuation lifts, refuges, and trained assistance.
- "Inclusive design" principles, supported by **BS 8300** and equality legislation, require fire strategies to explicitly consider **persons with reduced mobility (PRM)**.

Fire Strategy Implication: Evacuation modelling must account for **slower travel speeds, staff-assisted evacuation, and use of refuges or lifts**.

Group Behaviour

- People rarely behave as isolated individuals. **Families, friendship groups, and co-workers** often evacuate together, sometimes at the pace of the slowest member.
- In workplaces, occupants may follow **staff-led instructions** rather than alarms alone.
- In large assemblies (stadiums, concert halls, transport hubs), crowd dynamics such as **bottlenecking, arching at exits, and herding** become critical.

Fire Strategy Implication: Evacuation modelling should include **group behaviours** and **crowd dynamics**, not just individual flows. Strategies must also ensure **exit distribution** prevents dangerous congestion.

Panic vs Orderly Evacuation

- The idea that people commonly "panic" in fires is largely a myth. Research shows that most evacuations are orderly if communication is clear and staff are present.
- Panic-like behaviour may occur in **extreme crowding or perceived entrapment**, particularly where exits are not obvious.
- Key influences:
    - **Clear signage and wayfinding**.
    - **Voice alarm systems** providing instructions.
    - Presence of **trained staff or wardens**.

Fire Strategy Implication: Strategies must include **clear communication systems** and **management procedures** to guide occupants and avoid confusion.

Human Behaviour in Modelling

- Evacuation software (e.g., Building Exodus, Pathfinder, STEPS, MassMotion) allows engineers to assign:
    - Pre-movement times.
    - Walking speeds (differentiated by age and mobility).
    - Route choices, including preference for familiar exits.
    - Grouping behaviours.
- However, models are only as reliable as their inputs. If unrealistic assumptions are used, outputs may misrepresent actual safety.

Fire Strategy Implication: Fire engineers must calibrate models using **empirical data** (from drills, research, or case studies) to reflect realistic occupant behaviour.

Summary

Human behaviour is as important as fire behaviour in determining safety outcomes. A credible fire strategy must consider:

- **Recognition and response delays** (pre-movement time).
- **Mobility limitations** and inclusive evacuation provisions.
- **Group behaviour** and crowd dynamics.
- The balance between **orderly evacuation and panic**, shaped by communication and management.

By applying **realistic behavioural assumptions**, strategies ensure that evacuation models are not overly optimistic and that safety provisions reflect how people actually respond in fire emergencies.

3.8 Integration with Building Design

Fire engineering is most effective when integrated early in the design process. Key interfaces include:

- **Architecture** – spatial layouts, open plan designs, atriums.
- **Structural Engineering** – load-bearing capacity under fire.
- **Mechanical and Electrical Systems** – HVAC and smoke control systems.
- **Management Systems** – evacuation planning, staff training, maintenance regimes.

This multidisciplinary integration ensures fire safety is embedded, not added as an afterthought.

3.9 Fire Service Intervention

A core principle is ensuring that fire service access, facilities, and timings are accounted for. Strategies must consider:

- Arrival and set-up time for appliances.
- Firefighting shafts, lobbies, and lifts.
- Hose lengths and pressure requirements.

- Water supplies and hydrants.
- Safe working conditions within the building.

Case Study: King's Cross Underground Fire (1987, London, UK)

The **King's Cross Underground fire** is one of the most significant UK case studies demonstrating the role of human behaviour in fire.

- **Incident Overview**
    - A small fire started beneath a wooden escalator at King's Cross station.
    - Initially it appeared minor, with limited visible flames. Many staff and passengers underestimated the severity.
    - After approximately 15 minutes, the fire underwent a sudden **flashover (the "trench effect")**, producing a fireball that travelled up the escalator shaft into the ticket hall.
    - 31 people were killed and over 100 injured.
- **Human Behaviour Factors**
    - **Recognition delay**: Early signs were ignored; passengers assumed smoke was a normal part of train operations.
    - **Staff response**: Some staff delayed calling the fire brigade, believing the fire was minor and manageable.
    - **Pre-movement time**: Many passengers continued using escalators and trains, not realising evacuation was necessary.
    - **Communication gaps**: Lack of a clear public announcement system meant passengers were not given timely instructions.
- **Lessons for Fire Strategies**
    - Fire strategies must not assume immediate recognition of danger. **Pre-movement delays** can be significant.
    - **Staff training and clear communication systems** are critical. Voice alarms and evacuation procedures reduce confusion.
    - Fire engineering analysis must include **scenarios where occupants remain in place for extended periods** before evacuating.
    - Infrastructure designs must consider how smoke moves in **vertical shafts** and how delayed evacuation can trap people in untenable conditions.

This case reshaped UK fire safety policy, leading to the **Fire Precautions (Sub-surface Railway Stations) Regulations 1989**, mandating improved staff training, communication, and materials testing.

Additional Example: The Station Nightclub Fire (2003, Rhode Island, USA)

- A pyrotechnics display ignited flammable foam insulation.
- Despite visible flames, many occupants initially attempted to exit via the **familiar main entrance**, creating a bottleneck.
- 100 people died, many within minutes due to smoke and toxic gases.
- Demonstrates how **exit familiarity, crowd herding, and poor distribution of egress routes** can overwhelm evacuation plans.

Key Insight for Fire Strategies Both case studies show that **human behaviour can undermine even well-designed fire safety systems**. Strategies must:

- Anticipate **recognition and response delays.**
- Provide **redundant and clearly signed exits**.
- Incorporate **voice alarms and trained staff** to guide occupants.
- Model **realistic behaviours** (use of familiar exits, group dynamics, delayed movement).

3.10 Summary

Fire safety engineering provides the **scientific and analytical foundation** for modern fire strategies. While prescriptive codes offer valuable baseline solutions, it is the application of engineering principles that allows strategies to be **tailored, flexible, and robust**.

In this chapter, we have explored the **fundamentals of fire behaviour** — from ignition and growth through flashover, full development, and decay — and how these stages shape conditions within buildings. We examined the central role of **heat release rate (HRR)** as the defining parameter of fire severity, dictating smoke production, tenability, and structural loading.

We then considered the two dominant approaches to fire safety: the **prescriptive approach**, which provides certainty and simplicity, and the **performance-based approach**, which uses modelling and analysis to justify complex or innovative designs. In practice, most fire strategies employ a **hybrid**

model, applying prescriptive rules where appropriate and performance-based solutions where necessary.

Engineering tools and methods — such as **fire modelling, evacuation modelling, structural fire analysis, and quantitative risk assessment** — form the practical toolkit that enables engineers to develop evidence-based strategies. These tools are applied within **fire engineering scenarios** representing credible worst cases, ensuring resilience against a wide range of possible fire events.

Crucially, this chapter also highlighted that **human behaviour in fire is as important as fire behaviour itself**. Occupant recognition, response delays, mobility limitations, group dynamics, and communication all profoundly influence evacuation outcomes. Case studies such as the **King's Cross Underground fire (1987)** and **The Station Nightclub fire (2003)** remind us that without realistic assumptions about human behaviour, even well-designed strategies can fail.

Key Message

- Fire safety engineering is about **integration**:
- Fire science explains how fires ignite, grow, and spread.
- Engineering tools model smoke, evacuation, and structural performance.
- Human behaviour studies provide realistic assumptions about occupant response.
- All of these are combined into a **coherent fire strategy** that satisfies both legal requirements and practical safety objectives.

Looking Ahead

Having established the **principles of fire safety engineering**, the next chapter will turn to the **structure and contents of a fire strategy document**. It will provide a practical roadmap showing how the ideas explored in Chapter 3 are translated into a clear, defensible, and usable strategy that serves designers, regulators, building managers, and emergency responders.

Chapter 4

Structure and Contents of a Fire Strategy Document

A fire strategy is only effective if it is **structured, clear, and comprehensive.** The way information is presented is as important as the technical content, because regulators, designers, contractors, fire authorities, and building managers all rely on the document for different purposes.

A well-prepared fire strategy should be **logical, transparent, and traceable**, showing not only what measures are in place but also the reasoning behind them. Typically, a fire strategy is structured into the following key sections:

1. Introduction and Administrative Details
2. Description of the Building
3. Design Conditions and Assumptions
4. Compliance Basis and Deviations
5. Means of Escape
6. Fire Protection Systems
7. Firefighting Access and Facilities
8. Management and Maintenance
9. Appendices and Supporting Evidence

Each section plays a vital role in demonstrating how legal obligations are met and how safety is achieved in practice.

4.1 Introduction and Administrative Details

The opening section of a fire strategy document establishes the **framework, accountability, and context** for the entire report. Without this, the document risks becoming ambiguous or difficult to use in future. Regulators, fire authorities, building managers, and even courts may refer back to a fire strategy years after it was written, so a clear introduction ensures that its **purpose, authorship, scope, and supporting evidence** are transparent and traceable.

Purpose of the Document

The introduction should explicitly state the **purpose for which the fire strategy has been prepared**, which can vary depending on the project stage:

- **Design Approval** – prepared during concept or detailed design stages to demonstrate compliance with Building Regulations Part B and relevant standards. Often submitted to Building Control or an Approved Inspector.
- **Regulatory Submission** – produced to satisfy planning or licensing authorities, insurers, or Fire and Rescue Services.
- **Construction Phase** – used by contractors to interpret design intent and install fire protection measures correctly.
- **Handover / As-Built** – updated to reflect the completed building, forming the baseline document for ongoing management.
- **Retrospective Assessment** – commissioned for existing or historic buildings that were built without a formal strategy, or where changes in use require a new compliance review.

By stating the purpose clearly, all stakeholders understand whether the document represents **design intent, construction detail, or operational reference**.

Authorship and Commissioning

Accountability is fundamental. The document should record:

- **Name(s) of author(s)** – including professional qualifications and affiliation.
- **Organisation** – the fire engineering consultancy or responsible party producing the strategy.
- **Commissioning client** – who requested the document (developer, owner, landlord, managing agent).
- **Version control and revision history** – including dates, changes made, and approvals.

This ensures **ownership and responsibility** are clear. In the event of later disputes or updates, stakeholders can confirm who prepared the strategy, for whom, and under what assumptions.

Regulatory Framework

The introduction should cite the **statutory and technical references** upon which the strategy is based. Typical inclusions are:

- **Building Regulations 2010 (as amended)** – with specific reference to **Part B – Fire Safety**.
- **Approved Document B (Volumes 1 & 2)** – the most common prescriptive guidance.
- **British Standards** – including BS 9999 (non-residential), BS 9991 (residential), BS 7974 (performance-based engineering), and others as relevant.
- **European and International Standards** – such as BS EN 13501 (reaction to fire classification), BS EN 12101 (smoke control), or ISO 23932 (fire safety engineering).
- **Other guidance** – such as NFPA codes, insurer requirements, or local fire service design guides, where applicable.

By making this explicit, the fire strategy shows that its recommendations are **grounded in recognised standards and statutory duties**. It also clarifies the **boundary between prescriptive compliance and performance-based justifications**, an important distinction during approvals.

Supporting Documents and References

No fire strategy stands alone; it is part of a **wider design package**. The introduction should list all **supporting documents** that informed the strategy, for example:

- Architectural and engineering drawings (plans, elevations, sections).
- Structural and mechanical specifications.
- Fire engineering calculations, modelling reports, and test certificates.
- Previous fire risk assessments, planning submissions, or heritage reports.
- Manufacturers' data sheets for critical fire safety products.

This creates a **traceable record** of the evidence base, ensuring consistency between the strategy and the wider project documentation.

Clarity and Traceability

The key purpose of this section is to ensure the document remains **usable and defensible**, even years after its production. A strategy that does not identify its purpose, author, standards, or evidence risks:

- Misinterpretation during construction or operation.
- Rejection at regulatory review.
- Difficulty during refurbishment, where design intent cannot be traced.
- Legal challenges, where accountability cannot be established.

By contrast, a well-structured introduction provides **clarity, continuity, and confidence** to all parties who rely on the strategy.

4.2 Description of the Building

The fire strategy must begin by establishing a clear and comprehensive **description of the building**. This section provides the **baseline context** upon which the strategy is interpreted, ensuring that regulators, fire authorities, designers, and building managers all understand the physical and functional environment in which fire safety measures will operate.

A well-written description of the building avoids ambiguity, creates consistency across project documentation, and allows future assessors (for example, during refurbishment or retrospective risk assessments) to understand the assumptions on which the strategy was based.

Site and Surroundings

The first element of the building description should address the **wider context of the site**, including:

- **Location** – geographic position, urban or rural setting, and any planning designations (heritage sites, conservation areas, etc.).
- **Neighbouring risks** – proximity to other buildings, hazardous industries, fuel storage, or transport infrastructure.
- **Access routes** – main roads, emergency vehicle access points, site constraints (e.g., narrow streets, railway lines, rivers).
- **Topography** – slope, gradient, or flood risk, which may affect fire appliance access or evacuation strategies.

This context helps determine external fire spread risks, boundary conditions, and fire service intervention assumptions.

Building Form and Height

The physical characteristics of the building must be clearly defined:

- **Number of storeys** – including basements, mezzanines, and rooftop plant rooms.
- **Vertical arrangement** – high-rise, mid-rise, or low-rise; whether floors are interconnected by atria or voids.
- **Special spaces** – large-volume compartments (e.g., warehouses, exhibition halls), atria, malls, or auditoria.
- **Extensions and refurbishments** – any recent or planned alterations that affect compartmentation or escape routes.
- **Footprint and layout** – open-plan vs cellular, single building vs complex of blocks.

This information is critical to assess escape distances, firefighting access, and structural fire resistance requirements.

Occupancy and Use

The **use of the building and the nature of its occupants** directly influence the fire strategy:

- **Residential** – houses, flats, student accommodation, care homes.
- **Commercial** – offices, retail, entertainment venues.
- **Healthcare** – hospitals, clinics, care facilities, where phased evacuation or defend-in-place strategies may apply.
- **Assembly** – stadiums, theatres, airports, places of worship, where large crowds require careful evacuation analysis.
- **Mixed-use** – increasing common in urban developments, combining residential, retail, and commercial uses.

This section should also note **occupant characteristics**:

- Maximum occupancy loads.
- Age and mobility distribution (children, elderly, disabled).
- Staffing levels and role of trained personnel in evacuation.
- Hours of use (daytime vs overnight occupancy).

These factors shape assumptions about evacuation strategy, alarm systems, and management responsibilities.

Construction and Fire Resistance

Finally, the building description must outline the **construction method and fire resistance characteristics**:

- **Structural frame** – concrete, structural steel, timber, or composite.
- **Fire resistance class** – e.g., REI 60 (loadbearing, integrity, insulation for 60 minutes) under Eurocodes.
- **Compartmentation strategy** – compartment sizes, subdivision of floors, presence of fire-resisting walls and floors.
- **External walls and roofs** – materials, fire performance classifications (BS EN 13501 reaction-to-fire ratings, combustibility, spread of flame).
- **Interior finishes** – surface spread of flame ratings of linings.
- **Special structural features** – such as transfer structures, long spans, or lightweight construction, which may behave differently under fire.

This information underpins later sections of the strategy, particularly means of escape, fire protection systems, and firefighting facilities.

Why This Section Matters

The description of the building provides the **baseline understanding** needed to interpret the rest of the fire strategy. For regulators, it ensures that the measures proposed are proportionate to the risks. For fire services, it provides essential pre-incident planning information. For future building managers, it documents the assumptions that must be preserved to keep the strategy valid.

A vague or incomplete description can lead to **misinterpretation, under-protection, or regulatory rejection**, while a clear and comprehensive description builds confidence and establishes the foundation for a robust strategy.

4.3 Design Conditions and Assumptions

Every fire strategy is underpinned by a set of **design conditions and assumptions** that shape its recommendations. These assumptions must be clearly stated at the outset so that regulators, designers, building managers, and future assessors understand the **context within which the strategy was prepared**.

If not made explicit, these assumptions can become points of contention later — for example, if the actual occupant load exceeds design values, if local fire service response is slower than anticipated, or if material changes alter fire load densities. By documenting them clearly, the strategy provides a **benchmark against which future changes and fire risk assessments can be measured**.

Occupant Load

Occupancy is one of the most critical assumptions in fire strategy development.

- **Maximum numbers** – The design strategy must state the maximum number of people expected, either by regulation (e.g., floor space factors in Approved Document B or BS 9999) or by client specification.
- **Occupant distribution** – Spread of people across floors, compartments, or zones must be defined, as this affects evacuation modelling and exit sizing.
- **Types of occupants** – Age, mobility, and vulnerability (e.g., children in schools, elderly in care homes, passengers in airports).
- **Staffing levels** – Assumptions about trained staff available to lead evacuation, especially in hospitals, prisons, or assembly venues.

Why it matters: If real occupant numbers exceed assumptions, escape times may become unsafe. This is why occupant load forms a baseline for both evacuation strategy and ongoing management controls.

Fire and Rescue Service Response

The effectiveness of fire service intervention depends on **local resources and response times**. The strategy must record:

- **Proximity of nearest fire station** – and expected turnout time.
- **Appliance access** – roads, turning circles, obstructions, and hard standings.
- **Water supplies** – location and flow rates of hydrants, tanks, or alternative supplies.
- **Fire service facilities within the building** – firefighting shafts, firefighting lifts, rising mains.
- **Operational assumptions** – e.g., that the fire service can intervene within 10 minutes of alarm activation.

Why it matters: Overestimating fire service capability may lead to unsafe reliance on intervention. Under the Building Safety Act and the "design for resilience" principle, strategies must avoid assumptions that cannot be guaranteed in practice.

Fire Load Intensity and Location

The type and distribution of combustible material within a building defines the **severity of potential fires**. Assumptions must cover:

- **Material type** – e.g., office furniture, retail stock, hospital equipment, industrial chemicals.
- **Fuel load density** – expressed in MJ/m^2, often benchmarked against guidance such as Eurocodes or BS 7974 examples.
- **Compartment distribution** – whether fuel loads are evenly spread or concentrated in certain zones (e.g., warehouses, archives, plant rooms).
- **Special hazards** – kitchens, laboratories, IT server rooms, workshops, or storage of flammable liquids.

Why it matters: Fire load assumptions directly influence design fires, heat release rates, structural fire resistance, and smoke control design. If actual fuel loads exceed design assumptions, strategies may be invalidated.

Special Requirements

In addition to life safety, many buildings have **secondary objectives** that shape the fire strategy. These should be stated explicitly at the design stage:

- **Business continuity** – ensuring that critical operations can resume quickly after a fire (data centres, financial institutions, utilities).
- **Heritage protection** – safeguarding historic fabric and artefacts, often requiring enhanced suppression or compartmentation.
- **High-value assets** – museums, galleries, archives, or infrastructure with irreplaceable contents.
- **Resilience objectives** – ensuring that essential facilities such as hospitals, transport hubs, or government buildings remain operational during and after fire incidents.

Why it matters: Strategies that only address minimum life safety may be inadequate if the client or insurer requires resilience, continuity, or property protection.

Why Stating Assumptions Explicitly Matters

Documenting design conditions and assumptions serves several purposes:

- **Transparency** – regulators can see the basis for design decisions.
- **Benchmarking** – fire risk assessments during occupation can compare actual use against design assumptions.

- **Accountability** – if assumptions change (e.g., increased occupant load, altered layouts), it is clear whether the strategy requires updating.
- **Golden Thread compliance** – under the Building Safety Act 2022, assumptions must be recorded to preserve the safety case across a building's lifecycle.

Summary

This section of the fire strategy records the **assumed context** of the building: occupant load, fire service response, fire load intensity, and special objectives. By making these assumptions explicit, the strategy becomes:

- Easier to interpret at approval stage.
- More resilient to disputes.
- A clear baseline for ongoing risk assessment and management.

4.4 Compliance Basis and Deviations

A fire strategy must set out clearly the **basis of compliance** — i.e., how the proposed design satisfies the functional requirements of the Building Regulations and any other applicable standards. It must also highlight and justify any **deviations from prescriptive codes**, showing that alternative approaches achieve an equivalent or greater level of safety.

This section is critical because it provides regulators, fire authorities, insurers, and clients with a **transparent explanation of how the design complies with statutory obligations**. Without it, the strategy risks rejection, challenge, or misinterpretation.

Prescriptive Compliance

Prescriptive design remains the **most straightforward route to compliance**, and many aspects of building fire safety are fully addressed through guidance documents such as:

- **Approved Document B (ADB)** – statutory guidance for England.
- **BS 9999** – risk-based code of practice for fire safety in non-residential buildings.
- **BS 9991** – specific to residential buildings such as flats, care homes, and student housing.

When these codes are followed directly, the strategy should:

- Cite the **specific clauses, tables, or figures** used.
- Confirm that the design parameters (travel distances, door widths, fire resistance periods) meet the prescribed values.
- Reference any supporting evidence, such as manufacturers' test reports or third-party certifications.

Example: *"The escape stair width of 1,200 mm complies with the minimum width requirement of Approved Document B Volume 2, Table 3.1 for an occupancy of 200 persons."*

Minor Deviations

Even in straightforward buildings, **small departures from prescriptive guidance** often occur. These should not be hidden or minimised but instead highlighted and justified clearly.

- **Typical Minor Deviations**
    - Slightly extended travel distances.
    - Reduced door widths where historic constraints apply.
    - Small increases in compartment sizes.
    - Minor shortfalls in stair widths or lobby dimensions.
- **Justification Approaches**
    - Reference **risk-based codes** (e.g., BS 9999, which allows risk adjustments).
    - Demonstrate compensatory measures (e.g., sprinklers, additional detection, enhanced signage).
    - Provide comparative analysis to show negligible risk impact.

Example: *"Travel distance on the first floor is 21 m for office in one direction escape, exceeding the 18 m permitted under ADB Table 2.1. This is justified under BS 9999 (risk profile 2, unsprinklered), which permits 25 m under equivalent conditions."*

Alternative / Performance-Based Solutions

For complex or innovative designs, prescriptive compliance alone may be impossible. In these cases, **performance-based fire engineering** is applied, typically following **BS 7974 and its supporting documents**.

- **Methodology**
  - Define credible **fire scenarios** (fast-growing, slow-growing, ventilation-controlled, etc.).
  - Establish **design fires** (heat release rate, growth rate, fuel load).
  - Conduct **modelling and analysis**:
    - Smoke spread (zone models or CFD).
    - Evacuation (Exodus, Pathfinder, STEPS, MassMotion).
    - Structural performance (finite element modelling, parametric fire curves).
  - Compare outputs against **tenability and safety criteria**.
- **Principle of Equivalence**
  - Demonstrate that the engineered solution provides a level of safety **at least equal to** that intended by prescriptive guidance.
  - Clearly document assumptions, inputs, results, and conclusions.

Example: "A 35 m travel distance is proposed in the atrium office. Evacuation modelling (Exodus) indicates a Required Safe Egress Time (RSET) of 3.5 minutes. Smoke modelling (Smartfire) shows that tenable conditions (visibility > 10 m, temperature < 60°C) are maintained for 7 minutes (ASET), demonstrating compliance with functional requirement B1 of the Building Regulations."

Why Transparency Matters

A clear statement of compliance and deviations ensures that:

- Regulators can **verify equivalence** without unnecessary queries.
- Fire and Rescue Services understand the **operational implications**.
- Designers and contractors know exactly which standards to follow.
- Building managers and future assessors can check whether **assumptions remain valid**.

Hidden or poorly justified deviations often lead to **delays, rejections, or disputes**. Transparency not only builds trust with regulators but also strengthens the legal defensibility of the strategy.

Summary

This section of the fire strategy must:

- Identify where prescriptive compliance has been achieved.

- Clearly highlight and justify **minor deviations**.
- Provide robust evidence for **performance-based solutions**.

By adopting the principle of **equivalence** and documenting decisions transparently, the strategy becomes both **compliant and defendable**, satisfying regulators today and remaining traceable for future use.

4.5 Means of Escape

The provision of **safe means of escape** is central to life safety in every fire strategy. This section must describe in detail how occupants are expected to leave the building, or reach a place of relative safety, in the event of fire. It should also demonstrate that the chosen strategy satisfies the **functional requirement B1 of the Building Regulations**:

"The building shall be designed and constructed so that there are appropriate provisions for the early warning of fire, and appropriate means of escape in case of fire from the building to a place of safety outside the building, capable of being safely and effectively used at all material times."

This assurance is provided through a combination of **escape philosophy, physical escape routes, supporting systems, and management procedures**.

Escape Strategy

The **evacuation philosophy** must be clearly stated, as it underpins the design of exits, alarms, and management procedures. Common strategies include:

- **Simultaneous evacuation**
    - All occupants evacuate at once when the alarm sounds.
    - Most common in small-to-medium buildings such as schools, offices, retail units.
- **Phased evacuation**
    - Occupants are evacuated floor by floor, usually starting with the fire floor and those above.
    - Typical in high-rise offices, where immediate evacuation of all occupants would cause congestion.
- **Progressive horizontal evacuation (PHE)**
    - Used in hospitals and care facilities. Occupants are moved horizontally into a fire-resisting compartment, often with staff assistance, before further evacuation.

- **Stay-put policy**
    - Applied in certain residential buildings (e.g., blocks of flats), where compartmentation allows occupants not directly affected by fire to remain in their flats.
    - Requires robust fire-resisting construction, effective compartmentation, and management procedures.

The chosen strategy must be justified in relation to building type, occupancy, and fire service response.

Stair and Exit Provision

The number, type, and rating of **escape stairs and exits** must be documented:

- **Number of stairs** – determined by occupant load, building height, and redundancy requirements.
- **Fire rating** – stair enclosures typically designed to achieve 60 minutes (office buildings) or 120 minutes (tall buildings).
- **Separation** – stairs must be positioned to provide independent routes (remoteness).
- **Exit capacity** – width and capacity calculations should demonstrate compliance with ADB, BS 9999, or evacuation modelling results.
- **Final exits** – sufficient in number and width, leading to open air and clear dispersal areas.

Example: *"Three protected stairs are provided, each with 1,200 mm clear width, sufficient to accommodate the calculated occupant load of 1,800 persons (600 per stair). Enclosures are fire-resisting to 120 minutes in accordance with BS EN 13501."*

Travel Distances

Travel distance is the maximum distance an occupant must travel to reach a place of safety. Strategies should:

- Record measured distances.
- Compare against prescriptive values in ADB or BS 9999.
- Where exceeded, provide performance-based justification (e.g., by evacuation modelling, enhanced detection/sprinklers).

Example: *"Travel distance on the first floor is 22 m for office in one direction escape, exceeding ADB Table 2.1, 18 m limit for unsprinklered offices. This is justified under BS 9999 (risk profile 2) which permits 25 m, with additional mitigation provided by an L1 fire alarm system."*

Doors and Hardware

All doors on escape routes must be described, with emphasis on their fire safety function:

- **Swing direction** – doors must open in the direction of escape where occupancy exceeds 60 persons.
- **Self-closing devices** – required for fire doors forming part of a compartment or stair enclosure.
- **Panic bars and fastenings** – final exits serving large numbers should comply with BS EN 1125 (panic exit devices).
- **Vision panels and ironmongery** – to support safe egress and fire integrity.

Signage and Emergency Lighting

Escape routes must be visible and usable under emergency conditions:

- **Signage** – in accordance with BS 5499 and ISO 7010. Must be illuminated, visible even in smoke, and clearly indicate alternative routes.
- **Emergency lighting** – designed to BS 5266 and BS EN 1838, providing adequate illumination (typically 1 lux along escape routes, higher at exits and call points).
- **Wayfinding** – critical in large or complex buildings; illuminated directional signs or low-mounted lighting may be required in smoke-prone areas.

Alarm Systems

The fire detection and alarm system must be described, with reference to **BS 5839**:

- **System category** – e.g., L1 (maximum life protection), L2, L3, L4, L5.
- **Zoning** – building divided into zones for alarm identification.
- **Audibility** – minimum sound levels (65 dB(A) general, 75 dB at bedheads in sleeping areas).
- **Voice alarms** – increasingly used in large assemblies, transport hubs, and shopping centres, as they improve clarity and reduce panic compared to bells or sirens.
- **Integration** – with suppression, smoke control, lifts, and fire service notification.

Why This Section Matters

This section provides regulators with assurance that **occupants can escape safely without external assistance**, the cornerstone of UK fire safety law. It also ensures that:

- Exit capacity is matched to occupant load.
- Evacuation strategy is realistic and appropriate to the building type.
- Escape routes remain protected long enough for use.
- Supporting systems (lighting, alarms, signage) enable effective evacuation under emergency conditions.

Summary

The **means of escape** section of the fire strategy demonstrates how the building's design ensures safe evacuation. By documenting the evacuation philosophy, stair and exit provision, travel distances, door hardware, signage, lighting, and alarm systems, the strategy provides a **comprehensive and defensible account of life safety measures.**

4.6 Fire Protection Systems

A fire strategy must provide a comprehensive record of all **active and passive fire protection systems** incorporated into the building. These measures work together to prevent fire spread, control smoke, and ensure that occupants can evacuate and firefighters can intervene safely.

The description of fire protection systems should detail not only their **design intent and standards of reference**, but also their **operational interdependencies** — for example, how detection systems activate suppression, or how compartmentation supports smoke control performance. This ensures that building managers and regulators understand how the design translates into practical safety outcomes.

4.6.1 Passive Fire Protection Systems

Passive systems form the **backbone of fire safety**, controlling fire and smoke spread through robust construction measures:

- **Compartmentation**
    - Fire-resisting walls and floors divide the building into manageable compartments.

    - Sizes must comply with ADB, BS 9999, or be justified by performance-based design.
    - Special attention should be paid to **basements, atria, and large open spaces**, where compartmentation may be modified.
- **Fire doors**
    - Essential to maintain compartmentation integrity.
    - Fire resistance ratings (e.g., FD30, FD60, FD120) should be specified and aligned with the compartment boundary.
    - Doors should be self-closing, fitted with smoke seals, and tested to BS 476 or BS EN 1634.
- **Cavity barriers and fire stopping**
    - Prevent hidden fire spread within voids, cavities, and service penetrations.
    - Details should specify material type, rating, and installation standards.
- **Structural fire protection**
    - Application of fire-resistant coatings, sprays, boards, or concrete encasement to maintain stability.
    - Fire resistance periods (30, 60, 90, 120 minutes) must be consistent with strategy requirements.
    - Should reference relevant standards, e.g., BS EN 13381 for fire resistance testing.

4.6.2 Active Fire Protection Systems

Active systems are designed to **detect, suppress, or control fire development**:

- **Automatic sprinkler systems**
    - Designed to BS EN 12845 (commercial) or BS 9251 (residential).
    - Provide rapid fire suppression or control, protecting occupants, structure, and business continuity.
    - The strategy should specify system type (e.g., wet pipe, dry pipe, pre-action), coverage areas, and design density.

- **Water mist systems**
  - Used in sensitive environments (e.g., heritage, IT rooms, museums) where water damage must be minimised.
  - Standards: BS 8458 (domestic/residential) and BS 8489 (commercial/industrial).
- **Gaseous suppression systems**
  - CO_2, inert gases (IG-55, IG-541), or chemical agents (FM-200, Novec 1230).
  - Applied in critical spaces such as server rooms, control centres, archives.
  - Must comply with BS EN 15004.
- **Smoke and heat control systems**
  - Designed to BS EN 12101 suite of standards.
  - Includes smoke vents (natural or powered), smoke curtains, pressurisation of stair cores, and extraction fans.
  - Must ensure smoke is managed to keep escape routes tenable and assist firefighting.

4.6.3 Detection and Alarm Systems

Detection and alarm systems are the **trigger mechanism** for many other fire safety provisions. The strategy should define:

- **Category of system** – in accordance with BS 5839 (e.g., L1 for maximum life protection, L2/L3 for partial coverage, L4/L5 for specific risks).
- **Types of detectors** – smoke, heat, flame, aspirating (VESDA), or multi-sensor.
- **Zoning and audibility** – alarms must be audible above background noise levels; voice alarms should be considered for large public buildings.
- **Integration** – alarms must interface with sprinklers, smoke control, lifts, and fire service alerting.
- **Redundancy and reliability** – dual power supplies, monitoring, and fault reporting.

4.6.4 Linking Design to Management

It is not enough for a fire strategy to list the systems; it must also explain:

- **Maintenance obligations** – e.g., weekly fire alarm tests, monthly emergency lighting checks, annual sprinkler servicing.
- **Management responsibilities** – the "Responsible Person" (as defined by the Regulatory Reform Fire Safety Order) must ensure systems remain functional.
- **Golden Thread compliance** – records of systems must be kept updated throughout the building lifecycle, forming part of the digital safety case.

Summary

This section of the fire strategy records the **layers of protection** provided by passive and active systems:

- **Passive protection** (compartmentation, doors, barriers, structural stability) prevents uncontrolled spread.
- **Active protection** (sprinklers, suppression, smoke control) limits fire growth and assists evacuation.
- **Detection and alarm systems** ensure rapid response and integration with other measures.

By linking **design intent** to **management requirements**, the fire strategy ensures these systems remain effective throughout the building's life.

4.7 Firefighting Access and Facilities

A fire strategy must address the **operational needs of the Fire and Rescue Service (FRS)**. Even the most carefully designed escape and protection systems assume that firefighters will ultimately attend, contain, and extinguish the fire. If a building does not provide the necessary facilities for safe intervention, occupants may be placed at higher risk and property damage may be severe.

This section therefore sets out how the building supports firefighting operations, from external access for appliances to internal firefighting facilities and information systems.

4.7.1 Appliance Access

- **Vehicle access roads** – In accordance with Approved Document B (ADB) and Manual for Fire Service Vehicles, buildings must provide access routes of sufficient width, gradient, and strength for fire appliances.

- **Turning facilities** – Turning circles or hammerheads may be required where reversing distances exceed 20 m.
- **Hard standings** – Load-bearing capacity must accommodate high-axle-weight appliances; surface finishes should remain usable in all weather conditions.
- **Access to all facades** – Particularly important for tall or large buildings, where aerial appliances may be needed for firefighting or rescue.

Example: *"A 3.7 m wide access road encircles the building, with turning facilities at the north and south ends. Hard standings at the east façade are designed to support 17 tonne appliances, providing aerial ladder access to upper floors."*

4.7.2 Hydrants and Water Supplies

- **Proximity** – Hydrants should typically be within 90 m of fire appliance access points.
- **Flow rates** – Supplies must meet local water authority requirements, often 1,500 – 2,000 litres/min for commercial sites.
- **Redundancy** – For large or high-risk developments, multiple hydrants or on-site water storage (tanks, lagoons, reservoirs) may be required.
- **Alternative supplies** – In rural or remote sites, static tanks or dry hydrants may substitute for mains supplies.

Example: *"Two hydrants are located on the west service road, providing a combined flow of 2,200 litres/min at 1 bar. An additional 90,000-litre static water tank is provided for redundancy."*

4.7.3 Internal Firefighting Facilities

Modern buildings must provide internal facilities that allow firefighters to intervene effectively, especially in high-rise or complex layouts:

- **Firefighting shafts** – Consist of a protected stair, firefighting lobby, and rising main; required in tall buildings or those with deep basements.
- **Firefighting lifts** – Provided within shafts where floors are above 18 m or below 10 m from fire service access level, enabling crews to reach the fire quickly with equipment.

- **Rising mains** – Dry or wet risers installed in shafts to allow hose connection on each floor.
- **Access lobbies** – Provide smoke separation between stair and accommodation, allowing safe hose deployment.

Example: *"Two firefighting shafts are located in the north and south cores. Each contains a 1,100 mm stair, 60 m/min firefighting lift, and wet riser with outlets on every floor."*

4.7.4 Smoke Control for Firefighters

Smoke is the greatest hazard to firefighting crews. The strategy must detail systems provided to maintain tenable conditions:

- **Pressurised stairwells** – Maintain a smoke-free escape and firefighting route.
- **Firefighting lobbies** – Serve as safe bridgeheads for hose deployment.
- **Smoke extraction systems** – Natural or mechanical smoke ventilation in basements, atria, or car parks.
- **Automatic controls** – Systems should be clearly linked to the fire alarm for reliable activation, with manual override available to the FRS.

Example: *"Pressurisation fans maintain a positive pressure of 50 Pa in firefighting stairs. Basement smoke extract fans achieve 10 air changes per hour, maintaining visibility above 10 m for 20 minutes."*

4.7.5 Information for Fire Crews

The effectiveness of firefighting is enhanced by clear and reliable building information:

- **Plans and signage** – Firefighting access plans, riser locations, hydrants, and escape routes should be displayed at entrances (e.g., "premises information plates").
- **Control rooms / panels** – Fire alarm panels, sprinkler control valves, and smoke control panels must be readily accessible and clearly labelled.
- **Pre-incident information (PII)** – High-risk or unusual sites should provide information packs to the local FRS, covering hazards (chemicals, fuel storage), layouts, and emergency contacts.

Example: *"The main entrance contains a Premises Information Plate (PIP) identifying firefighting shafts, hydrants, and control panels. A secure control room adjacent to reception houses alarm, smoke control, and sprinkler panels."*

Why This Section Matters

Without adequate firefighting access and facilities:

- Occupant evacuation may be compromised if fire crews cannot intervene quickly.
- Fires may spread beyond their compartment of origin.
- Firefighters may be exposed to unacceptable risks.
- Insurance costs and liability exposure may increase.

A clear and detailed account of firefighting provisions provides confidence that the building can be safely managed in the event of fire.

Summary

This section of the fire strategy demonstrates how the building supports safe and effective firefighting by providing:

- **External access** for appliances.
- **Reliable water supplies** for suppression.
- **Internal facilities** such as shafts, lifts, and rising mains.
- **Smoke control measures** to maintain safe conditions.
- **Building information** for operational crews.

By addressing these needs, the strategy maximises both the safety of occupants and the protection of property.

4.8 Management and Maintenance

A fire strategy is only effective if it is **properly managed, maintained, and periodically reviewed**. Design measures and engineered systems can only achieve their intended purpose if occupants and building managers understand their role, procedures are rehearsed, and fire protection measures are kept in full working order.

This section of the strategy must therefore set out **management responsibilities, maintenance obligations, and procedures for review**, ensuring the document remains valid for the lifetime of the building.

4.8.1 Emergency Procedures

Clear and rehearsed emergency procedures are vital to ensure effective evacuation when fire occurs. The fire strategy should reference:

- **Fire drills** – Regular drills must be carried out, with records maintained. Frequency should reflect building type (e.g., at least annually for residential blocks, quarterly or more for workplaces and schools).
- **Staff training** – Training should cover alarm response, evacuation leadership, fire extinguisher use, and assisting vulnerable persons.
- **Personal Emergency Evacuation Plans (PEEPs)** – Individual plans for occupants who may need assistance, such as wheelchair users or those with sensory impairments.
- **Visitor management** – Systems must account for contractors, visitors, and others unfamiliar with the premises.
- **Coordination with Fire and Rescue Service** – Building management should ensure crews are familiar with site layout and key systems (e.g., during familiarisation visits).

4.8.2 Maintenance Schedules

Fire protection measures are only reliable if they are maintained to relevant standards. The fire strategy should set out minimum requirements for inspection, testing, and servicing:

- **Fire alarms and detection systems** – tested weekly (user test) and serviced every six months, per BS 5839.
- **Emergency lighting** – monthly functional tests and annual duration tests, per BS 5266.
- **Sprinkler systems** – weekly pump checks, quarterly inspections, and annual servicing, per BS EN 12845 / BS 9251.
- **Smoke control systems** – routine inspections and annual testing in accordance with BS EN 12101.
- **Portable extinguishers** – annual servicing in accordance with BS 5306.
- **Fire doors and compartmentation** – regular inspection to ensure self-closing devices, seals, and integrity remain intact.

A **maintenance log** should be kept on site (and digitally under the "golden thread" of information), available for inspection by regulators and insurers.

4.8.3 Management Responsibilities

Under the **Regulatory Reform (Fire Safety) Order 2005**, every non-domestic building must have a clearly identified **Responsible Person** (RP). This role must be explicitly identified within the fire strategy, alongside any delegated duties.

- **Responsible Person** – usually the employer, building owner, or managing agent. Accountable in law for fire safety management.
- **Duty holders** – e.g., facilities managers, fire wardens, health and safety officers.
- **Training and competence** – managers must ensure that those with fire safety responsibilities are trained and competent to carry out their roles.
- **Record-keeping** – evidence of drills, maintenance, and reviews must be retained to demonstrate compliance.

4.8.4 Updating the Fire Strategy

A fire strategy should be a **dynamic, living document**. It must not be filed away after completion but regularly reviewed and updated when conditions change. Triggers for review include:

- **Refurbishment or alterations** – any works that affect escape routes, compartmentation, fire loads, or protection systems.
- **Change of use** – new occupancies (e.g., converting offices to residential) or altered occupant profiles.
- **Significant incidents** – after a fire, near-miss, or serious false alarm.
- **Regulatory updates** – to reflect changes in Building Regulations, British Standards, or enforcement guidance.
- **Periodic review** – even without changes, the strategy should be reviewed at least every 3–5 years to ensure ongoing validity.

All revisions should be recorded in a **version control log**, ensuring traceability and accountability.

Why This Section Matters

History shows that **many fire tragedies arise not from poor initial design but from failures in management and maintenance**. Grenfell Tower (2017) and Lakanal House (2009) both highlighted how poorly maintained fire protection systems and inadequate management can have catastrophic consequences.

By embedding management and maintenance requirements into the fire strategy, the document becomes a **living framework** that continues to safeguard occupants, protect assets, and meet regulatory duties throughout the building lifecycle.

Summary

This section ensures the fire strategy does not end at design approval but continues into operation and long-term management. It documents:

- **Emergency procedures** – drills, training, PEEPs, and coordination with the FRS.
- **Maintenance schedules** – inspection and servicing of alarms, lighting, sprinklers, smoke control, extinguishers, and doors.
- **Management responsibilities** – identification of the Responsible Person and duty holders.
- **Updating protocols** – ensuring the strategy remains current after changes or incidents.

By linking **design intent** with **operational management**, the fire strategy is transformed into a sustainable, enforceable tool for safety.

4.9 Appendices and Supporting Evidence

The appendices form the **technical backbone** of a fire strategy. They provide the **supporting data, calculations, and documentation** that validate the conclusions of the main report. Without this evidence, the fire strategy risks being seen as a narrative rather than a defensible, engineering-based document.

This section ensures transparency, traceability, and confidence, allowing regulators, fire authorities, insurers, and building managers to review the underlying justification for each fire safety provision.

4.9.1 Floor Plans and Fire Drawings

Clear, accurate drawings are essential to illustrate the provisions described in the strategy. They should:

- Be to scale, legible, and consistent with the latest architectural issue.
- Identify **escape routes, stairs, exits, refuges, and assembly points**.
- Show the location of **compartment lines, fire doors, and cavity barriers**.
- Indicate **active systems** – alarm zones, sprinkler heads, smoke vents, firefighting shafts, rising mains, hydrants.
- Include **legends and north points** for clarity.

Fire strategies without aligned drawings are extremely difficult to interpret and risk rejection at regulatory stage.

4.9.2 Test Certificates and Product Data

Evidence of fire performance for materials, products, and systems must be included:

- **Fire resistance test reports** (to BS 476, BS EN 1364, BS EN 1634, or equivalent).
- **Reaction-to-fire classifications** (to BS EN 13501 series).
- **Third-party certifications** (e.g., LPCB, UL, BSI Kitemark).
- **Product data sheets** – confirming installation standards, compatibility, and limitations.

Including these certificates provides assurance that materials and systems are not only designed but also **proven under test conditions**.

4.9.3 Calculations and Analytical Evidence

For both prescriptive and performance-based approaches, supporting calculations should be appended:

- **Occupant load calculations** – based on floor area and use.
- **Exit capacity and travel distance calculations** – demonstrating compliance with ADB, BS 9999, or alternative justifications.
- **Structural fire resistance calculations** – using Eurocode methods or tabulated data.
- **Smoke control calculations** – reservoir sizing, vent areas, fan duty points.

These ensure that the numerical basis for design decisions is transparent and reproducible.

4.9.4 Modelling Results

Where performance-based analysis has been applied, outputs from engineering tools should be included:

- **CFD smoke modelling** (e.g., SmartFire, FDS) – graphical and tabular results showing temperature, visibility, CO concentration, and smoke layer interface.
- **Evacuation modelling** (e.g., Exodus, Pathfinder, STEPS) – movement times, congestion points, ASET vs RSET comparisons.
- **Structural analysis** (e.g., SAFIR, ANSYS) – demonstrating stability under fire exposure.

- **Risk assessments or QRA outputs** – where probabilistic methods have been used.

These results should be **summarised in the main body** of the strategy but included in full in the appendices for traceability.

4.9.5 Version Control Log

A robust fire strategy must include a **version history** documenting when and why changes were made:

- Date of issue.
- Revision number.
- Nature of amendment (e.g., updated drawings, revised occupant load).
- Author and approver.

This ensures that the document remains **traceable over time**, supporting the "golden thread" of information required by the Building Safety Act 2022.

4.9.6 Optional Supporting Documents

Depending on project type, additional appendices may include:

- **Fire risk assessments** (for existing buildings).
- **Emergency procedures** – evacuation policies, staff training records, PEEPs.
- **Correspondence with authorities** – fire service consultation responses, planning approvals, insurer requirements.
- **Heritage or conservation reports** – where special protection applies.

Why This Section Matters

The appendices are not "extras" but a **core part of the fire strategy**. They:

- Provide the **evidence base** for the conclusions in the main report.
- Enable regulators and insurers to verify compliance independently.
- Serve as a **reference library** for building managers during operation and maintenance.
- Ensure that future updates or refurbishments can be benchmarked against original design intent.

A fire strategy without robust appendices risks being viewed as **opinion** rather than **defensible analysis**.

Summary

This section of the fire strategy ensures that all claims made in the document are supported by **hard evidence**. Appendices typically include:

- **Fire drawings and floor plans**.
- **Certificates and product data**.
- **Calculations and modelling outputs**.
- **Version control logs**.

Together, these documents form the **technical backbone** of the fire strategy, transforming it from a descriptive narrative into a robust, transparent, and defensible framework for fire safety.

4.10 Summary

The structure of a fire strategy document is not arbitrary. Each section plays a defined role in:

- Establishing accountability.
- Describing the building context.
- Stating design assumptions.
- Demonstrating compliance and justifying deviations.
- Documenting escape, protection, and firefighting provisions.
- Outlining management responsibilities.

By following this structure, fire strategies become **clear, defensible, and practical documents** that serve as both a regulatory submission and a long-term management tool.

Each section plays a vital role in demonstrating how legal obligations are met and how safety is achieved in practice.

Chapter 5
Description of the Building

5.1 Introduction

The starting point of any fire strategy is a clear, accurate, and comprehensive **description of the building**. This establishes the context for all subsequent fire safety decisions. Without a robust description, it is impossible to judge the suitability of fire protection measures, evacuation provisions, or firefighting facilities.

This section should capture the **location, layout, construction, use, and occupancy profile** of the building. It acts as the "scene-setting" chapter that underpins every technical and management consideration in the fire strategy.

5.2 Location and Site Context

The **location of a building within its site** is a fundamental determinant of its fire safety provisions. The arrangement of surrounding access roads, adjacent structures, water supplies, and site-specific hazards directly influences the design and effectiveness of fire strategies. Failure to adequately consider site context can undermine otherwise well-designed internal fire protection measures.

5.2.1 Access for Fire and Rescue Services

- **Road Access and Hardstandings**
    - Fire appliances must be able to approach within prescribed distances of principal entrances, firefighting shafts, and rising main inlets.
    - Hardstanding areas must be designed to bear the weight of appliances (typically up to **26 tonnes gross vehicle weight**).
    - **Turning circles or hammerheads** are required for appliances in dead-end access roads.

- **Appliance Staging Areas**
    - Sufficient clear space must be provided for aerial appliances where high-rise access is necessary.
    - Clear signage and control of obstructions (bollards, street furniture, landscaping) must be enforced.
- **Proximity to Water Supplies**
    - Hydrants should normally be located **within 90 m of building entrances** (ADB guidance).
    - In remote or rural sites, on-site water tanks and fire pumps may be necessary.
    - Redundancy in water supply should be considered for high-value or high-occupancy buildings.

5.2.2 Proximity to Other Buildings

- **Boundary Distances**
    - Building separation distances are critical in limiting external fire spread. Guidance is provided in ADB, BS 9999, and BS EN 1991-1-2 (Eurocodes).
    - Where buildings are closer than prescribed separation distances, additional façade fire resistance or automatic suppression may be required.
- **Neighbouring Risks**
    - Fire spread across boundaries has historically been a major contributor to catastrophic urban fires.
    - The fire strategy should identify combustible external claddings, large, glazed areas, or openings facing adjacent properties that may accelerate fire transfer.

5.2.3 Site Features and Hazards

- **Fuel Stores and Hazardous Materials**
    - LPG cylinders, fuel depots, or chemical storage increase the hazard level and may require enhanced separation or blast-resistant design.

- **Car Parks**
  - Car parks, particularly multi-storey structures, represent concentrated fire loads due to fuel tanks and plastics.
  - Fire strategy should consider smoke extraction, ventilation, and compartmentation between car parks and occupied spaces.
- **Vegetation and Wildfire Risk**
  - In rural or peri-urban areas, wildfire risk should be considered, especially with climate change increasing severity of seasonal fires.
  - Fire breaks, clear perimeters, and defensible zones may be required.
- **Topography**
  - Sloping sites or raised terraces may complicate fire appliance access.
  - Cuttings, embankments, or restricted bridge loads can limit the routes available to the fire service.

5.2.4 Security, Assembly, and Business Continuity Considerations

- **Assembly Points**
  - Safe, clearly marked assembly points must be shown on the site plan, located far enough away to avoid exposure to radiant heat or falling debris.
- **Access Control and Security**
  - Gated developments must provide fire service override facilities to prevent delays.
- **Business Continuity**
  - For critical infrastructure, the location and resilience of site features (e.g., backup generators, IT hubs) should be included in fire strategy documentation.

5.2.5 Documentation and Best Practice

Best practice is to include a **site plan** in the fire strategy showing:

- The **building footprint**.
- Firefighting **access routes and hardstandings**.

- Location of **hydrants, tanks, or water supplies**.
- **Assembly points** for occupants.
- Neighbouring structures and boundary conditions.

Such documentation provides clarity to regulators, designers, and fire services, reducing disputes during approval and ensuring operational readiness in an emergency.

Summary

The site context determines whether the fire service can intervene effectively and whether external fire spread risks are adequately mitigated. By assessing **access, proximity to other buildings, hazards, and site features**, and documenting these on a clear plan, the fire strategy provides a robust foundation for life safety, property protection, and business resilience.

5.3 Building Form and Dimensions

The physical form and dimensions of a building establish the **fundamental parameters for fire safety analysis**. By documenting storeys, heights, basements, voids, and areas, the fire strategy creates the baseline against which compliance with Building Regulations, British Standards, and performance-based fire engineering can be assessed. These characteristics influence fire resistance requirements, means of escape, structural resilience, and firefighting access.

5.3.1 Number of Storeys

- **Above Ground**
    - The total number of storeys above ground determines fire safety provisions such as:
        - **Escape strategy** (e.g., phased evacuation in high-rise vs simultaneous in low-rise).
        - **Firefighting shafts and lifts** (mandatory above 18 m floor height).
        - **External fire service access** (aerial appliances limited to ~30 m reach).
    - Tall buildings (over **30 m**) require enhanced structural fire resistance, robust smoke control, and redundancy in escape routes.

Below Ground

- Basements present unique challenges due to lack of natural light, limited ventilation, and complex firefighting access.
- **Basements deeper than 10 m** typically require mechanical smoke ventilation and firefighting lifts.
- Special risks include car parks (fuel and smoke hazards) or plant rooms (electrical and gas services).

5.3.2 Basements

- The fire strategy must record:
  - **Number and depth** of basements.
  - **Use and occupancy** (e.g., storage, parking, leisure facilities).
  - **Connection to upper storeys** (protected stair cores, lifts, escalators).
- Requirements include:
  - Smoke ventilation compliant with **BS EN 12101**.
  - Fire resistance to prevent fire spread into upper floors.
  - Separate means of escape where basement occupancy exceeds prescribed limits.

5.3.3 Attics and Roof Spaces

- Roof voids, attics, or lofts must be clearly described, as they often conceal **hidden fire spread pathways**.
- **Habitable spaces** (e.g., loft conversions) require:
  - Protected stairs.
  - Escape windows or alternative exits.
- **Storage voids** may need cavity barriers and fire-stopping.
- Roof constructions incorporating **timber trusses or combustible insulation** require special consideration due to rapid fire spread risk.

5.3.4 Floor-to-Floor Heights

- Floor-to-floor and floor-to-ceiling heights directly affect:
  - **Evacuation modelling** (travel speeds on stairs are influenced by stair riser height).

  - **Firefighting access** (equipment reach and hose length limitations).
  - **Smoke layer stratification** in large spaces or atria.
- High floor-to-floor heights may require intermediate refuges, additional staircases, or enhanced smoke control.

5.3.5 Total Floor Area

- Both **gross internal floor area (GIFA)** and **footprint area** must be recorded.
- Implications:
  - Large floor plates increase **travel distances** to exits.
  - Fire resistance requirements scale with floor area and compartment size.
  - Sprinkler systems may become mandatory (e.g., sprinklered compartments >2,000 m² for shops, per ADB).
- For mixed-use developments, each functional area (e.g., retail, residential, office) must be broken down separately.

5.3.6 Interaction with Regulations

- The documented building form and dimensions inform:
  - **Part B of the Building Regulations** – structural fire resistance, travel distances, separation distances.
  - **BS 9999/9991** – escape capacity calculations based on occupancy and risk profiles.
  - **BS 7974 analyses** – input parameters for evacuation and smoke modelling.
- Accurate data at this stage ensures that subsequent calculations and justifications are defensible.

Summary

A clear description of **building form and dimensions** is essential for every fire strategy. By documenting storeys, basements, roof spaces, heights, and floor areas, the strategy establishes the physical framework that drives decisions on:

- **Fire resistance levels**.

- **Means of escape and evacuation times.**
- **Compartmentation and suppression systems.**
- **Firefighting access and intervention facilities.**

Without this baseline information, neither prescriptive compliance nor performance-based design can be robustly justified.

5.4 Occupancy and Use

The **intended use and occupancy characteristics of a building** are critical factors in fire safety design. They determine the level of fire risk, the evacuation strategy, and the type of active and passive fire protection measures required. A robust fire strategy must therefore clearly document the building's **primary use, mixed uses, high-risk areas, and occupant characteristics.**

5.4.1 Primary Use

The fire strategy should classify the building according to its primary function, as defined by **Approved Document B, BS 9999, BS 9991, and international codes such as NFPA 101**:

- **Residential** – houses, flats, care homes, student residences.
- **Commercial** – offices, retail stores, shopping centres.
- **Industrial** – manufacturing facilities, warehouses, logistics hubs.
- **Healthcare** – hospitals, clinics, care homes.
- **Educational** – schools, universities, nurseries.
- **Assembly and Recreation** – theatres, cinemas, stadiums, leisure centres.
- **Storage and Other** – archive facilities, cold stores, IT/data centres.

Each category carries unique fire risks: for example, **residential buildings** must prioritise night-time evacuation, while **industrial facilities** may require enhanced containment of hazardous processes.

5.4.2 Mixed Uses

Modern developments often combine multiple occupancies within a single structure.

- **Examples**:
    - Retail or commercial units at ground level with residential apartments above.

  - Transport hubs incorporating shopping centres, offices, and hotels.
- **Implications**:
  - Mixed uses increase **complexity of evacuation** — different groups may require different strategies (e.g., phased vs simultaneous).
  - **Fire separation** between occupancies is critical (ADB requires compartmentation between different use classes).
  - Shared services (ducts, lifts, atria) must be carefully protected to prevent cross-occupancy fire spread.

5.4.3 High-Risk Areas

Certain areas present higher ignition potential, fuel load, or operational hazard. These must be explicitly identified in the fire strategy:

- **Kitchens** – grease and oil-based fires with high flame spread potential.
- **Laboratories** – presence of flammable chemicals, pressurised gases, or specialist equipment.
- **Plant rooms** – electrical switchgear, boilers, fuel storage.
- **IT/data rooms** – high fire load density in cabling and electronics, requiring suppression such as gaseous systems.
- **Car parks** – concentrations of fuel and plastics within vehicles, Electric Vehicle Charging Points.

Design measures for such areas may include **automatic suppression**, enhanced **compartmentation**, and **specialist detection systems**.

5.4.4 Occupant Characteristics

Understanding who occupies the building, and how they use it, are essential to safe evacuation. The fire strategy must describe:

- **Maximum Occupant Load**
  - Calculated according to relevant guidance (e.g., persons per m^2 in ADB or BS 9999).
  - For example, office accommodation may use **1 person per 6 m^2**, while assembly areas may assume **1 person per 0.5 m^2**.

- **Vulnerable Occupants**
    - Presence of elderly, disabled, children, or those with reduced mobility.
    - Requirement for **refuges, evacuation lifts, and Personal Emergency Evacuation Plans (PEEPs)**.
- **Patterns of Occupancy**
    - **Continuous** (e.g., residential, healthcare).
    - **Intermittent** (e.g., schools, theatres).
    - **24/7 operation** (e.g., hospitals, airports, transport hubs).
    - Recognition of **peak loads** (e.g., shopping centres at weekends, stadiums during events).

5.4.5 Implications for Fire Strategy Design

The documented occupancy and use data directly influence:

- **Evacuation planning** – whether simultaneous, phased, or progressive horizontal strategies are required.
- **Active protection systems** – sprinklers, gaseous suppression, detection type and zoning.
- **Passive protection** – fire resistance levels, compartment sizes, and boundary conditions.
- **Management and training** – staff procedures, drills, and responsibilities.

Summary

The **occupancy and use profile** of a building defines its fire risk. A well-prepared fire strategy must identify:

- The building's **primary function**.
- Any **mixed uses** and their interactions.
- **High-risk areas** requiring enhanced measures.
- The **number, type, and vulnerability** of occupants.
- Typical and peak **patterns of use**.

Only with this information can evacuation plans, fire protection systems, and management procedures be tailored to provide **safe, effective, and inclusive fire safety solutions**.

5.5 Construction and Materials

The construction method and materials of a building have a direct impact on its **fire resistance, compartmentation integrity, and potential for fire spread.** A fire strategy must therefore provide a clear description of the structural frame, floors, roofs, walls, and finishes, together with the compartmentation strategy that underpins life safety and property protection.

5.5.1 Structural Frame

- **Types of frame**:
  - **Steel** – requires applied fire protection (spray coatings, board encasement, intumescent paint) to achieve resistance periods of 30, 60, 90, or 120 minutes as required by Building Regulations. Susceptible to rapid heating and loss of strength if unprotected.
  - **Reinforced Concrete** – inherently fire resistant, but spalling risk must be addressed. Often used in high-rise buildings to provide robustness.
  - **Timber** – increasingly common in modern methods of construction (MMC). Structural timber chars predictably, but combustibility raises additional concerns in high-rise and residential use. CLT (cross-laminated timber) requires careful consideration of compartmentation and fire testing data.
  - **Hybrid or Composite** – steel–concrete composites, or mixed timber and concrete solutions, which require combined fire performance assessments.
- **Fire resistance rating of primary elements**:
  - Required minimums set out in **Approved Document B (ADB), BS EN 1992/1993 (Eurocodes), and BS EN 13501-2**.
  - High-rise or high-risk occupancies typically require **120 minutes resistance** for loadbearing structures.

5.5.2 Floors and Roofs

- **Floors**:
  - **Concrete slabs** – robust fire resistance, good for multi-storey buildings.

- **Timber joists** – common in residential properties; may require upgrading (e.g., plasterboard protection).
- **Composite decking** – widely used in commercial developments; fire performance depends on slab thickness and protection of steel decking.

- **Roofs**:
  - **Flat roofs** – may incorporate insulation layers; combustibility of roof build-ups must be assessed against **BS EN 13501-5 (roof external fire performance)**.
  - **Pitched/trussed roofs** – timber trusses can lead to rapid collapse; cavity barriers essential to limit fire spread.
  - **Green roofs** – may include combustible planting layers; guidance requires control of fire breaks and irrigation systems to reduce spread.

5.5.3 External Walls

- **Materials and cladding systems**:
  - External wall performance is a **critical issue post-Grenfell Tower (2017)**.
  - Systems must comply with **Regulation 7(2)**, requiring materials in relevant buildings (high-rise residential, hospitals, care homes, student housing) to be **A2-s1,d0 or better (non-combustible)**.
  - Systems must be tested to **BS 8414** where performance-based justification is sought.
- **Risk of fire spread**:
  - Combustible cladding, inappropriate cavity barriers, or poor workmanship can enable vertical fire spread.
  - Windows, balconies, and service penetrations require robust detailing to prevent breaches.

5.5.4 Internal Linings and Finishes

- **Surfaces** on walls, ceilings, and floors must be controlled to limit fire growth rate.
- **Reaction-to-fire classification** is governed by **BS EN 13501-1**:

- Class A1/A2 = non-combustible (e.g., plasterboard, stone, concrete).
- Class B–D = limited to medium combustibility (treated timber, some plastics).
- Class E–F = high combustibility (untreated timber, textiles, foams).

- **ADB provisions**:
  - Escape routes must use Class 0 (National Class) or Class B-s3,d2 (European Class) linings.
  - Assembly and residential buildings require stricter limits.

5.5.5 Compartmentation Strategy

Compartmentation is the **primary defence against fire and smoke spread**, maintaining escape route tenability and containing damage.

- **Vertical separation**: fire-resisting floors between storeys.
- **Horizontal separation**: compartment walls between flats, retail units, or industrial areas.
- **Fire resistance**: typically, 30–120 minutes depending on use and building height.
- **Cavity barriers**: required in concealed spaces (e.g., roof voids, wall cavities) to prevent unseen fire spread.
- **Special occupancies**: higher resistance often required for plant rooms, data centres, or hazardous material stores.

5.5.6 Documentation

The fire strategy should include:

- A **technical description** of materials and build-ups.
- Fire test reports, classification certificates (BS EN 13501), or third-party certification.
- Compartmentation drawings highlighting walls, floors, and cavity barrier lines.

Summary

The **construction and materials** of a building define how it will resist and contain fire. A well-prepared fire strategy must describe:

- The **structural frame** and its fire resistance.
- The performance of **floors, roofs, and external walls**.
- The **reaction-to-fire classification of internal linings**.
- The **compartmentation strategy**, including cavity barriers and fire-resisting separations.

This information provides the technical foundation for assessing compliance with the Building Regulations, determining evacuation safety, and protecting both occupants and property.

5.6 Fire Resistance Classification

The classification of fire resistance is central to demonstrating compliance with the Building Regulations and ensuring that structural stability, compartmentation, and escape route integrity are maintained during fire. Fire resistance is expressed in terms of **duration (minutes)** for which elements of construction can withstand standardised fire exposure while maintaining:

- **Load-bearing capacity (R)**,
- **Integrity against flames/smoke passage (E)**, and
- **Thermal insulation (I)**.

The fire strategy must specify these requirements for each critical element of the building.

5.6.1 Load-bearing Structure

- **Purpose**: Prevent premature collapse during evacuation and firefighting operations.
- **Typical requirements (ADB)**:
    - Buildings up to 5 storeys: **30–60 minutes**.
    - Medium-rise (6–18 m): **60 minutes**.
    - High-rise (>18 m): **90–120 minutes**, depending on use.
- **Considerations**:
    - Car parks, storage, or industrial uses may require enhanced ratings due to higher fire loads.

  - Structural fire engineering (to **BS EN 1991-1-2** and **BS 7974-structural response**) can justify alternative ratings through performance-based design.

5.6.2 Compartment Floors and Walls

- **Purpose**: Contain fire and smoke within defined areas to protect escape routes and limit fire spread.
- **Requirements**:
  - **Between storeys**: typically **60–120 minutes** depending on height and occupancy.
  - **Between flats, retail units, or occupancies**: minimum **60 minutes**.
  - **Separating high-risk areas (e.g., plant rooms, kitchens)**: often **120 minutes**.
- **Cavity barriers and fire-stopping** must supplement walls and floors to maintain compartment integrity.

5.6.3 Doorsets and Glazing

- **Fire Doors**: Classified under **BS EN 1634-1** and denoted as FD30, FD60, FD90, or FD120.
  - **FD30S** = 30 minutes with smoke control.
  - **FD60S** = 60 minutes with smoke control.
- **Glazing**: Must be fire-resisting and tested to the same standards; typically used in corridors, lobbies, and stair enclosures.
- **Considerations**:
  - Doors in escape routes should be **self-closing** and fitted with **intumescent and smoke seals**.
  - Vision panels are acceptable if tested as part of the doorset.

5.6.4 Escape Routes

- **Objective**: Maintain tenability for the full evacuation period.
- **Requirements**:
  - Protected corridors and lobbies: **30–60 minutes** fire resistance.

    - Protected stairs: **60–120 minutes**, especially in high-rise or complex buildings.
    - Pressurisation and smoke control systems may reduce reliance on longer durations, but redundancy is essential.
- **Design principle**: Fire resistance must **exceed the evacuation time (RSET)**, providing a **safety margin**.

5.6.5 Derivation of Classifications

Fire resistance classifications may be determined by three approaches:

1. **Prescriptive (Approved Document B)**
    - Specifies minimum fire resistance based on height, use, and compartmentation.
    - Widely used for standard designs.
2. **Risk-Based (BS 9999)**
    - Adjusts fire resistance periods depending on occupancy risk profile, management quality, and building features.
    - Can reduce or increase requirements compared to ADB.
3. **Performance-Based (BS 7974)**
    - Uses fire engineering methods and scenario modelling to justify fire resistance.
    - *Example*: Using time–temperature curves (parametric fires) instead of the standard fire curve to evaluate real performance.

5.6.6 Documentation in the Fire Strategy

The strategy should include:

- A **schedule of fire resistance ratings** for all elements (structure, floors, walls, doors, glazing, escape routes).
- References to **test certificates, assessments, or third-party accreditation**.
- Justifications where performance-based or risk-based approaches deviate from prescriptive guidance.

Summary

Fire resistance classifications ensure that a building's **structure remains stable**, **compartments contain fire**, and **escape routes remain usable** for the duration of evacuation. By clearly setting out ratings for structure, floors, walls, doors, glazing, and escape routes — and showing whether these are derived from ADB, BS 9999, or BS 7974 analysis — the fire strategy provides a **transparent, defensible foundation** for compliance and safety.

5.7 Integration with Drawings and Appendices

A fire strategy must be supported by **drawings, schedules, and technical appendices** that illustrate and evidence the written content. Fire safety is inherently spatial, and without graphical representation, textual descriptions can be misinterpreted or lose detail during later stages of design, construction, or operation.

Drawings form a **bridge between the written strategy and physical reality**, ensuring that all stakeholders — from architects and engineers to contractors, regulators, and the Fire and Rescue Service — can interpret and apply the strategy consistently.

5.7.1 Essential Drawings

- **General Arrangement Plans (GA Drawings)**
  - Show the overall building layout (floor plans, elevations, sections).
  - Indicate room functions, occupancy types, and circulation routes.
  - Provide baseline context for interpreting fire strategy provisions.
- **Fire Compartmentation Drawings**
  - Highlight compartment walls, floors, cavity barriers, and protected shafts.
  - Differentiate fire resistance levels (e.g., 30, 60, 120 minutes).
  - Critical for demonstrating compliance with **Building Regulations Part B** and ensuring design intent is carried through to construction.

- **Escape Route Plans**
  - Show primary and secondary escape routes, staircases, refuges, and final exits.
  - Include travel distances, exit widths, and capacity calculations.
  - Where relevant, overlays may show **ASET vs RSET analysis zones** or evacuation modelling outputs.
- **Fire Service Access Diagrams**
  - Show fire appliance access routes, turning circles, hardstandings, and hydrants.
  - Identify rising main inlets, firefighting shafts, and control rooms.
  - Useful for consultation with Fire and Rescue Services at planning stage.

5.7.2 Supporting Schedules

Drawings should be supplemented with **schedules and tabulated data**, for example:

- **Fire doorset schedules** (FD30/FD60, with smoke seals, vision panels, self-closing devices).
- **Compartmentation schedules** (wall/floor type, fire resistance rating, tested details).
- **Fire protection system schedules** (sprinkler zoning, alarm categories, smoke control zones).

These ensure that detailed technical requirements are **clearly documented and traceable**.

5.7.3 Version Control and Referencing

To ensure integrity and traceability of the fire strategy:

- All drawings and appendices must be **referenced within the main text**, including drawing number, title, revision, and date.
- A **drawing register** should be included in the appendices to confirm which documents were current at the time of strategy issue.
- Later revisions must be recorded in the **version control log** to avoid ambiguity during design development, construction, or future refurbishments.

5.7.4 Digital Integration

Modern best practice increasingly requires fire strategy documentation to be digital and BIM-compatible:

- Fire safety drawings may be embedded within a **Building Information Model (BIM)**, allowing coordination across disciplines.
- Digital appendices ensure ongoing access for building operators and facilitate compliance with the **Golden Thread of Information** as mandated by the Building Safety Act 2022.

Summary

Drawings and appendices are the **visual and technical backbone** of a fire strategy. They provide clarity, traceability, and enforceability by showing exactly how written provisions translate into the physical building. Through **general arrangements, compartmentation drawings, escape route plans, fire service diagrams, and schedules**, a fire strategy becomes a **living reference document** that can guide design, construction, regulation, and operation.

5.8 Case Example (Illustrative)

To illustrate how building description underpins the fire strategy, consider the following **mixed-use development**:

- **Overall Form**
    - The scheme consists of a **20-storey residential tower** constructed above a **ground-floor retail podium**, with **two basement levels** dedicated to car parking and plant space.
    - The building sits on a confined city-centre site, with fire appliance access provided on two sides and a dedicated firefighting lobby and lift core.
- **Structure and Materials**
    - The tower employs a **reinforced concrete frame** with post-tensioned concrete floor slabs, providing inherent fire resistance and robustness.
    - **External walls** are formed with a **unitised cladding system**, incorporating **non-combustible panels (A2-s1,d0)** to comply with Regulation 7(2).

  - **Compartmentation** is achieved at every floor plate, with robust separation between the **retail podium** and the **residential tower**.
- **Occupancy and Use**
  - The residential tower accommodates up to **1,200 residents**, with an average of **60 persons per floor**.
  - The podium retail units attract transient occupancy, requiring shorter travel distances and clear exit provision.
  - Basements are designed for **300 cars**, with additional space allocated for mechanical plant and services.
- **Fire Safety Provisions**
  - **Basements**: Sprinkler protection (BS EN 12845) and mechanical smoke extraction (BS EN 12101-6).
  - **Residential tower**: A **stay-put evacuation policy** supported by compartmentation, with a simultaneous evacuation override for fire service use.
  - **Retail podium**: Simultaneous evacuation supported by a voice alarm system.
  - **Escape**: Two independent stair cores, each protected to 120 minutes fire resistance, with firefighting lifts integrated into firefighting shafts.
  - **Fire service access**: Rising mains (wet above 60 m), hydrants within 90 m, and a dedicated fire control room at ground level housing the alarm panel, smoke control interfaces, and communication systems.

Commentary

This type of **structured description** sets a clear foundation for the technical fire safety design. By recording:

- The **building form and occupancy**,
- The **materials and construction methods**,
- The **compartmentation principles**, and
- The **active and passive fire safety provisions**,

The fire strategy creates a **baseline reference** against which compliance can be tested, design assumptions validated, and future modifications benchmarked.

5.9 Summary

The description of the building is the **foundation chapter** of a fire strategy. It sets out where the building is located, how it is constructed, what it is used for, and who will occupy it. By documenting this information comprehensively and linking it to drawings, the strategy ensures that all subsequent fire safety measures are tailored to the actual risks of the building.

The next chapter will address **design assumptions and fire scenarios**, showing how the building description informs the selection of fire safety measures.

Chapter 6

Design Assumptions and Fire Scenarios

6.1 Introduction

Every fire strategy is underpinned by a set of **design assumptions and fire scenarios**. These assumptions form the **baseline conditions** against which all fire safety measures are tested, whether through prescriptive compliance routes or performance-based engineering analysis. They provide clarity on what the design is expected to achieve, the hazards it must address, and the limits within which it can be considered safe.

In prescriptive design, assumptions are often implicit — for example, Approved Document B assumes certain occupant densities, travel distances, and fire growth rates. In performance-based design, however, these assumptions must be **made explicit and justified**, since the strategy often depends on engineered alternatives supported by modelling.

Why Assumptions Matter

The importance of documenting assumptions lies in three areas:

1. **Compliance and Defensibility**
    - Regulators and Building Control must be able to see exactly which conditions the design addresses.
    - Insurers and clients need confidence that the assumptions are realistic and cover credible worst-case scenarios.
2. **Clarity for Future Management**
    - Once a building is occupied, managers need to know what was assumed at design stage — e.g., maximum occupant load, fire load densities, or response times.
    - If these assumptions are exceeded in practice, the fire strategy may no longer be valid.

3. **Robustness of Performance-Based Design**
    - In engineered designs, assumptions determine the inputs to **smoke modelling, evacuation modelling, and structural fire analysis**.
    - Unrealistic assumptions (e.g., assuming zero pre-movement time in evacuation) can make a design appear safe on paper but unsafe in reality.

Risks of Poorly Defined Assumptions

If assumptions are not explicit or are unrealistic, they can lead to:

- **Regulatory rejection** – because the strategy does not clearly show equivalence to prescriptive guidance.
- **Unsafe conditions** – if, for example, actual occupant numbers exceed the assumed load.
- **Legal vulnerability** – in the aftermath of a fire, if it is shown that design assumptions were incomplete, outdated, or unsubstantiated.
- **Invalidation of the strategy** – particularly if refurbishment or change of use alters the building context without updating the original assumptions.

The Grenfell Tower Inquiry has highlighted how **unrealistic assumptions about compartmentation performance and evacuation behaviour** contributed to systemic failure. Similarly, in other international cases, incorrect assumptions about smoke control performance or fire service response times have undermined otherwise well-engineered strategies.

The Role of Fire Scenarios

Alongside assumptions, fire strategies rely on **design fire scenarios**. These scenarios define the types of fire events that the strategy must be robust against. They include:

- **Fast-growing fires** – testing early warning and evacuation effectiveness.
- **Slow-growing fires** – assessing detection and management response.
- **Ventilation-controlled fires** – testing smoke control and firefighting strategies.
- **Localised or shielded fires** – in plant rooms, kitchens, or storage areas.
- **Multiple fire scenarios** – in very large or complex developments.

Selecting credible scenarios ensures that the fire strategy covers both **typical and worst-case events**, avoiding both under- and over-design.

Golden Thread of Assumptions

Under the **Building Safety Act 2022**, design assumptions must now form part of the "golden thread" of information that follows a building throughout its life. This means:

- Assumptions must be documented at design stage.
- They must be accessible to regulators, managers, and occupants.
- They must be updated if the building undergoes change.

This approach ensures that fire strategies remain **living, practical documents**, rather than static reports that quickly lose relevance.

Looking Ahead in This Chapter

The remainder of this chapter will examine in detail:

- The **types of design assumptions** that must be recorded (occupancy, fire load, fire service response, management level).
- How **design fire scenarios** are selected and justified.
- The methods used to derive **design fires** (heat release rates, t-squared fires, fuel load densities).
- How assumptions are validated through **modelling, standards, and case studies**.
- The importance of ensuring assumptions remain valid during the building lifecycle.

6.2 Occupancy Load

The **expected number of occupants** in a building is one of the most critical assumptions in any fire strategy. Occupancy load directly determines the required **capacity of escape routes, exits, and stairs**, and it influences almost every other aspect of life safety design, including alarm audibility, evacuation times, and fire service intervention planning.

A credible fire strategy must therefore demonstrate not only how the occupant load has been calculated, but also that it is **realistic, transparent, and robust enough** to account for peak conditions and changes in use over time.

Why Occupant Load Matters

- **Escape route capacity and width** – Exit and stair sizes are based on the number of people they must accommodate, usually expressed in mm per person.
- **Alarm and communication systems** – Larger occupant loads may require voice alarms, phased evacuation, or zoned alarm systems.
- **Fire service planning** – Fire crews base their operational response on the likely number of people inside a building, affecting search, rescue, and firefighting strategies.
- **Evacuation modelling** – Occupant numbers feed directly into RSET (Required Safe Egress Time) calculations.

If occupant load is underestimated, escape routes may become dangerously congested. If overestimated, the design may become unnecessarily expensive or impractical.

6.2.1 Methods of Calculation

Occupant load is typically derived using one of the following methods:

- **Approved Document B (ADB)**
  - Provides floor space factors (m^2 per person) depending on building use.
  - *Example*: office areas use 6 m^2/person; assembly spaces use 0.5 m^2/person.
- **BS 9999 (2017)**
  - Offers more flexibility, with risk-based adjustments for management level, fire growth rate, and detection systems.
  - *Example*: well-managed office buildings may justify lower occupant densities than prescriptive ADB values.
- **Functional use approach**
  - For specific uses such as cinemas, lecture theatres, or sports venues, capacity may be set by **seating count**, not floor area.
  - *Example*: 300 fixed seats = 300 persons maximum occupancy.
- **Performance-based design**
  - Where ADB or BS 9999 do not reflect real conditions, fire engineers may use observed or modelled occupancy data (e.g., shopping centres, transport hubs).

6.2.2 Maximum vs Typical Occupancy

It is important to distinguish between:

- **Maximum theoretical occupancy** – The figure calculated using standards (always used for compliance).
- **Typical operational occupancy** – The number usually present during day-to-day operation.

For example:

- An office designed for 500 people may typically house only 350 staff, but fire safety must still assume the **maximum load**.
- A cinema may have a calculated load of 300 people, but this figure is only reached during weekend evenings.

The strategy should use the **maximum figure** for design purposes, while recognising typical variation in management and operational planning.

6.2.3 Peak Occupancy and Special Conditions

Certain occupancies experience significant **peaks in use**, which must be accounted for:

- **Assembly spaces** – stadiums, theatres, arenas, where sudden crowd surges can occur.
- **Retail environments** – shopping centres during sales or holiday periods.
- **Transport hubs** – train stations and airports at peak travel hours.
- **Educational facilities** – lecture theatres with timetabled peak sessions.

Where occupancy is highly variable, strategies may require **dynamic management controls** (e.g., limiting access, live monitoring, or event-day procedures).

6.2.4 Vulnerable and Diverse Occupants

In addition to numbers, the **characteristics of occupants** must be considered:

- **Mobility-impaired persons** – requiring refuges, evacuation lifts, or assisted escape.
- **Children and elderly** – slower travel speeds and higher dependence on staff.
- **Language and cultural differences** – influencing wayfinding and compliance with instructions.

- **Transient populations** – hotels, airports, tourist attractions, where occupants are unfamiliar with the building.

Fire strategies must ensure that **evacuation times and systems are based on realistic assumptions** about these populations, not idealised averages.

6.2.5 Case Examples

- **Nightclub Fire (Cocoanut Grove, 1942, USA)**
  - Occupant load exceeded legal capacity several times over; escape routes were overwhelmed.
  - 492 deaths. Shows the fatal consequences of exceeding calculated occupancy.
- **Modern Cinemas**
  - Design occupancy = number of fixed seats (e.g., 300 seats = 300 people).
  - Management may record lower typical use but must always plan for full capacity.
- **Stadiums and Arenas**
  - Occupant load in the tens of thousands. Escape strategy must include crowd modelling and controlled dispersal, not just raw exit width calculations.

Summary

Occupant load is one of the **most fundamental assumptions** in fire strategy development. It affects escape capacity, alarms, evacuation modelling, and firefighting planning. A robust strategy must:

- Use recognised calculation methods (ADB, BS 9999, or functional counts).
- Assume maximum design occupancy while acknowledging operational variations.
- Consider peak periods and vulnerable groups.
- Provide transparent justification of numbers for regulators and managers.

A miscalculation of occupant load can undermine the entire fire strategy. By contrast, a clear and defensible assumption provides the foundation for safe, resilient, and compliant design.

6.3 Nature of Building Use and Activities

The **intended use and activities within a building** are fundamental determinants of fire safety measures. It is not enough to know how many people are present; their purpose, behaviour, mobility, and familiarity with the building also influence how evacuation and fire protection systems must be designed. A fire strategy must therefore carefully document the building's use and activity profile, considering both **normal operation** and **peak or exceptional conditions**.

6.3.1 Continuous vs Intermittent Use

- **Continuous-use buildings**
  - *Examples*: residential apartments, care homes, hotels, hospitals.
  - Occupants may be present **24/7**, including during night-time hours when alertness is lower and pre-movement times are longer.
  - Fire strategies must prioritise **early warning**, robust compartmentation, and management procedures that account for sleeping risk.
- **Intermittent-use buildings**
  - *Examples*: offices, schools, commercial workshops.
  - Typically occupied only during the day or fixed periods.
  - Strategies may focus on **simultaneous evacuation** with shorter pre-movement times, but must still allow for occasional out-of-hours use (e.g., cleaning staff, night shifts).

Why it matters: Sleeping occupancies generally require higher levels of fire protection and alarm audibility than daytime-only occupancies.

6.3.2 Static vs Dynamic Activity

- **Static activity environments**
  - *Examples*: factories with fixed workstations, libraries, call centres.
  - Occupants remain in known locations, making evacuation modelling more predictable.
  - Detection systems may focus on fixed fire loads.
- **Dynamic activity environments**
  - *Examples*: airports, railway stations, shopping centres.

    - Occupants are constantly moving, unfamiliar with the layout, and evacuation routes must accommodate **unpredictable flows**.
    - Dynamic environments often require **voice alarm systems**, enhanced wayfinding signage, and multiple wide exits to disperse crowds safely.

Why it matters: Static environments allow design to be based on predictable occupant positions, while dynamic environments demand **flexibility and redundancy** in escape routes.

6.3.3 Public Access vs Controlled Access

- **Public access buildings**
    - *Examples*: stadiums, retail malls, leisure centres.
    - Occupants are often **unfamiliar** with escape routes.
    - Require clear signage, high levels of emergency lighting, intuitive wayfinding, and robust alarm audibility.
    - Security and crowd management may also be considerations (e.g., preventing bottlenecks).
- **Controlled access buildings**
    - *Examples*: industrial sites, research facilities, data centres.
    - Occupants are usually trained staff who are familiar with evacuation procedures.
    - Fire strategies may rely more heavily on staff action, with targeted suppression systems for specific hazards.

Why it matters: The level of public access directly influences whether evacuation can rely on **trained responses** or must cater for **untrained and unfamiliar occupants**.

6.3.4 Vulnerable Occupants

Certain building uses involve populations with **reduced mobility or special needs**:

- **Healthcare settings** – patients may be bed-bound or reliant on staff assistance. Strategies must include **progressive horizontal evacuation (PHE)** and compartmentation.

- **Care homes** – elderly residents may require evacuation aids (chairs, sleds) and extended evacuation times.
- **Schools** – children may respond unpredictably; staff-led evacuation is essential.
- **Custodial environments** – prisons and secure facilities must balance fire safety with security, often requiring defend-in-place strategies.

Why it matters: Vulnerable occupants cannot evacuate independently, so fire strategies must **integrate management procedures, staff resources, and specialised equipment** into design assumptions.

6.3.5 Influence on Fire Safety Systems

The nature of building use affects requirements across multiple safety systems:

Detection – sleeping occupancies require enhanced alarm audibility; industrial facilities may require flame or heat detection where smoke is common.

Evacuation planning – phased evacuation may suit high-rise offices, while simultaneous evacuation is essential for schools.

Suppression – residential care homes often require sprinklers (BS 9251), while data centres may require gaseous suppression systems.

Smoke control – transport hubs may require large-scale smoke extraction to protect crowds during phased dispersal.

6.3.6 Case Examples

- **King's Cross Underground Fire (London, 1987)**
  - A transport hub with dynamic passenger flows. Inadequate smoke control and lack of evacuation clarity contributed to 31 deaths.
  - Demonstrates the challenge of designing for **dynamic, transient populations**.
- **Residential High-Rise (Stay-Put vs Evacuate)**
  - The Grenfell Tower tragedy showed how a stay-put strategy relies completely on robust compartmentation. When that failed, occupants were left without an appropriate evacuation option.

    - Shows the importance of aligning **strategy to actual building use and risk profile**.
- **Hospitals**
    - Fires in healthcare settings highlight the need for **progressive horizontal evacuation**, staff training, and compartmentation robust enough to hold fire for extended evacuation times.

Summary

The **nature of building use and activities** is as critical as occupant numbers. A fire strategy must account for:

- Whether use is **continuous or intermittent**.
- Whether activity is **static or dynamic**.
- Whether access is **public or controlled**.
- Whether occupants are **vulnerable or self-reliant**.

These distinctions shape detection, evacuation, suppression, and management requirements, ensuring that the fire strategy is tailored to the **real-world use of the building** rather than generic assumptions.

6.4 Fire and Rescue Service Assumptions

Fire strategies must clearly state the **assumptions made about Fire and Rescue Service (FRS) intervention**, since this directly affects how long a building's fire safety systems must operate independently. It is unsafe to assume immediate intervention, or that specialist resources will always be available. Instead, a fire strategy should set realistic, transparent assumptions about attendance, capabilities, and constraints.

6.4.1 Response Times

- **Urban settings**
    - In many UK cities, first appliance attendance is typically within **5–10 minutes** of call receipt.
    - This is consistent with national performance data, but local variations exist.
- **Suburban and rural settings**
    - Response times may be significantly longer due to greater travel distances or on-call staffing.

    - Delays of **15–20 minutes** are not unusual, particularly in remote areas.
- **Design implications**
    - Buildings in areas with longer expected response times must rely more heavily on **passive fire protection, sprinklers, and smoke control systems.**
    - Fire strategies should not assume intervention will occur before life safety is secured.

6.4.2 Firefighting Capabilities

Not all fire services have the same equipment, staffing, or specialist resources. Strategies should record assumptions about:

- **Aerial appliances**
    - In many services, aerial ladder platforms or turntable ladders may only be available in certain stations, adding to response time.
    - Reliance on external rescue from height should therefore be avoided in strategy design.
- **Specialist rescue teams**
    - Urban Search and Rescue (USAR), hazardous materials teams, or high-rise response crews may not be locally based.
    - Buildings requiring such intervention (e.g., tunnels, chemical plants) must have **robust internal systems** until specialists arrive.
- **Crew size and continuity**
    - Initial appliances may carry 4–6 firefighters; large or complex incidents require reinforcement.
    - Fire strategies should therefore assume **limited manpower during the early stages** of an incident.

6.4.3 Access Constraints

Even when response is prompt, physical access may hinder effective intervention. Strategies should note:

- **Road layouts** – narrow streets, cul-de-sacs, or weight-limited bridges may delay appliances.
- **Traffic congestion** – particularly in urban centres, where peak-time response may be slowed.
- **Security barriers** – gates, bollards, rising barriers, or hostile vehicle mitigation (HVM) features.
- **Site complexity** – campuses, airports, or industrial estates may require extensive internal navigation before crews reach the incident location.

Mitigation includes:

- **Dedicated fire appliance routes and hardstandings**.
- **Automatic opening systems for barriers** linked to alarms.
- **On-site water supplies** where hydrant access is unreliable.

6.4.4 Duration of Independent System Performance

The key purpose of defining FRS assumptions is to determine **how long passive and active systems must perform without intervention**:

- **Escape systems** (stairs, lobbies, refuges) must remain usable until evacuation is complete.
- **Compartmentation** must resist fire spread long enough to contain the incident before fire crews arrive.
- **Smoke control systems** must maintain tenable conditions for evacuation and initial firefighting.
- **Suppression systems** (sprinklers, mist, gas) must operate reliably to control or contain fire until crews can take over.

This is sometimes expressed as the **critical independence time** — the period during which the building must protect life unaided.

6.4.5 Case Examples

- **Grenfell Tower (2017)**
    - Fire spread far faster than assumed; FRS could not prevent escalation despite rapid response.
    - Highlights that **reliance on fire service intervention cannot substitute for robust passive protection**.

- **Rural warehouse fires**
    - Large industrial fires in remote areas often outstrip FRS water supply and arrival times.
    - Demonstrates the need for **on-site water storage and independent suppression**.
- **Hospitals and care homes**
    - Evacuation is prolonged; strategies assume **FRS assistance with staged evacuation**, but design must allow for safe refuge and compartmentation until crews arrive.

Summary

Fire strategies must be explicit about assumptions relating to the Fire and Rescue Service. These include:

- **Response times** – shorter in urban areas, longer in rural.
- **Capabilities** – aerial appliances, crew numbers, specialist teams.
- **Access constraints** – road layouts, congestion, barriers, site complexity.

The role of the fire service is to **support and reinforce** safety, not to provide the primary means of escape. Strategies must therefore ensure that **passive and active systems can function independently** long enough to safeguard occupants until intervention is possible.

6.5 Fire Load Intensity

The **fire load intensity** of a building represents the amount of combustible energy available for release in a fire. It is typically expressed in **megajoules per square metre (MJ/m²)** and provides a quantitative measure of the potential severity of a fire within a compartment. Fire load is a cornerstone of fire engineering because it informs:

- The likely **heat release rate (HRR)** of a fire.
- The **duration of fire resistance** required for structural elements.
- The sizing of **suppression and smoke control systems**.
- The credibility of **design fire scenarios** used in modelling.

6.5.1 Definitions

- **Average Fire Load (f1)**
    - The mean energy density of combustible contents distributed across the entire floor area of a building.

- Provides a generalised measure of risk used in code-based design.

Peak Local Fire Load (f2)

- Concentrated combustibles in specific rooms or areas, leading to significantly higher energy densities.
- Examples include **archives, server rooms with plastics and cabling, laboratories with solvents**, or **warehouses with high-storage racks**.
- Peak fire loads often drive worst-case scenario modelling, since localised intensity can exceed compartment design assumptions.

6.5.2 Typical Fire Load Values

Fire load varies significantly depending on building use:

- **Residential**: 400–800 MJ/m² (furniture, textiles, plastics).
- **Offices**: 500–1,000 MJ/m² (paper, electronics, furniture).
- **Retail**: 800–1,200 MJ/m² (stock density, packaging).
- **Libraries/Archives**: 2,000–3,000 MJ/m² (high paper content).
- **Industrial/Warehouses**: up to 4,000–6,000 MJ/m², depending on storage materials.
- **Laboratories**: highly variable, with localised extreme fire loads due to chemicals.

6.5.3 Factors Influencing Fire Load Intensity

- **Type of combustibles** – plastics, foams, and chemicals release more energy per kilogram than wood or paper.
- **Arrangement of combustibles** – compact shelving or high-rack storage increases intensity compared to dispersed loads.
- **Ventilation** – oxygen availability governs whether a fire is fuel-controlled or ventilation-controlled.
- **Compartment size** – larger compartments with high fire loads may sustain longer fully developed fires.
- **Housekeeping and management** – poor storage practices can create unintended high fire load zones.

6.5.4 Fire Load Assessment in Fire Strategies

Strategies should include an explicit statement of assumed fire loads, supported by:

- **Reference standards**: Eurocode 1 (EN 1991-1-2) provides methods for calculating fire loads in different occupancies.
- **Surveys and inventories**: In high-risk premises (museums, archives), physical audits may be required.
- **Worst-case assumptions**: Where precise data is unavailable, conservative values should be applied.
- Fire load assessments support:
- **Structural fire engineering** – determining whether structures need 30, 60, 90, or 120 minutes of fire resistance.
- **Suppression design** – sizing sprinkler densities or gaseous suppression agent volumes.
- **Scenario definition** – selecting credible design fires for CFD and evacuation modelling.

6.5.5 Example

- An **office building** with open-plan workspaces may have an **average fire load** of 600 MJ/m^2.
- However, a **server room** within the same building may present a **local fire load** of 3,000 MJ/m^2 due to plastics and cabling.
- The fire strategy must therefore consider both **average building-wide values** and **local peak hazards**, ensuring suppression and compartmentation are robust enough to handle extremes.

6.5.6 Case Examples

- **Summerland Fire, Isle of Man (1973)**
    - High plastic fire loads in cladding and internal finishes drove rapid fire growth, overwhelming evacuation.
    - Demonstrates the consequences of underestimating combustible materials in design.

- **Warehouse Fires**
    - Several UK warehouse fires (e.g., Atherstone-on-Stour, 2007) escalated due to very high fire loads, showing the importance of aligning suppression systems with real storage conditions.

Summary

Fire load intensity is a **fundamental assumption** in fire strategy development. By distinguishing between **average fire load (f1)** and **peak local fire load (f2)**, strategies can:

- Quantify the severity of potential fires.
- Design suppression and smoke control systems to credible worst-case conditions.
- Ensure structural fire resistance is robust.
- Support defensible design fire scenarios for modelling.

Failure to consider both average and peak fire loads risks creating strategies that work in theory but fail under real fire conditions.

6.6 Location of Fire Loads

The **location of combustible materials** is as critical as their quantity. A moderate fire load in the wrong place — for example, beside an escape stair or ventilation intake — may present a far greater hazard than a high fire load in a well-compartmented storeroom. Fire strategies must therefore identify where fire loads are situated within the building and in its surroundings and assess how their placement influences risk.

6.6.1 Internal Hazards

Certain internal spaces often contain disproportionately high fire loads or ignition risks:

- **Storage rooms** – concentrated quantities of paper, packaging, clothing, or flammable stock can generate high-intensity fires.
- **Kitchens and catering facilities** – presence of cooking oils, fryers, and gas appliances introduces both ignition sources and high-energy fuels.
- **Plant rooms** – electrical switchgear, transformers, fuel tanks, and lubricants can drive localised fires.

- **Laboratories and workshops** – solvents, compressed gases, and reactive chemicals create complex hazards requiring tailored suppression strategies.
- **Archives, libraries, and data centres** – high density of paper, plastics, or cabling leads to significant localised fire loads with prolonged burning duration.

Implication: Internal hazard areas should be clearly identified, compartmented with appropriate fire resistance, and provided with targeted suppression or detection systems.

6.6.2 External Hazards

Fire risk is not limited to the inside of the building. External fire loads can significantly influence safety and must be documented:

- **Refuse stores** – bins and skips close to walls can allow fire to spread cladding or windows.
- **Car parks** – vehicles contain plastics, fuel, and batteries (increasingly lithium-ion), which can produce intense fires.
- **Adjacent buildings** – high fire load neighbours (warehouses, fuel depots, industrial sites) may present radiated or direct flame impingement risks.
- **Landscaping features** – wooden decking, outdoor furniture, or combustible façade materials can act as ignition pathways.

Implication: Separation distances, boundary conditions, and external wall construction must all be assessed in relation to these hazards.

6.6.3 Proximity to Escape Routes and Critical Systems

The greatest risk arises when high fire loads are located **close to escape or life-safety provisions**:

- **Near escape stairs and corridors** – fires in lobbies or adjacent rooms may compromise primary evacuation routes.
- **Adjacent to refuges or evacuation lifts** – hazardous locations nearby can render these unusable for vulnerable occupants.
- **Near ventilation intakes** – smoke or toxic gases may be drawn deep into a building if intakes are located near refuse stores, loading bays, or car parks.

- **Around structural supports** – if high fire loads surround critical columns or transfer structures, premature structural failure may occur.

Implication: Fire strategies should ensure that hazardous storage or ignition sources are **separated from critical life safety systems** by compartmentation, distance, or protective enclosures.

6.6.4 Case Examples

- **Grenfell Tower (2017, UK)**
  - External combustible cladding acted as a continuous vertical fire load adjacent to escape routes, undermining the stay-put strategy.
  - Demonstrates the catastrophic impact of external fuel placement.
- **Kings Cross Underground Fire (1987, UK)**
  - A small ignition under an escalator ignited accumulated grease and debris in a critical escape route, leading to 31 deaths.
  - Shows the importance of addressing "hidden" fire loads in proximity to escape systems.
- **Warehouse fires**
  - Storage of high fire load goods (plastics, aerosols) adjacent to external boundaries has caused fire spread to neighbouring properties.

6.6.5 Design and Management Controls

To mitigate risks from fire load location, strategies should include:

- **Compartmentation** – high fire load areas separated by rated walls and doors.
- **Suppression systems** – sprinklers, gaseous, or mist systems in high-risk rooms.
- **Safe storage policies** – refuse kept at safe distances; flammable liquids stored in cabinets.
- **Access planning** – fire appliance access routes kept clear of external hazards.
- **Ventilation design** – intakes located away from potential smoke or flame sources.

Summary

The **placement of fire loads** can significantly magnify risks. A fire strategy must therefore identify:

- **Internal hazard areas** (storage, kitchens, plant rooms, labs, archives).
- **External hazards** (refuse, car parks, neighbouring risks).
- **Proximity to escape routes, refuges, ventilation, and structure.**

By addressing both **where combustibles are located** and **how they interact with life-safety systems**, strategies ensure that localised hazards do not undermine overall fire safety.

6.7 Fire Growth and Development Scenarios

To evaluate the robustness of a fire strategy, it is not enough to consider a "generic fire." Instead, designers must test building performance against a **range of credible design fire scenarios** that reflect both typical and worst-case conditions. These scenarios provide input for smoke modelling, evacuation analysis, suppression design, and structural fire engineering.

By defining a spectrum of fire behaviours, the strategy demonstrates resilience, ensuring that safety objectives are met even when conditions deviate from the idealised case.

6.7.1 Fast-Growing Fires

- **Description**
  - Characterised by rapid heat release rate (HRR) growth, often following a t^2 fast or ultra-fast growth curve.
  - Common where plastics, foams, hydrocarbons, or accelerants are involved.
- **Implications**
  - Tests the speed of **detection and alarm systems**, as delay in warning can be critical.
  - Challenges **evacuation times**, particularly in assembly spaces with large crowds.
  - Evaluates **sprinkler and suppression system effectiveness** under rapidly escalating conditions.
- **Example**: A fire in a retail store display of polyurethane furniture reaching flashover in under 5 minutes.

6.7.2 Slow-Growing Fires

- **Description**
  - Fires with long smouldering or low-intensity stages before open flaming.
  - Often caused by electrical faults, overheated cabling, or smouldering upholstery.
- **Implications**
  - Tests the effectiveness of **early detection systems**, particularly smoke and aspirating detectors.
  - May generate large quantities of toxic gases (CO, HCN) before flames or heat trigger alarms.
  - Challenges **management procedures**, as fires may smoulder undetected during low-occupancy periods (e.g., night-time in offices or hotels).

Example: A smouldering sofa in a flat, producing toxic smoke long before flames appear.

6.7.3 Ventilation-Controlled Fires

- **Description**
  - Fires in sealed or partially ventilated rooms, where oxygen availability rather than fuel load limits combustion.
  - May remain suppressed until a window breaks or a door is opened.
- **Implications**
  - Tests **smoke movement modelling**, since incomplete combustion produces large smoke volumes.
  - Raises risks of **backdraft or flashover** upon sudden reintroduction of air.
  - Evaluates robustness of **compartmentation and firefighting tactics**.
- **Example**: A basement plant room fire with closed doors and limited ventilation, producing dense smoke and sudden flashover risk when firefighters enter.

6.7.4 Compartment Fires

- **Description**
    - Fires confined to a single fire-resisting compartment, assumed to remain within its boundaries for a specified fire resistance period.
- **Implications**
    - Used to test the **integrity of compartmentation** (walls, floors, fire doors).
    - Validates that occupants in adjacent areas remain safe.
    - Forms the basis for most prescriptive designs in ADB and BS 9999.
- **Example**: A kitchen fire contained within a fire-rated shaft or residential flat.

6.7.5 Multi-Compartment Fires

- **Description**
    - Fires that breach compartmentation, whether through structural failure, penetrations, or external fire spread (e.g., façade ignition).
- **Implications**
    - Tests **redundancy of fire safety measures** — e.g., if one stair becomes compromised, are alternative escape routes adequate?
    - Evaluates the role of **sprinklers, cavity barriers, and external wall systems** in preventing fire spread.
    - Critical in high-rise or complex buildings where reliance on compartmentation is central to evacuation strategy.
- **Example**: The Grenfell Tower fire (2017) where flames spread externally across compartments, undermining the stay-put evacuation strategy.

6.7.6 Integration of Scenarios in Design

Fire strategies should not rely on a single scenario but must:

- Define a **suite of scenarios** covering slow, fast, ventilation-controlled, and multi-compartment events.
- Apply each scenario to test specific systems (e.g., alarms, smoke control, structural fire resistance).

- Justify scenario selection against **building use, occupancy profile, and fire load data.**

Performance-based fire engineering (BS 7974) provides structured methods for defining and analysing these scenarios using **probabilistic and deterministic approaches.**

6.7.7 Case Examples

- **King's Cross Fire (1987, UK)** – a smouldering ignition under an escalator developed slowly but then transitioned rapidly to flashover, demonstrating the need to test both slow and fast growth scenarios.
- **Grenfell Tower (2017, UK)** – highlighted the consequences of **multi-compartment and external fire spread** invalidating evacuation strategies.
- **Basement car park fires** – often ventilation-controlled, producing heavy smoke that tests extraction systems.

Summary

Defining **credible fire growth and development scenarios** is essential for a defensible fire strategy. By considering:

- **Fast-growing fires** (sprinkler and evacuation stress tests).
- **Slow-growing fires** (early detection challenges).
- **Ventilation-controlled fires** (smoke and backdraft hazards).
- **Compartment fires** (validating containment).
- **Multi-compartment fires** (testing redundancy).

fire strategies ensure resilience across the spectrum of potential fire events.

6.8 Evacuation Assumptions

The success of any fire strategy ultimately depends on how **people respond to alarms and move to safety**. Assumptions about occupant behaviour form the foundation of evacuation design and must therefore be explicit, realistic, and defensible. Overly optimistic assumptions — such as immediate response to alarms or uniform mobility — can lead to designs that appear adequate on paper but fail in practice.

When evacuation modelling is used, these assumptions become **inputs to the Required Safe Egress Time (RSET)** calculation and must be documented, justified, and cross-checked against real-world evidence.

6.8.1 Alarm Response Time

- **Definition**
    - The delay between an alarm sounding and occupants beginning to move.
    - Includes the time to recognise the alarm, gather belongings, and decide to act.
- **Typical values**
    - *Offices*: 1–3 minutes (staff may first confirm if the alarm is real).
    - *Residential (night-time)*: up to 5 minutes or more, due to waking delay.
    - *Assembly spaces*: often rapid if alarms are clear and crowd behaviour is guided.
- **Implications**
    - Longer response times increase RSET, reducing safety margins.
    - Strategies must assume **realistic pre-movement delays**, not instantaneous evacuation.

6.8.2 Movement Speed

- **Factors affecting speed**
    - **Age** – children and elderly move more slowly.
    - **Ability** – mobility impairments reduce speed, sometimes requiring assistance.
    - **Congestion** – movement slows significantly in crowds and at door bottlenecks.
    - **Environment** – smoke reduces visibility, increasing delays; stair descent is slower than level walking.
- **Reference data**
    - Typical walking speeds:
        - Healthy adults: 1.2–1.5 m/s.

    - Elderly occupants: 0.8 m/s or less.
    - Stair descent: 0.5–0.7 m/s.
- **Implications**
    - Evacuation modelling must incorporate **realistic variability in speeds**, not a single average value.
    - Crowd dynamics (arch formation at exits, merging flows) must also be reflected.

6.8.3 Use of Refuge Areas

- **Purpose**
    - Provide protected waiting areas for mobility-impaired occupants until assisted evacuation can occur.
    - Required in multi-storey buildings where evacuation lifts or staff assistance may be delayed.
- **Design implications**
    - Refuge areas must be within **fire-resisting construction**, with communication systems to alert staff or firefighters.
    - Strategies must specify the **expected occupancy of refuges** and the procedure for clearing them during incidents.
- **Assumption requirement**
    - The fire strategy must clarify **who uses refuges, how long they will remain there, and how they will be evacuated further**.

6.8.4 Staff-Assisted Evacuation

- **Where required**
    - Hospitals, care homes, schools, and custodial environments.
- **Implications**
    - The speed of evacuation is determined not only by occupant mobility but by the **availability and training of staff**.
    - Progressive horizontal evacuation (PHE) relies on staff moving patients into compartments before full building evacuation.
    - Strategies must therefore include **staffing levels, training commitments, and realistic time allowances**.

Example: In a hospital ward of 30 patients, assumptions must include the number of staff on duty at night and the time required for assisted movement.

6.8.5 Documentation and Justification in Modelling

If evacuation modelling is used (e.g., Exodus, Pathfinder, STEPS, MassMotion), assumptions must be:

- **Documented** – clearly listed in the fire strategy.
- **Justified** – supported by guidance (BS 7974-6, SFPE Handbook, empirical data).
- **Variable** – models should use distributions of pre-movement times and speeds, not uniform values.
- **Tested under multiple scenarios** – to check robustness against optimistic and pessimistic assumptions.

6.8.6 Case Examples

- **Station Nightclub Fire (2003, USA)**
  - Over 400 people attempted to use the main entrance despite multiple exits being available.
  - Assumptions of even distribution across exits would have been unrealistic.
  - Result: 100 fatalities due to bottleneck congestion.
- **Grenfell Tower (2017, UK)**
  - A "stay-put" strategy assumed compartmentation would protect occupants indefinitely.
  - When this assumption failed, there was no alternative evacuation plan, leading to catastrophic consequences.
- **Hospital Fires (Global examples)**
  - Staff shortages during night shifts often undermine evacuation assumptions.
  - Demonstrates the importance of basing assumptions on **real staffing patterns**, not idealised ones.

Summary

Evacuation assumptions are essential to credible fire strategies. They must address:

- **Alarm response times** – realistic delays in occupant reaction.
- **Movement speeds** – accounting for age, ability, and congestion.
- **Use of refuges** – for mobility-impaired occupants.
- **Staff-assisted evacuation** – where vulnerable occupants rely on trained staff.

By documenting and justifying these assumptions, strategies ensure that evacuation models and designs reflect **real-world behaviour**, not idealised scenarios.

6.9 Environmental and External Conditions

Fire strategies cannot be developed in isolation from their **environmental context**. Conditions outside the building — such as weather, power supply, and water availability — can significantly influence the effectiveness of fire safety systems and firefighting operations.

Documenting these assumptions ensures that safety measures are designed with **realistic constraints** in mind and that backup provisions are identified where external factors are unreliable.

6.9.1 Wind Direction and Speed

- **Influence on smoke movement**
  - Wind can drive smoke plumes into escape routes, atriums, or across external facades.
  - Stack effect in tall buildings can be compounded by wind pressure, altering smoke control assumptions.
- **Impact on external fire spread**
  - Fires involving cladding or roof structures may spread more rapidly under strong wind.
  - Openings such as windows or vents exposed to prevailing winds can accelerate flame impingement.
- **Design implications**
  - Smoke control design in atriums, tunnels, or open concourses must consider **worst-case wind conditions**.
  - Fire service access strategies should note exposure risks when appliances are positioned downwind of smoke plumes.

6.9.2 Power Supply Assumptions

- **Reliability of mains power**
    - Many life-safety systems — alarms, emergency lighting, smoke control, suppression pumps — depend on electrical supply.
    - Assumptions must be clear on whether mains power can be relied upon during fire conditions.
- **Backup systems**
    - Critical systems must have **secondary power supplies**, such as batteries for alarms/emergency lighting or diesel generators for smoke extract fans and sprinkler pumps.
    - Fire strategies should state the expected **autonomy period** of backup supplies (e.g., 3 hours for emergency lighting, 30 minutes for smoke extract).
- **Risks**
    - Fires in plant rooms, switch rooms, or cable basements can disable primary power unexpectedly.
    - Strategies must include **redundancy and separation** of electrical supplies to avoid single points of failure.

6.9.3 Water Supply Assumptions

- **Availability of hydrants**
    - Strategies must confirm whether hydrants are provided on-site or in the street, and their flow/pressure capacity.
    - Where hydrants are distant or unreliable, on-site tanks may be required.
- **Suppression system demands**
    - Sprinkler and mist systems require significant and sustained water flows.
    - Water storage tanks, break tanks, or dual supplies (town main plus tank) may be necessary to ensure reliability.
- **Redundancy**
    - Where suppression is critical (e.g., in high-rise, heritage, or industrial storage), at least two independent water supplies are often recommended.

- Rural sites with no hydrant coverage must provide **dedicated on-site storage** sized to sustain firefighting for the required duration.

6.9.4 Wider Environmental Considerations

In addition to wind, power, and water, other environmental conditions may affect fire strategies:

- **Seismic activity** – in international contexts, earthquakes may damage fire systems before ignition.
- **Flooding** – water ingress can disable electrical systems, pumps, or lower-level escape routes.
- **Temperature extremes** – freezing conditions can compromise sprinklers and hydrants if not adequately heated or insulated.

Although less common in the UK, these factors may be highly relevant in international or specialist projects.

6.9.5 Case Examples

- **Montreal Olympic Stadium (1987)** – wind-driven smoke in the velodrome highlighted the importance of accounting for wind in smoke control design.
- **Grenfell Tower (2017, UK)** – external fire spread was accelerated by wind exposure along the façade.
- **Warehouse fires in rural UK** – firefighting operations delayed due to inadequate hydrant provision, emphasising the importance of on-site water storage.

Summary

Environmental and external conditions are crucial design assumptions that must be explicitly documented in every fire strategy. These include:

- **Wind direction and speed** – influencing smoke behaviour and external fire spread.
- **Power supply reliability** – with clear provision for backup and redundancy.
- **Water supply assumptions** – ensuring hydrants, tanks, and suppression systems can operate effectively.

By identifying and addressing these external dependencies, fire strategies ensure that **safety systems remain resilient** even under adverse environmental conditions.

6.10 Sensitivity and Conservatism

Every fire strategy relies on **assumptions about fire growth, occupant behaviour, and system performance**. These assumptions are inherently uncertain, which means that the way they are framed has a direct impact on the **safety and cost-effectiveness** of the design.

The key principle is that assumptions should be **conservative but realistic**:

- **Conservative enough** to ensure life safety is never compromised.
- **Realistic enough** to avoid unnecessary cost, space constraints, or operational restrictions.

6.10.1 The Risks of Overly Optimistic Assumptions

Assumptions that underestimate risk may make a design appear adequate on paper but unsafe in reality. Examples include:

- Assuming all occupants begin evacuating immediately when alarms sound, ignoring realistic **pre-movement delays**.
- Assuming sprinklers will always operate effectively, without considering possible failure or maintenance issues.
- Assuming compartmentation will remain intact indefinitely, even under extreme fire load conditions.

Consequences: Optimistic assumptions can lead to **unsafe evacuation models, undersized stairs or exits, and insufficient suppression capacity**, leaving occupants at risk.

6.10.2 The Risks of Excessive Conservatism

Conversely, excessively conservative assumptions may lead to **overdesign** and wasted resources:

- Designing stairs or exits far larger than realistically required, consuming valuable floor area.
- Mandating suppression systems where risk could reasonably be managed with passive measures.

- Setting unrealistic fire resistance periods (e.g., requiring 240 minutes when 120 minutes would suffice under codes).

Consequences: Over-conservatism can make buildings **unnecessarily costly**, reduce design flexibility, and in some cases, discourage innovative architectural solutions.

6.10.3 Achieving the Balance

The balance between optimism and conservatism can be achieved by:

- **Referencing recognised standards** (ADB, BS 9999, BS 7974, SFPE).
- **Using empirical data** from case studies and evacuation research.
- **Aligning with management capability** — if a building has robust management, certain risk factors can be reduced; where management is weak, more conservative assumptions are justified.
- **Applying safety factors** — ensuring a margin between Required Safe Egress Time (RSET) and Available Safe Egress Time (ASET).

6.10.4 Sensitivity Testing

A powerful method to ensure assumptions are defensible is **sensitivity testing**:

- Running evacuation models with varied pre-movement times, speeds, and exit usage.
- Modelling smoke spread under different ventilation and wind conditions.
- Testing suppression system reliability by considering single-point failures.

Example: If an evacuation model shows safety is achieved with a 1.5-minute alarm delay, testing scenarios with 3–4 minutes ensures that the design remains safe even if occupants are slower than expected.

6.10.5 Case Examples

- **King's Cross Underground Fire (1987)**
    - Management assumed escalator fires would remain small and localised, an overly optimistic view that proved fatal.

- **High-rise office design (modern practice)**
  - Designers often run **multiple sensitivity tests** for stair capacity and evacuation times, ensuring that even slower-than-average evacuation speeds remain safe.
- **Grenfell Tower (2017)**
  - The stay-put policy was based on an assumption of indefinite compartmentation integrity. When this failed, there was no safety margin or alternative evacuation option.

Summary

Design assumptions must find the **right balance**:

- Overly optimistic assumptions = unsafe.
- Overly conservative assumptions = wasteful.
- Sensitivity testing ensures assumptions are **robust and defensible** across a range of conditions.

By applying conservatism where life safety is critical, and flexibility where innovation is possible, fire strategies can be both **safe and practical**.

6.11 Case Example (Illustrative)

To illustrate how design assumptions and fire scenarios are brought together, consider a **multi-storey office tower with mixed-use ground floors**.

Building Context

- **Type**: 30-storey office tower with ground-floor retail and a restaurant.
- **Location**: Dense urban centre with surrounding high-rise buildings.
- **Occupancy**: Designed as an open-plan office environment with associated support spaces.

Occupancy Assumptions

- **Calculation basis**: 1 person per **6 m² net floor area** (as per Approved Document B for offices).
- **Peak design load**: **1,800 occupants** at full occupancy.
- **Typical occupancy**: 1,200–1,400 occupants during weekdays, but design is based on the maximum figure.

- **Special conditions**: Restaurant peak occupancy adds 200 people during lunch/evening periods.

Fire and Rescue Service Assumptions

- **Response time**: First appliance assumed within **8 minutes** (urban average, confirmed by liaison with local FRS).
- **Access constraints**: Only one principal appliance access road due to neighbouring developments; aerial appliance access limited to front elevation.
- **Implication**: Fire protection and smoke control must provide independence for at least 10 minutes before external intervention can be relied upon.

Design Fire Scenarios Considered

1. **Fast-Growing Fire**
    - *Location*: Restaurant kitchen at ground floor.
    - *Fuel*: Cooking oils, plastics, and packaging (fast t^2 growth rate).
    - *Purpose*: To test sprinkler effectiveness, alarm response, and protection of escape routes serving retail and office lobbies.
2. **Slow-Growing Fire**
    - *Location*: Upper-level office (21st floor).
    - *Fuel*: Electrical fault leading to smouldering cabling and furnishings.
    - *Purpose*: To test early detection systems (aspirating smoke detectors) and ensure management response is sufficient even before open flaming develops.
3. **Ventilation-Controlled Fire**
    - *Location*: Basement car park.
    - *Fuel*: Vehicle fire involving plastics and hydrocarbons.
    - *Purpose*: To evaluate smoke extraction design and firefighter access.

Evacuation Assumptions

- **Alarm response time**: **2 minutes** (based on staff training and typical office delay).

- **Occupant speed**: Average of **1.2 m/s** on level ground, **0.7 m/s** on stairs.
- **Stair provision**: 6 fire-protected stairs distributed around the floor plate.
- **Refuge strategy**: Evacuation lifts and protected refuge lobbies provided for mobility-impaired occupants.
- **Management factor**: Regular fire drills ensure staff are familiar with escape routes, reducing pre-movement uncertainty.

Smoke Control and System Assumptions

- **Atrium smoke control**: Designed to maintain tenable conditions for at least 10 minutes in the event of a fast-growing fire at ground level.
- **Pressurised stairs**: Maintain positive pressure of 50 Pa to prevent smoke ingress.
- **Power supply**: Emergency generators support smoke fans and sprinkler pumps for up to 2 hours.
- **Water supply**: Dual-feed sprinklers supplied by both town mains and 200 m^3 on-site storage.

Integration of Assumptions

These assumptions form the **basis for evacuation time calculations and system performance checks**:

- Evacuation modelling demonstrates that **Required Safe Egress Time (RSET)** is less than **Available Safe Egress Time (ASET)** under all scenarios, with a minimum safety factor of 1.5.
- Stair capacity analysis confirms that the six stairs are sufficient to evacuate the full design load within calculated RSET.
- Smoke modelling confirms that pressurised stairs and atrium extract maintain tenable conditions until fire service arrival.

Why This Case Matters

This example demonstrates how a **fire strategy must bring together occupancy, fire load, fire service response, evacuation behaviour, and environmental assumptions into a coherent framework**. By explicitly stating these assumptions, the strategy is:

- **Transparent** – regulators and fire services can see the design basis.
- **Defensible** – assumptions are grounded in recognised guidance and sensitivity-tested.

- **Maintainable** – building managers know what the design relied upon, enabling them to monitor compliance during operation.

6.12 Summary

Design assumptions and fire scenarios form the **analytical backbone** of every fire strategy. They define the conditions under which fire safety measures must perform, ensuring that evacuation routes, suppression systems, smoke control, and firefighting facilities are designed against **credible risks** rather than generic models.

Throughout this chapter, we have seen how assumptions must be:

- **Explicit** – clearly documented so regulators, managers, and future assessors can understand the design basis.
- **Realistic** – reflecting true occupancy, human behaviour, fire service capabilities, and environmental conditions.
- **Conservative but balanced** – safeguarding life safety without imposing unnecessary cost or restricting design innovation.

Key considerations include:

- **Occupancy load** – establishing the maximum number of people the building may contain, with attention to peaks, vulnerable groups, and operational variations.
- **Nature of building use** – differentiating between continuous or intermittent occupation, public or controlled access, and the needs of vulnerable populations.
- **Fire and Rescue Service assumptions** – recognising limits of response times, access, and specialist capabilities, and ensuring systems perform independently until intervention.
- **Fire load intensity and location** – quantifying both average and localised hazards, and identifying critical placements near escape routes or ventilation systems.
- **Fire growth scenarios** – testing performance against fast, slow, ventilation-controlled, compartmentalised, and multi-compartment fires.
- **Evacuation behaviour** – accounting for alarm delays, movement speeds, refuges, and staff-assisted evacuation.

- **Environmental factors** – including wind, power, and water supply reliability, which can affect system performance.
- **Sensitivity and conservatism** – validating assumptions through sensitivity testing and applying margins of safety.

The **case example** of a multi-storey office tower demonstrated how these factors combine into a defensible, transparent, and practical fire strategy that regulators can approve, and building managers can maintain.

Ultimately, assumptions provide not just technical inputs but also the **"golden thread" of accountability**, ensuring that fire safety decisions are traceable across a building's lifecycle. If assumptions change — for example, through refurbishment, increased occupancy, or altered management procedures — the strategy must be updated accordingly.

Looking Ahead

Having established the analytical basis through design assumptions and scenarios, the next chapter turns to one of the **core pillars of fire safety**: **Means of Escape**. Here, the focus shifts to the physical and procedural measures that allow all occupants to evacuate safely, building directly on the assumptions outlined in this chapter.

Chapter 7
Means of Escape

7.1 Introduction

The **provision of safe and effective means of escape (MoE)** is the **cornerstone of every fire strategy**. All other fire safety measures — from compartmentation and suppression systems to smoke control and firefighting access — are ultimately designed to support this fundamental objective: ensuring that occupants can move from a place of danger to a place of ultimate safety before fire and smoke conditions become untenable.

Means of escape applies to **all building types and occupancies**, from single-family dwellings to complex high-rise mixed-use developments. The principle is universal: occupants must be able to escape **without reliance on external rescue**, such as the fire and rescue service, although external intervention may play a supporting role.

Key aspects underpinning this principle include:

- **Self-rescue without assistance** – routes must be designed so that people can escape unaided, with special provisions made for those with reduced mobility.
- **Predictability and reliability** – escape routes must be available, unobstructed, and protected, even under adverse fire conditions.
- **Time as a critical factor** – escape design must ensure that the **Required Safe Egress Time (RSET)** is always less than the **Available Safe Egress Time (ASET)**, providing a margin of safety.
- **Integration with wider fire strategy** – MoE provisions cannot be considered in isolation. Their success depends on compartmentation, fire resistance, active systems (alarms, sprinklers, smoke control), and management arrangements.

This chapter will examine the **principles, requirements, and strategies** that underpin means of escape design. It will explore:

- The **legal and regulatory framework** (Building Regulations, Approved Document B, BS 9999/9991).
- **Core design principles**, including travel distances, exit capacities, and separation of escape routes.
- **Evacuation strategies** (simultaneous, phased, zoned, defend-in-place, progressive horizontal).
- **Supporting measures**, such as refuges, signage, and emergency lighting.
- **Human behaviour considerations**, recognising that real-world evacuation rarely follows idealised models.
- **Case examples** to illustrate both successful designs and lessons learned from failures.

By the end of this chapter, readers will understand how means of escape provisions are **developed, justified, and integrated** within a wider fire strategy, and why they remain the most critical component of safeguarding life in fire emergencies.

7.2 Legal and Regulatory Framework

The design of **means of escape (MoE)** in the UK is underpinned by a well-established legal and regulatory framework. This framework provides the balance between **mandatory functional requirements** and **guidance routes to compliance**, enabling designers to adopt either prescriptive or performance-based approaches depending on the complexity of the building.

7.2.1 Building Regulations Part B

- The **Building Regulations 2010 (as amended)** set out the **functional requirements** relating to fire safety under **Requirement B1**: *"The building shall be designed and constructed so that there are appropriate provisions for the early warning of fire, and appropriate means of escape in case of fire from the building to a place of safety outside the building, capable of being safely and effectively used at all material times."*
- Part B is **mandatory** - every building must demonstrate that its means of escape provisions achieve this overarching objective.
- However, the Regulations themselves do not prescribe exact distances, widths, or strategies — these are instead provided in **statutory guidance documents** such as Approved Document B.

7.2.2 Approved Document B (ADB)

- **Volume 1 (dwellings)** and **Volume 2 (buildings other than dwellings)** provide **prescriptive guidance** for means of escape.
- Key aspects include:
    - Maximum **travel distances** (e.g., 18 m for dead-end corridors, 45 m for two-way escape routes in offices).
    - Minimum **exit widths** e.g. 5 mm per person for door width serving more than 220 persons,
    - Number and separation of **escape stairs and exits**.
    - Specific requirements for **protected routes, refuges, and smoke control**.
- ADB is the most common compliance pathway, but its prescriptive nature may limit flexibility in modern, complex, or innovative designs.

7.2.3 BS 9999 – Risk-Based Design

- **BS 9999:2017 – Code of Practice for Fire Safety in the Design, Management and Use of Buildings** provides a **risk-based framework**.
- It adjusts MoE requirements according to:
    - Occupant characteristics (mobility, familiarity, alertness).
    - Building risk profile (low, medium, high).
    - Fire safety management quality (good, average, poor).
- This allows for greater **flexibility in travel distances, escape capacities, and exit provision**, where supported by strong management procedures.
- BS 9999 is often used for **large, complex buildings** where prescriptive ADB measures would be too restrictive.

7.2.4 BS 9991 – Residential-Specific Provisions

- **BS 9991:2024** focuses on **residential occupancies** (flats, maisonettes, student accommodation, sheltered housing, care homes).
- It addresses:
    - **Stay-put strategies** in high-rise residential towers.
    - **Progressive horizontal evacuation** in care and healthcare environments.

- Enhanced measures for **vulnerable occupants**, including refuges, evacuation lifts, and assisted evacuation procedures.

- BS 9991 is particularly relevant in the **post-Grenfell context**, where residential evacuation and fire resistance are under intense scrutiny.

7.2.5 Prescriptive, Performance-Based, or Hybrid Approaches

- **Prescriptive (ADB)** – suitable for conventional, low-complexity buildings where standard travel distances, widths, and exits are sufficient.
- **Performance-Based (BS 7974 framework)** – used where innovative designs (e.g., atria, open-plan layouts, supertall towers) require fire engineering analysis of evacuation times (ASET vs RSET), smoke movement, and occupant behaviour.
- **Hybrid Approaches** – most strategies use prescriptive guidance for standard areas and performance-based justification for complex or non-standard features.

7.2.6 International Influence

Although this section focuses on UK practice, it is worth noting that international codes such as **NFPA 101 (Life Safety Code)** and the **International Building Code (IBC)** influence many UK-based projects, particularly for clients with international portfolios. Strategies may need to demonstrate compliance with both UK and overseas frameworks.

Summary

The regulatory framework for means of escape combines:

- The **mandatory functional requirement** under Part B of the Building Regulations.
- **Prescriptive guidance** in ADB.
- **Flexible, risk-based codes** (BS 9999, BS 9991).
- **Performance-based fire engineering** where necessary.

A fire strategy must clearly state which compliance pathway is being followed, justify any deviations, and ensure that escape provisions are **both compliant and defensible**.

7.3 Principles of Means of Escape

The principles of means of escape form the foundation of all fire strategy design. They ensure that **every occupant**, regardless of age, mobility, or familiarity with the building, can reach a place of ultimate safety in the event of fire. These principles are embedded in the **Building Regulations Requirement B1**, which requires that "appropriate means of escape… capable of being safely and effectively used at all material times" are provided.

7.3.1 Availability and Adequacy of Escape Routes

- Escape routes must be **unobstructed, continuously available, and clearly defined**.
- They must be sized to match the **calculated occupant load**, ensuring that exit widths and stair capacities allow timely evacuation without dangerous congestion.
- Escape must be possible from **all parts of the building**, including remote rooms, basements, and upper floors, with no person more than the prescribed travel distance from a protected route.

7.3.2 Escape Without External Assistance

- Occupants must be able to evacuate **without relying on the Fire and Rescue Service** for rescue.
- This principle reflects lessons from historic fires (e.g., Summerland 1973, Grenfell Tower 2017), where delayed external intervention led to significant loss of life.
- Firefighters may provide assistance to vulnerable occupants, but the building design must not assume that such intervention is the **primary escape route**.

7.3.3 Inclusivity – Accounting for All Users

- Escape provisions must consider the needs of **vulnerable occupants**, including:
    - People with reduced mobility or sensory impairments.
    - Children, the elderly, or those unfamiliar with the building.

- Provisions include:
  - **Refuge areas** within protected lobbies or stair enclosures.
  - **Evacuation lifts** (BS 9999 and BS EN 81-76).
  - **Evacuation aids** such as chairs or sleds in healthcare and residential environments.
  - **Personal Emergency Evacuation Plans (PEEPs)** for individuals who need assistance.
- This principle aligns with the **Equality Act 2010**, ensuring that escape is not discriminatory and is accessible to all.

7.3.4 Tenability of Escape Routes

- Escape routes must remain safe — **free of intolerable smoke, heat, and flames** — for the full duration of evacuation.
- Tenability is achieved through:
  - **Compartmentation** to delay fire spread into escape routes.
  - **Fire-resisting construction** of stairs, lobbies, and corridors.
  - **Smoke control systems** (natural or mechanical ventilation, stair pressurisation, atrium smoke extraction).
  - **Fire doors** fitted with smoke seals and self-closers.
- Evacuation safety is assessed by comparing:
  - **Available Safe Egress Time (ASET)** – the time until escape routes become untenable.
  - **Required Safe Egress Time (RSET)** – the time needed for occupants to evacuate.
- A **safety margin** must always exist (ASET > RSET).

7.3.5 Integration with Wider Fire Strategy

The success of means of escape provisions depends on their integration with the building's **overall fire safety design**, including:

- Early **detection and alarm systems**.
- Effective **compartmentation and suppression**.
- Clear **management procedures and staff training**.
- Ongoing **maintenance** of escape routes and doors.

Summary

The principles of means of escape are clear: routes must be **available, accessible, inclusive, and tenable**, enabling every occupant to reach a place of safety without reliance on external rescue. These principles provide the benchmark against which all detailed escape designs are tested, whether through prescriptive compliance, risk-based approaches, or fire engineering analysis.

7.4 Escape Route Design

Designing escape routes requires balancing **regulatory requirements, human behaviour, and practical building layouts**. The objective is to ensure that occupants can reach a place of safety **within the available safe egress time (ASET)** and without excessive congestion or exposure to smoke. Key parameters include travel distance, exit width and capacity, number of exits, and separation of escape routes.

7.4.1 Travel Distance

- Defined as the **maximum distance occupants must travel from any point within a building to reach a place of relative safety** (protected route, stair, or final exit).
- **Approved Document B (ADB)** prescribes limits, for example:
  - **Dead-end corridors** in offices: maximum **18 m**.
  - **Two-direction escape routes** in offices: maximum **45 m**.
  - Shorter limits apply in higher-risk occupancies (e.g., assembly or healthcare).
- **BS 9999** allows extended travel distances where management quality is high, risk profiles are favourable, and building features provide enhanced resilience (e.g., sprinklers, superior detection systems).
- **Fire engineering analysis (BS 7974)** can further justify bespoke travel distances by comparing **ASET vs RSET** using smoke modelling and evacuation simulations.

7.4.2 Exit Width and Capacity

- Exit widths must be sufficient to allow **all occupants to escape before tenability is lost**.

- **ADB guidance**:
    - **5 mm per person** for door widths serving more than 220 persons.

Examples:

- A 1,000-person assembly hall requires at least **5 m of exit width** in total.
- This may be provided as multiple smaller exits provided they are independent.

- **BS 9999 adjustments**:
    - Risk profile and management factors may allow reductions or require increases in exit widths.
- **Performance-based design**: Evacuation models (e.g., Pathfinder, STEPS) can demonstrate whether proposed exits achieve acceptable evacuation times.

7.4.3 Number of Exits

- **Single escape routes** are only acceptable in **limited, low-risk scenarios**, such as:
    - Small single-storey offices (<60 m^2).
    - Certain residential dwellings with compliant layouts.
- **Most multi-occupancy buildings** require **at least two independent escape routes**.
- Larger, more complex buildings (e.g., stadiums, airports) require multiple exits distributed across the building to prevent congestion and ensure redundancy if one route is compromised.

7.4.4 Separation of Exits

- Independent exits must be located **sufficiently far apart** to ensure that a **single fire event cannot compromise all routes simultaneously**.
- **The 45° rule** is commonly applied:
    - Where two exits are required, they should be separated by an angle of at least **45°**, measured from the most remote point within the space.

- For **larger or more complex buildings**, fire engineering analysis may be used to demonstrate alternative separation principles, particularly where atria, open plans, or interconnected levels are present.

7.4.5 Additional Design Considerations

- **Door swings**: must open in the direction of escape if serving >60 persons.
- **Obstructions**: escape routes must be clear of furniture, storage, or security barriers.
- **Lighting and signage**: escape routes must be visible under emergency lighting conditions (BS 5266 / BS EN 1838).
- **Security systems**: exit doors with access control must fail-safe to ensure escape at all times.

Summary

Escape route design is one of the most **quantifiable elements** of a fire strategy. By applying limits on **travel distance, exit width, number of exits, and separation**, and by adjusting these using **BS 9999 or BS 7974 analysis**, designers can demonstrate that occupants will be able to evacuate safely, even under worst-case conditions.

7.5 Evacuation Strategies

Different building types, occupancies, and risk profiles require different approaches to evacuation. The selection of an evacuation strategy is one of the most important decisions in fire strategy development, as it dictates the design of escape routes, alarm systems, fire resistance requirements, and management procedures.

7.5.1 Simultaneous Evacuation

- **Definition**: All occupants evacuate the building immediately when the alarm sounds.
- **Typical Applications**:
    - Small, low-rise buildings such as offices, shops, and community centres.
    - Residential dwellings and smaller blocks of flats where travel distances are short and exit provision is straightforward.

- **Advantages**:
  - Simple to design and manage.
  - Requires only basic alarm and communication systems.
- **Limitations**:
  - May not be feasible for large or complex buildings where congestion and long travel distances increase RSET.

7.5.2 Phased Evacuation

- **Definition**: Occupants leave in stages, starting with those closest to the fire, followed by progressively more remote areas.
- **Typical Applications**:
  - High-rise office towers, large atrium buildings, and some mixed-use complexes.
- **Design Requirements**:
  - Fire-resisting structure to contain fire for sufficient time.
  - Voice alarm systems to direct occupants.
  - Management procedures for coordination.
- **Advantages**:
  - Reduces congestion on stairs.
  - Allows continued use of unaffected areas during minor incidents.
- **Limitations**:
  - Requires strict management and robust fire resistance.
  - Occupants further from the fire may delay action, assuming they are not at immediate risk.

7.5.3 Zoned Evacuation

- **Definition**: Evacuation is triggered in **specific zones** (floors, compartments, or smoke control sectors), usually based on detector or alarm activation.
- **Typical Applications**:
  - Shopping centres, transport hubs, airports, and large public assembly venues.

- **Design Requirements**:
  - Reliable, zoned fire detection and alarm systems.
  - Clear communication to prevent unnecessary evacuation of unaffected areas.
- **Advantages**:
  - Limits disruption by focusing on at-risk zones.
  - Avoids unnecessary evacuation of very large occupancies.
- **Limitations**:
  - Risk of miscommunication if zones are poorly defined.
  - Requires careful coordination with smoke control and firefighting strategies.

7.5.4 Defend-in-Place

- **Definition**: Occupants remain within their compartment of origin, which is designed to resist fire and smoke spread for a defined period. External rescue or fire service intervention may be required.
- **Typical Applications**:
  - **Hospitals and care homes** where evacuation is impractical for high-dependency occupants.
  - **Residential towers** with stay-put policies, provided compartmentation is robust.
- **Design Requirements**:
  - High levels of compartmentation and fire resistance.
  - Smoke control to prevent vertical or horizontal spread.
  - Staff training to reassure and manage occupants.
- **Advantages**:
  - Reduces stress on vulnerable occupants.
  - Avoids the risks associated with mass evacuation in complex environments.
- **Limitations**:
  - Places strong reliance on passive protection systems.
  - Can become unsafe if fire breaches compartmentation (as highlighted by Grenfell Tower).

7.5.5 Progressive Horizontal Evacuation (PHE)

- **Definition**: Occupants move horizontally from one fire-resisting compartment into an adjacent compartment on the same level. Evacuation continues progressively until occupants reach ultimate safety.
- **Typical Applications**:
    - **Healthcare facilities**, **care homes**, and **secure institutions** (e.g., prisons) where occupants cannot self-evacuate quickly.
- **Design Requirements**:
    - Fire compartments of sufficient size and resistance (usually 60 minutes or more).
    - Refuge points with staff support.
    - Clearly signed progressive routes.
- **Advantages**:
    - Reduces risk for dependent occupants who cannot immediately leave the building.
    - Allows controlled, staged evacuation with staff assistance.
- **Limitations**:
    - Requires substantial compartmentation.
    - Longer-term management responsibility for patient or resident safety.

7.5.6 Strategy Selection

The appropriate evacuation strategy depends on:

- **Building type and size**.
- **Occupancy profile** (independent vs vulnerable occupants).
- **Management resources**.
- **Integration with other fire safety measures** (compartmentation, alarms, suppression).

The fire strategy must explicitly **state and justify** the chosen strategy, referencing prescriptive guidance or performance-based analysis where relevant.

Summary

Evacuation strategies range from **simple simultaneous evacuation** in small buildings to **complex staged, zoned, or defend-in-place approaches** in large, high-rise, or healthcare occupancies. Selecting and documenting the correct approach ensures that evacuation is both achievable and defensible under realistic fire scenarios.

7.6 Refuge Areas and Assisted Evacuation

Not all occupants are able to self-evacuate in the event of a fire. People with **mobility impairments, sensory disabilities, medical conditions, or cognitive challenges** may require additional assistance. A comprehensive fire strategy must therefore include provisions for **refuge areas, assisted evacuation methods**, and **personal evacuation planning**, ensuring that escape is possible for everyone.

This approach aligns with:

Building Regulations Part B1, which requires that escape is available to all occupants.

The **Equality Act 2010**, which prohibits discrimination in building access and use.

Best practice guidance in **BS 9999**, **BS 9991**, and **BS EN 81-76** (evacuation lifts).

7.6.1 Refuge Areas

Refuge areas provide **temporary safe spaces** for people who cannot immediately use stairs or exits.

- **Location and Design**:
  - Usually located within **protected stair lobbies** or **fire-resisting compartments**.
  - Must provide **at least 30 minutes fire and smoke protection** (often FD30S-rated construction).
  - Should be large enough to accommodate wheelchairs without obstructing circulation.
- **Facilities**:
  - Equipped with **two-way communication systems** (e.g., Emergency Voice Communication Systems, EVCs) compliant

with **BS 5839-9**, allowing direct contact with building control or fire services.
  - Clearly signed and accessible.
- **Role in Strategy**:
  - Refuge areas are **not final exits** but interim spaces until assisted evacuation can be arranged.

7.6.2 Evacuation Aids

Specialised equipment may be required to assist people in leaving the building safely:

- **Evacuation Chairs** – portable devices for moving mobility-impaired persons down staircases.
- **Evacuation Sleds or Mats** – particularly in healthcare settings, where bedbound patients may need to be moved quickly.
- **Evacuation Lifts** – lifts specifically designed for evacuation (not ordinary passenger lifts), in compliance with **BS EN 81-76**.
  - Require fire-protected lobbies, dual power supply, and communication facilities.
  - Increasingly important in **high-rise residential buildings** post-Grenfell.
- **Management Responsibility**:
  - Staff must be trained in the use of evacuation aids, and drills should confirm that evacuation can be achieved within required timescales.

7.6.3 Personal Emergency Evacuation Plans (PEEPs)

- **Definition**: A PEEP is an **individualised evacuation plan** tailored to the specific needs of a person with a disability or other special requirement.
- **Content**:
  - Identifies the person's typical location(s) in the building.
  - Specifies the refuge area or evacuation lift to be used.
  - Details the staff roles and equipment required to assist.

- **Implementation**:
  - Must be prepared for staff, residents, or frequent building users.
  - For transient populations (e.g., hotels), **Generic Emergency Evacuation Plans (GEEPs)** may be prepared instead.
- **Legal Context**:
  - Following legal challenges post-Grenfell, the requirement for PEEPs in high-rise residential buildings has become a central policy debate in UK fire safety.

7.6.4 Integration with the Wider Strategy

- Refuge and assisted evacuation provisions must be **integrated into the overall escape strategy**, ensuring they are not treated as separate or optional features.
- The fire strategy should specify:
  - **Number and location** of refuge areas.
  - **Type of evacuation aids** available.
  - **Procedures and responsibilities** for staff.
  - **Testing and maintenance** of equipment and communication systems.

Summary

Provision for **refuge areas and assisted evacuation** ensures that fire strategies are **inclusive, legally compliant, and practical**. By combining **protected refuge spaces, evacuation aids, and PEEPs**, designers and building managers can provide assurance that **all occupants — including those with reduced mobility or special needs — have a safe and realistic means of escape**.

7.7 Signage and Emergency Lighting

Escape routes are only effective if occupants can **quickly identify and navigate them** during a fire. Smoke, darkness, and stress impair visibility and decision-making, making **signage and emergency lighting** essential to maintaining safe egress. Together, they provide the **visual, tactile, and auditory cues** that guide people — including those with sensory impairments — to safety.

7.7.1 Escape Signage

- **Requirements**:
  - Exit routes must be marked with **illuminated signs** showing the location and direction of travel.
  - Symbols and pictograms must comply with **BS EN ISO 7010** and **BS 5499**, ensuring international consistency.
- **Types of Signage**:
  - **Exit Signs** – above final exits and intermediate doors.
  - **Directional Arrows** – indicating the quickest route to an exit (left, right, up, or down).
  - **Fire Safety Signs** – for equipment (extinguishers, alarms, hydrants) as required by the **Health and Safety (Safety Signs and Signals) Regulations 1996**.
- **Design Considerations**:
  - Signs must be visible in both **normal and emergency lighting conditions**.
  - **Maintained or non-maintained illuminated signs** should be selected based on building type and occupancy profile.
  - Mounting height and positioning should ensure visibility even in **crowded or smoke-filled conditions**.

7.7.2 Emergency Lighting

- **Purpose**: To maintain visibility of escape routes and critical equipment during **power failure** or smoky conditions.
- **Standards**: Governed by **BS 5266-1** and **BS EN 1838**, which define performance criteria.
- **Requirements**:
  - Minimum illuminance of **1 lux** along the centre line of escape routes.
  - Higher levels (**5 lux or more**) in high-risk task areas (e.g., plant rooms, kitchens).
  - Coverage for stairwells, ramps, intersections, refuges, and final exits.
  - Maximum response time of **0.5 seconds** after power failure.

- **System Types:**
    - **Self-contained emergency luminaires** with integral batteries.
    - **Central battery systems** serving multiple luminaires, common in large or complex buildings.
- **Testing and Maintenance:**
    - Must include routine functional testing (monthly short test, annual full-duration test).
    - Records kept as part of the fire safety logbook.

7.7.3 Wayfinding Aids for Inclusivity

Modern fire strategies recognise the need for escape provisions to be **accessible to all users**, not only those with good sight or hearing. Additional aids include:

- **Tactile Indicators** – raised floor surfaces or handrail markings for visually impaired occupants.
- **Audible Systems** – voice alarm systems that provide spoken instructions (BS 5839-8), especially valuable in large or complex buildings.
- **Photoluminescent Markings** – low-level guidance strips on walls, floors, or stairs that remain visible in heavy smoke.
- **Inclusive Design** – alignment with the **Equality Act 2010** ensures non-discriminatory provision of escape information.

7.7.4 Integration with the Fire Strategy

Signage and emergency lighting are not standalone features but are integrated with:

- **Escape route design** – ensuring all protected routes are clearly identifiable.
- **Management responsibilities** – routine inspections, cleaning, and lamp replacement.
- **Fire drills and staff training** – ensuring occupants recognise and respond to the guidance provided.

Summary

Signage and emergency lighting transform escape routes into **usable and reliable lifelines** during fire emergencies. By ensuring compliance with **BS EN ISO 7010, BS 5499, BS 5266, and BS EN 1838**, and by incorporating inclusive wayfinding aids, the fire strategy provides for safe, intuitive, and equitable evacuation for all building occupants.

7.8 Maintaining Tenability of Escape Routes

The design of escape routes is only effective if those routes remain **tenable** for the full duration of evacuation. Tenability refers to the ability of occupants to move through escape paths without exposure to conditions that would cause harm — including excessive heat, smoke density, toxic gases, or flame spread.

To achieve this, escape routes must be **protected both passively and actively**, ensuring that **Available Safe Egress Time (ASET)** always exceeds **Required Safe Egress Time (RSET)** by an appropriate safety margin.

7.8.1 Compartmentation

- **Purpose**: To prevent the spread of fire and smoke into protected routes.
- **Methods**:
    - **Fire-resisting walls and floors** separating escape stairs, corridors, and lobbies from adjacent compartments.
    - **Cavity barriers** in voids to prevent unseen smoke and flame spread.
    - Minimum fire resistance periods typically **30, 60, or 120 minutes** depending on building height and risk profile.
- **Standards**: Must comply with **Approved Document B**, **BS EN 1364 (walls/partitions)**, and **BS EN 13501-2 (classification)**.

7.8.2 Smoke Control Systems

- Smoke is the greatest threat to tenability, often more dangerous than flames. Systems are therefore required to limit smoke ingress into escape routes:
    - **Stair Pressurisation Systems** – maintain higher air pressure in escape stairs, preventing smoke entry. Designed under **BS EN 12101-6**.

  - **Automatic Opening Vents (AOVs)** – at the head of stairs or in lobbies to vent smoke.
  - **Mechanical Extract Fans** – used in basements or atria where natural venting is insufficient.
  - **Smoke Curtains** – temporary barriers that direct smoke towards vents or away from escape paths.
- *Reliability is critica*l: systems must have **dual power supplies** and be subject to **regular testing and maintenance.**

7.8.3 Protected Lobbies and Shafts

- **Definition**: Intervening spaces that provide a buffer between risk areas and escape routes.
- **Design Requirements**:
  - **Double-door protection** – requiring occupants to pass through two sets of self-closing doors before reaching the escape stair.
  - **Positive pressure systems** may be applied to further enhance smoke protection.
  - **Firefighting lobbies** often double as refuge areas, supporting both occupant escape and fire service access.
- **Standards**: Typically designed to **60-minute fire resistance** with FD60S doors.

7.8.4 Fire-Resisting Doorsets

- **Role**: Act as the weakest link in many escape routes, making their specification and installation critical.
- **Performance**:
  - Doorsets rated FD30 or FD60 (30 or 60 minutes), depending on building type and height.
  - Fitted with **intumescent strips** and **cold smoke seals** to prevent leakage.
  - Self-closing devices ensure doors remain shut under fire conditions.
- **Glazing**: Vision panels must be tested as part of the doorset to the relevant classification.

- **Maintenance**: Regular inspection for gaps, warping, or missing seals is essential to maintain protection.

7.8.5 Integration and Management

Maintaining tenability is not solely about design but also about **ongoing management**:

- Escape routes must be **kept clear** of storage, obstructions, or alterations that compromise fire resistance.
- Smoke control systems require **routine testing** to confirm performance.
- Fire doors must be subject to **regular inspection** and immediate repair if compromised.

Summary

Maintaining the tenability of escape routes ensures that occupants can evacuate **before conditions become life-threatening**. Through a combination of **compartmentation, smoke control, protected lobbies, and fire-resisting doorsets**, supported by robust management practices, the fire strategy ensures that escape routes remain safe and effective under credible fire scenarios.

7.9 Human Behaviour Considerations

While prescriptive codes and engineering models often assume that occupants will respond **immediately and rationally** to alarms, real-world fire incidents show that human behaviour is far more complex. Understanding these behaviours is vital to designing escape strategies that are **realistic, inclusive, and resilient**.

7.9.1 Pre-Movement Time and Delays

- Occupants rarely evacuate immediately after hearing an alarm.
- **Delays may occur** due to:
    - Attempting to confirm whether the alarm is genuine ("alarm verification").
    - Collecting belongings or securing valuables.
    - Completing tasks before leaving (common in workplaces).
- Research shows average pre-movement times can range from **30 seconds in high-risk environments** to **several minutes in offices, hotels, or residential buildings**.

7.9.2 Route Choice and Wayfinding

- **Unfamiliar visitors** often default to the route they entered, even when signed alternatives exist.
- **Crowd dynamics** can exacerbate bottlenecks at main entrances while secondary exits remain underused.
- Wayfinding behaviour is influenced by:
    - **Signage visibility** and clarity.
    - **Environmental cues** such as smoke flow or the behaviour of others.
    - **Knowledge of the building** (staff and residents evacuate differently from visitors).

7.9.3 Congestion and Flow Limitations

- Stairheads, lobbies, and doorways often become **bottlenecks**, slowing evacuation.
- Research into **flow rates** suggests:
    - Maximum sustainable flow through a standard exit is **~80–100 persons per minute per metre of width** under ideal conditions.
    - Flow is reduced under smoke, heat, or poor visibility.
- Evacuation modelling (e.g., Pathfinder, STEPS) can simulate congestion to test whether exit capacities are realistic.

7.9.4 Panic and Social Behaviour

- Contrary to popular belief, **panic is rare** in fire evacuations.
- More common behaviours include:
    - **Social affiliation** – people tend to evacuate in groups (families, co-workers).
    - **Role of leadership** – trained staff or authoritative figures can significantly improve evacuation efficiency.
    - **Uncertainty or miscommunication** – unclear or conflicting instructions can create disorder that resembles panic.

7.9.5 Vulnerable Occupants and Special Needs

- Behavioural assumptions must account for people who:
    - Have mobility or sensory impairments.
    - Are elderly, young children, or non-native speakers unfamiliar with instructions.
    - May be reluctant to evacuate (e.g., hospital patients, prisoners).
- Strategies must include **refuges, assisted evacuation provisions, and clear communication methods** tailored to these groups.

7.9.6 Implications for Fire Strategies

- **Escape design must allow for delays** in response and movement.
- **Exit capacities must be validated** against realistic flow rates, not just theoretical maxima.
- **Voice alarm systems** (BS 5839-8) and trained staff are often more effective than bells or sirens in prompting evacuation.
- Evacuation modelling should be calibrated with **behavioural research data**, not idealised assumptions.

Summary

Human behaviour in fire is shaped by psychology, environment, and social context. A fire strategy must recognise that:

- Occupants will often **delay evacuation**.
- Many will attempt to leave via **familiar entry routes**.
- **Congestion and bottlenecks** are inevitable in large buildings.
- **Clear communication and management** are more effective than relying on idealised assumptions.

By incorporating behavioural science and evacuation research, fire strategies become more **credible, inclusive, and effective**, ensuring that escape routes and procedures perform under real-world conditions.

7.10 Case Example (Illustrative)

To demonstrate how means of escape provisions are applied in practice, consider the fire strategy for a **large multi-storey shopping centre**:

- **Escape Provision**
    - Each retail level is served by **three independent exit staircases**, positioned to comply with the **45° separation rule**, ensuring no single fire can compromise all routes.
    - **Travel distances** are limited to a maximum of **30 m**, well within the limits of Approved Document B for assembly occupancies.
    - Stairs are enclosed in **120-minute fire-resisting construction**, with FD60S doorsets providing smoke protection.
- **Occupancy Load and Exit Capacity**
    - The calculated **design occupancy** is **2,400 persons** across the shopping levels.
    - Applying the standard ADB calculation of **5 mm per person of door width serving more than 220 persons**,
    - This is distributed across stair cores, with each staircase capable of serving 800 people at an acceptable flow rate.
- **Evacuation Strategy**
    - A **simultaneous evacuation** strategy is adopted, suitable for this high-occupancy public building.
    - To ensure clarity and minimise pre-movement delays, a **voice alarm system** compliant with **BS 5839-8** provides spoken instructions rather than relying on bells or sirens.
- **Provisions for Vulnerable Occupants**
    - **Refuge areas** are provided adjacent to each stair core, sized to accommodate mobility-impaired occupants.
    - Each refuge is equipped with an **emergency voice communication system** (**EVC**), linked to the fire control room, enabling coordination of assisted evacuation.
- **Smoke Control and Tenability**
    - A **mechanical smoke ventilation system** is installed in the central atrium, designed in accordance with **BS EN 12101**.
    - The system is capable of maintaining a tenable environment for at least **30 minutes**, ensuring that evacuation can be completed before conditions deteriorate.

    - Pressurisation fans protect escape stairs, preventing smoke ingress.
- **Integration with Firefighting Facilities**
    - A dedicated **fire control room** at ground level houses the alarm panel, smoke control interfaces, and CCTV monitoring.
    - Fire service access is provided on three sides of the building, with hydrants within 90 m of each entrance.

Commentary

This case demonstrates how a fire strategy integrates:

- **Escape design** (travel distances, exit widths, number and separation of routes).
- **Occupant management** (voice alarms, refuges, EVC systems).
- **Engineering provisions** (smoke control and tenability analysis).
- **Fire service facilities** (control room and access).

By combining prescriptive measures (travel distances, exit widths) with performance-based analysis (smoke control maintaining 30-minute tenability), the strategy ensures a robust, defendable, and inclusive approach to occupant safety.

7.11 Summary

The provision of **means of escape** is the single most critical element of a fire strategy. It represents the ultimate purpose of all fire safety measures: to ensure that every occupant — regardless of age, mobility, or familiarity with the building — can reach a place of ultimate safety before conditions become untenable.

This chapter has demonstrated that effective means of escape relies on:

- **A strong regulatory framework** – Building Regulations Part B, Approved Document B, BS 9999, and BS 9991 provide the foundation for prescriptive and risk-based approaches.
- **Core design principles** – escape routes must be available, accessible, independent of fire service intervention, and tenable for the duration of evacuation.

- **Quantifiable design parameters** – including travel distance limits, exit width and capacity calculations, the number and separation of exits, and door and stair design.
- **Evacuation strategies tailored to use and occupancy** – ranging from simultaneous evacuation in small premises to phased, zoned, defend-in-place, or progressive horizontal evacuation in complex and vulnerable occupancies.
- **Inclusive provisions for vulnerable occupants** – through refuges, evacuation lifts, evacuation aids, and Personal Emergency Evacuation Plans (PEEPs).
- **Supporting systems** – signage, emergency lighting, and wayfinding aids that ensure escape is possible even in low visibility or high-stress situations.
- **Maintenance of tenability** – achieved through compartmentation, smoke control, protected lobbies, and fire-resisting doorsets.
- **Realistic human behaviour assumptions** – acknowledging pre-movement delays, route familiarity, congestion, and the influence of communication and leadership.
- **Case applications** – illustrating how these principles combine in real projects, such as the shopping centre example where prescriptive and engineered measures ensure both safety and practicality.

Looking Ahead

Means of escape provisions do not exist in isolation. Their effectiveness depends on the building's ability to **contain fire and smoke long enough** for evacuation to succeed. The next chapter will therefore examine **Compartmentation and Passive Fire Protection** — the structural backbone of fire safety — which ensures that escape routes remain viable and that fire does not spread unchecked between building areas.

Chapter 8

Compartmentation and Passive Fire Protection

8.1 Introduction

Passive fire protection (PFP) forms the **first line of defence** in building fire safety. Unlike active systems — such as sprinklers, alarms, or smoke control — which rely on detection, power, or mechanical activation, passive systems are **built into the structure of the building itself**. Their role is to contain fire and smoke at their source, protect structural stability, and preserve escape routes long enough for occupants to evacuate and for the fire service to intervene.

At the heart of passive fire protection lies **compartmentation**: the division of a building into fire-resisting sections designed to resist fire spread. By preventing rapid horizontal or vertical fire growth, compartmentation limits the size of any single fire event, reduces risk to life, and prevents catastrophic property loss.

8.1.1 Why Passive Fire Protection Matters

- **Reliability**: Passive measures are always "on." They do not depend on detection, power, or human intervention.
- **Life Safety**: By restricting the movement of fire and smoke, they protect occupants in unaffected areas and preserve the tenability of escape routes.
- **Property Protection and Resilience**: Limiting fire spread helps prevent total building loss and supports business continuity.
- **Firefighter Safety**: Well-compartmented buildings provide firefighters with defined zones to operate within, improving control and reducing operational risk.
- **Regulatory Compliance**: Building Regulations Part B, BS 9999, and BS EN standards all place heavy emphasis on compartmentation as the foundation of fire safety.

8.1.2 Core Elements of Passive Fire Protection

While compartmentation is central, PFP encompasses a wider range of measures, including:

- **Fire-resisting walls and floors** – maintaining separation between compartments.
- **Fire doorsets** – controlling smoke and flame spread through openings.
- **Cavity barriers and fire-stopping** – sealing voids, ducts, and service penetrations.
- **Structural fire protection** – protecting steel, concrete, or timber frames from premature collapse.
- **Fire-resisting glazing and partitions** – maintaining visibility and openness while providing separation.

8.1.3 Integration with the Fire Strategy

Compartmentation and PFP must be designed and documented as part of the fire strategy, ensuring:

- Escape routes are **protected and remain tenable**.
- **Active fire safety systems** are supported — sprinklers and alarms may fail if compartments are compromised.
- **Fire service facilities** can function effectively within a safe operational window.
- The building is resilient to **refurbishment, fit-out changes, and misuse**, which often compromise compartmentation in practice.

8.1.4 Scope of This Chapter

This chapter explores the **principles, applications, and best practice** for compartmentation and passive fire protection, including:

- The **regulatory framework** governing compartmentation.
- Principles of **fire and smoke separation**.
- Fire resistance ratings and testing standards.
- Fire doors, glazing, and cavity barriers.
- Common pitfalls in design and maintenance.
- Case examples where compartmentation succeeded — or failed — in controlling fire spread.

Summary

Passive fire protection is the **silent guardian** of fire safety, always present, always working, and fundamental to the success of the overall fire strategy. By embedding compartmentation and structural protection into the very fabric of a building, designers and operators provide the stability and containment that safeguard lives, protect property, and support firefighting operations.

8.2 Principles of Compartmentation

Compartmentation is the **central principle of passive fire protection**. By dividing a building into discrete, fire-resisting sections, it prevents uncontrolled fire and smoke spread, protects occupants during evacuation, and preserves the building's structural stability for firefighting operations.

8.2.1 Objectives of Compartmentation

The key objectives are to:

- **Contain fire within its compartment of origin**
  - Prevents rapid fire growth into adjacent spaces.
  - Limits the maximum size of a single fire event.
- **Protect escape routes and refuges**
  - Ensures stairs, lobbies, and protected corridors remain tenable during evacuation.
  - Supports evacuation strategies such as defend-in-place or progressive horizontal evacuation.
- **Safeguard structural elements**
  - Fire-resisting compartments protect columns, beams, and loadbearing walls, reducing the risk of **premature collapse**.
  - Provides firefighters with confidence that they can operate safely for a defined duration.
- **Reduce risk of external fire spread**
  - Compartmentation also extends to **external walls and boundaries**, reducing the chance of fire spreading to neighbouring buildings.

8.2.2 Design Considerations

1. **Size and Extent of Compartments**
   - Compartment sizes are limited by Building Regulations and guidance documents (ADB, BS 9999).
   - Typical maximum sizes:
     - **7,000 m²** for industrial buildings without sprinklers for not more than 18m and 2000 m² when more than 18m (ADB).
     - Significantly larger compartments may be permitted with sprinklers and engineered solutions.
   - In healthcare, residential, or high-risk occupancies, smaller compartments (e.g., 500–750 m²) may be required to support safe evacuation.
2. **Fire Resistance Ratings**
   - Structural and separating elements are rated in minutes (typically **30, 60, 90, or 120+ minutes**).
   - Ratings cover:
     - **Integrity (E)** – resistance to flames and hot gases.
     - **Insulation (I)** – limiting temperature rise on the unexposed face.
     - **Loadbearing capacity (R)** – structural performance under fire.
   - Higher ratings are generally required in **high-rise, healthcare, or high-occupancy buildings**.
3. **Continuity of Fire Resistance**
   - Fire-resisting elements must form a **continuous barrier** without weak points.
   - Particular attention is needed at:
     - **Junctions** (wall-to-floor, wall-to-roof, wall-to-wall).
     - **Service penetrations** (ducts, pipes, cables).
     - **Voids and cavities**, where fire or smoke can bypass barriers.
   - Correct installation of **fire-stopping, cavity barriers, and intumescent seals** is essential.

8.2.3 Integration with Fire Strategy

Compartmentation is not an isolated measure but underpins:

- **Means of escape** – protected corridors, lobbies, and stairs.
- **Evacuation strategies** – progressive horizontal evacuation in healthcare relies on robust compartment walls.
- **Firefighting operations** – enabling safe, staged fire suppression.
- **Property protection** – preventing entire buildings or complexes from being lost.

8.2.4 Challenges and Risks

- **Poor workmanship** (gaps around penetrations, missing cavity barriers).
- **Post-construction alterations** (e.g., new service runs breaching compartments).
- **Inadequate documentation** – making it unclear whether barriers are continuous.
- **Maintenance gaps** – fire-stopping is often neglected compared to active systems.

Summary

Compartmentation is the **bedrock of passive fire protection**, with objectives spanning life safety, structural protection, and resilience. Its effectiveness depends on careful design of compartment size, fire resistance ratings, and continuity, supported by **rigorous installation, inspection, and maintenance**.

8.3 Fire Resistance Ratings

Fire resistance ratings define the **duration for which a building element can resist fire exposure** while maintaining its intended function. Compartment walls, floors, structural frames, and doorsets are all required to achieve specific fire resistance periods depending on the **use, height, and risk profile of the building**. These ratings provide confidence that escape routes will remain tenable, compartmentation will hold, and structural collapse will not occur prematurely.

8.3.1 Performance Criteria

When tested under standard fire conditions (e.g., **BS EN 1363-1**), building elements are assessed against three key performance criteria:

- **Structural Stability (R – Resistance)**
  - The ability of a loadbearing element (e.g., beams, columns, floor slabs) to continue carrying its design load without collapse during fire exposure.
  - Critical in multi-storey buildings where premature failure could cause progressive collapse.
- **Integrity (E)**
  - The ability of a separating element (e.g., wall, floor, door) to prevent the passage of **flames and hot gases** that could ignite materials on the unexposed side.
  - Integrity failure is typically identified when sustained flaming or gap openings occur during fire testing.
- **Insulation (I)**
  - The ability to restrict **temperature rise** on the unexposed side, protecting occupants and preventing fire spread by heat conduction.
  - Maximum permitted temperature rise is usually **140°C average** or **180°C at any single point** above ambient.

Together, these criteria are expressed using the **REI classification system** in accordance with **BS EN 13501-2**.

8.3.2 Common Classifications

- **Fire Doors (FD ratings)**
  - Classified as **FD30, FD60, FD90, or FD120**, representing the number of minutes of fire resistance.
  - FD ratings typically cover **integrity and (sometimes) insulation**, but not structural loadbearing.
  - Example: **FD60S** = 60 minutes fire resistance with smoke seals.
- **REI Ratings (BS EN system)**
  - Expressed as a triplet, e.g., **REI 60** = 60 minutes loadbearing capacity (R), integrity (E), and insulation (I).
  - Used for walls, floors, roofs, and other structural or compartment elements.

- **Specialist Classifications**
  - **R only**: Loadbearing capacity only (e.g., structural frames protected by intumescent paint).
  - **EI only**: Integrity and insulation only (non-loadbearing partitions, doors, glazing).
  - **RE only**: Where insulation is not required but flame spread prevention is essential (rare cases).

8.3.3 Determining Required Durations

The required fire resistance depends on:

- **Building Height**
  - Tall buildings (≥18 m, ≥30 m, or ≥60 m thresholds in ADB) require longer fire resistance (often 90 or 120 minutes).
- **Occupancy Profile**
  - Higher occupancies (e.g., assembly, healthcare, residential towers) demand greater protection than small, low-risk workplaces.
- **Use and Function**
  - Hospitals and care homes typically require 60–120 minutes due to progressive evacuation strategies.
  - Single-storey warehouses may only need 30 minutes if unoccupied during night hours.
- **Structural System**
  - Concrete inherently provides longer fire resistance than unprotected steel, which requires intumescent coatings or encasement.

8.3.4 Testing and Certification

- Fire resistance performance is determined through **standard fire resistance tests**:
  - **BS EN 1364 / 1365** – non-loadbearing and loadbearing walls, floors, and roofs.
  - **BS EN 1634** – fire doors and shutters.

    - **BS EN 13381** – applied fire protection systems (e.g., intumescent paints).
- Products must carry **third-party certification** (e.g., BWF-CERTIFIRE for doors) to ensure tested performance is replicated in practice.
- Installation quality and ongoing maintenance are as critical as the initial test evidence.

8.3.5 Integration into Fire Strategy

The fire strategy must clearly specify:

- The **fire resistance ratings** required for structural and compartment elements.
- The **standards of testing and certification** applied.
- The rationale for durations chosen (e.g., ADB compliance, BS 9999 risk-based adjustments, BS 7974 performance-based justification).
- Integration with **means of escape timelines** to ensure RSET remains within available protection.

Summary

Fire resistance ratings provide a **quantifiable measure** of how long building elements can withstand fire conditions. Whether expressed as **FD ratings for doors** or **REI classifications for structures**, these ratings form the backbone of compartmentation design. By ensuring that walls, floors, doors, and frames maintain integrity, insulation, and structural stability for appropriate durations, fire strategies provide confidence that escape, containment, and firefighting objectives can be achieved.

8.4 Internal Fire Separation

Internal fire separation is a core component of **compartmentation** and passive fire protection. By subdividing a building into fire-resisting sections, designers limit the spread of fire and smoke, protect escape routes, and provide safe refuges for occupants. Internal separation is particularly critical in **multi-occupancy and high-dependency buildings**, where evacuation may be delayed or staged.

8.4.1 Applications of Internal Fire Separation

- **Between Occupancies**
  - Flats, hotel rooms, and offices must be separated from each other by fire-resisting walls and floors.
  - Prevents a fire in one unit compromising safety in another.
  - Central to the "stay put" strategy in high-rise residential buildings, where each flat is a protected compartment.
- **Between Risk Areas**
  - High-risk rooms (plant rooms, kitchens, laboratories, storage spaces, IT/data centres) must be enclosed in fire-resisting construction.
  - Protects escape routes and low-risk spaces from high fire load or ignition sources.
  - Often requires **enhanced resistance** (60–120 minutes) compared with general-use compartments.
- **Between Storeys**
  - Floor slabs act as fire-resisting barriers between levels, stopping vertical fire and smoke spread.
  - Any penetrations (services, risers, ducts) must be properly **fire-stopped**.
  - In multi-storey buildings, compartment floors are critical to prevent rapid vertical fire spread.
- **Service Risers and Shafts**
  - Shafts carrying utilities (electrical, mechanical, plumbing) are natural fire and smoke pathways.
  - These must be enclosed in fire-resisting construction and **fire-stopped at each floor level**.
  - Access doors to risers should be rated (FD30/FD60) and fitted with smoke seals.

8.4.2 Sub-Compartmentation in Healthcare and High-Dependency Buildings

- In hospitals, care homes, and other high-dependency occupancies, **progressive horizontal evacuation (PHE)** requires **sub-compartmentation**:
  - Large fire compartments are divided into **smaller sub-compartments**, each typically providing 30–60 minutes fire resistance.
  - Occupants are moved horizontally from the fire-affected sub-compartment into an adjacent protected area on the same level.
  - Evacuation is therefore achieved in stages, without immediate reliance on stairs or external exits.
  - This strategy reduces stress on vulnerable occupants and provides staff with manageable evacuation zones.

8.4.3 Design Considerations

- **Fire Resistance Ratings**: Typically 60 minutes for general separations; up to 120 minutes for high-risk areas.
- **Continuity of Barriers**: Walls and floors must extend to the underside of the structural slab or roof, avoiding gaps above suspended ceilings or below raised floors.
- **Penetrations**: Services passing through barriers must be sealed with tested fire-stopping systems (e.g., intumescent collars, fire-rated boards).
- **Doors and Access Points**: Any openings in separating elements must be fitted with fire doorsets of equivalent rating.
- **Testing Standards**: Construction and products must comply with BS EN 1364 (non-loadbearing walls), BS EN 1365 (loadbearing walls and floors), and BS EN 13501 classification.

8.4.4 Integration with Fire Strategy

Internal fire separation underpins:

- **Means of Escape** – ensuring that escape routes remain protected.
- **Evacuation Strategies** – enabling staged or progressive evacuation.
- **Firefighting Access** – providing defined zones where crews can operate safely.

- **Resilience** – limiting fire to small zones reduces business interruption and protects critical assets.

Summary

Internal fire separation is essential to preventing uncontrolled fire spread within buildings. By separating occupancies, isolating high-risk areas, maintaining compartment floors, and enclosing service risers, the fire strategy ensures that fire and smoke are contained long enough for evacuation and intervention. In healthcare and high-dependency occupancies, **sub-compartmentation** further supports progressive evacuation, safeguarding vulnerable occupants.

8.5 Fire-Stopping and Penetrations

Compartmentation only remains effective if all penetrations, voids, and junctions are **properly sealed with tested fire-stopping systems**. In practice, the most common failures in passive fire protection occur not in the design of walls and floors, but at the points where those elements are breached — by services, ducts, joints, or hidden cavities. A robust fire strategy must therefore set out the requirements for **fire-stopping, cavity barriers, and linear seals**, ensuring continuity of fire resistance throughout the building.

8.5.1 Service Penetrations

- **Typical Risks**: Pipes, cables, and ductwork passing through compartment walls and floors. If not sealed correctly, they act as conduits for fire and smoke spread.
- **Solutions**:
    - **Intumescent collars and wraps** – expand around plastic pipes when exposed to heat, sealing the opening.
    - **Fire-rated boards and mortars** – used for larger openings.
    - **Sealants and mastics** – for small gaps and joints, tested to BS EN 1366-3.
- **Requirements**:
    - Products must be tested as **systems**, not just components, and installed in accordance with certification.
    - Every service penetration should be **recorded and logged** in as-built documentation for later inspection.

8.5.2 Cavity Barriers

- **Purpose**: Hidden cavities in walls, floors, roofs, and cladding systems allow rapid, unseen fire spread.
- **Applications**:
    - Concealed voids above suspended ceilings.
    - Raised access floors.
    - Roof spaces and concealed wall cavities (especially in rainscreen cladding systems).
- **Design Standards**:
    - **ADB** requires cavity barriers at maximum spacing and around openings, junctions, and cavity edges.
    - **BS EN 1366-4** sets out test methods for cavity barriers.
- **Installation**:
    - Cavity Barriers must be **continuous, correctly fixed, and fitted tightly**.
    - Incorrect or missing cavity barriers have been a major factor in external fire spread, including the **Grenfell Tower disaster**.

8.5.3 Linear Joints and Junctions

- **Definition**: Points where compartment walls meet floors, ceilings, roofs, or other walls.
- **Risks**: Gaps or imperfections at these joints compromise compartmentation.
- **Solutions**:
    - Fire-rated sealants, tapes, or flexible joint systems tested to **BS EN 1366-4**.
    - Systems must allow for expected building movement (thermal or structural) without breaking the fire seal.
- **Best Practice**: Detail these junctions in design drawings, and verify during installation inspections.

8.5.4 Testing and Certification

- **Testing Standards**:
  - **BS EN 1366-3** – service penetrations.
  - **BS EN 1366-4** – linear joint seals and cavity barriers.
- **Certification**:
  - Products must be installed exactly as tested — substitution or improvisation voids certification.
  - Third-party accreditation schemes (e.g., LPCB, IFC Certification) should be used to verify products and installers.
- **Documentation**:
  - Fire strategies should include product references, certificates, and **as-built fire-stopping schedules** to aid future maintenance.

8.5.5 Common Failures in Practice

- Services installed after construction (retrofit cabling, IT, plumbing) left unsealed.
- Unauthorised trades removing or damaging barriers.
- "Foam-filled" gaps using non-rated materials instead of certified systems.
- Missing or poorly fitted cavity barriers in roof spaces and cladding systems.

Summary

Fire-stopping and penetration seals are critical to the success of compartmentation. By ensuring that **service penetrations, cavity barriers, and linear joints** are properly detailed, tested, certified, and documented, the fire strategy guarantees continuity of fire resistance throughout the building. Failures at these points are among the most common — and most dangerous — causes of fire spread, making rigorous specification and inspection essential.

8.6 Surface Finishes and Internal Linings

The choice of surface finishes and internal linings has a profound effect on **fire growth, smoke production, and tenability**. While structural elements and compartmentation limit the spread of fire between zones, the lining materials inside compartments can determine how quickly a fire develops, how much smoke is produced, and whether escape routes remain usable.

8.6.1 Role of Internal Linings in Fire Development

- **Ignitability** – some finishes (untreated timber, plastics, fabrics) ignite quickly and contribute to rapid fire spread.
- **Flame Spread** – linings can allow flames to travel rapidly across walls and ceilings, overwhelming detection and suppression systems.
- **Smoke and Toxic Gas Production** – low-quality or inappropriate finishes produce **dense, toxic smoke**, impairing visibility and increasing risk to occupants and firefighters.

8.6.2 Classification Systems

1. **BS 476 System (UK National Standard)**
    - **Class 0** – highest classification under BS 476 for surface spread of flame (typically required in escape routes).
    - **Class 1–3** – lower levels of performance; Class 3 acceptable only in low-risk areas.
2. **BS EN 13501-1 European Standard**
    - Provides harmonised classification of reaction to fire.
    - Ratings expressed as:
        - **Euroclass A1 / A2** – non-combustible (e.g., plasterboard, mineral wool).
        - **Euroclass B–F** – varying levels of combustibility.
        - **s1–s3** – smoke production (s1 = little smoke, s3 = substantial smoke).
        - **d0–d2** – flaming droplets/particles (d0 = none, d2 = significant).
    - Example: **B-s1,d0** = limited combustibility, very low smoke, no flaming droplets.

8.6.3 Fire Strategy Requirements

- **Escape Routes**
    - High-performance linings required: **Class 0 (BS)** or **B-s3,d2 (EN)** as a minimum.
    - Best practice: **B-s1,d0** for superior smoke performance.

- **General Areas**
  - Offices, classrooms, and low-risk spaces may permit **Class 3 (BS)** or **D-s2,d2 (EN)** materials.
  - Consideration must be given to **furniture and contents**, which often present higher fire loads than the linings themselves.
- **High-Risk Uses**
  - In **healthcare, sleeping accommodation, or assembly buildings**, higher performance is mandated: **Class 0 (BS) or B-s3,d2 (EN) or better** throughout.
  - Reduced smoke production (s1) is strongly recommended to preserve visibility for evacuation and reduce toxic exposure.

8.6.4 Examples of Surface Materials

- **Painted Plasterboard** – typically achieves **B-s1,d0**, making it suitable for most applications including escape routes.
- **Timber Panelling** – must be treated with **intumescent coatings or pressure-impregnation** to achieve Class 0 / Class B.
- **Fire-Rated Wall Laminates** – composite panels tested to **BS EN 13501-1**, commonly used in healthcare and commercial interiors.
- **Suspended Ceilings** – tiles must also be classified, especially in escape routes, to avoid hidden flame spread.

8.6.5 Design and Management Considerations

- **Consistency**: All surfaces along escape routes must achieve the same classification to prevent weak links.
- **Retrofitting**: Renovations or decorative finishes (e.g., new wall coverings, acoustic panels) must be checked for compliance.
- **Testing and Certification**: Products must carry valid test certificates to **BS EN 13501-1** or **BS 476**, and installation must match the tested specification.
- **Integration with Active Systems**: In some engineered solutions, slightly lower classifications may be acceptable if supported by sprinklers or smoke control.

Summary

Surface finishes and internal linings have a decisive influence on fire growth and smoke production. By specifying materials to **Class 0 / B-s3,d2 or better** in escape routes, adopting higher standards in high-risk occupancies, and ensuring products are properly certified and installed, fire strategies ensure that fires are contained, smoke hazards are reduced, and occupants are afforded maximum time for safe evacuation.

8.7 Pipe Insulation and Ductwork

Services such as **pipework and ventilation ducts** run throughout buildings, frequently penetrating compartment walls and floors. If not properly designed and protected, these systems can act as hidden pathways for **fire, smoke, and toxic gases**, undermining escape routes and allowing uncontrolled fire spread. The fire strategy must therefore clearly specify the requirements for **insulation materials, ductwork protection, and fire-stopping systems**.

8.7.1 Pipe Insulation

- **Hidden Hazards**: Many pipe insulations are made from combustible foams (e.g., polyethylene, phenolic, nitrile). While these improve thermal efficiency, they can contribute significantly to fire growth and smoke production.
- **Escape Routes**:
    - Only **non-combustible insulation** should be used, typically **mineral wool or Class A1/A2-s1,d0 materials** (as per BS EN 13501-1).
    - Ensures that stair cores, lobbies, and corridors remain free from hidden fire hazards.
- **Other Areas**:
    - Limited use of combustible insulation may be permitted in low-risk, non-escape areas, provided it is fully **enclosed in fire-resisting construction** or treated with intumescent coatings.
- **Penetrations**:
    - Pipe insulation must not bypass fire-stopping; service penetrations require **intumescent wraps or collars** that expand to seal gaps when insulation burns away.

8.7.2 Ductwork

- **Risk Pathways**: Ventilation systems can carry smoke and flame rapidly through compartments if left unprotected.
- **Fire Dampers**:
  - Required where ducts pass through **compartment walls and floors**.
  - Designed to close automatically under fire conditions, maintaining compartment integrity.
  - Must be tested to **BS EN 1366-2** and installed to manufacturer's instructions.
- **Smoke Control Dampers**:
  - Used in engineered smoke control systems; operate selectively to extract smoke or protect escape routes.
- **Fire-Resisting Ducts**:
  - In some cases (e.g., kitchen extract, smoke control), ducts themselves must be constructed of **fire-resisting materials** tested to **BS EN 1366-1 or -8**.
  - Common in high-rise or large commercial buildings where smoke control is critical.

8.7.3 Integration with Compartmentation

- **Fire-Stopping at Service Penetrations**:
  - All pipe and duct penetrations must be sealed with tested systems (e.g., intumescent mastic, fire boards, mortar seals) to maintain compartment rating.
- **Continuity of Protection**:
  - Fire resistance must be maintained along the entire route of the duct or pipework, not just at junctions.
- **Maintenance Access**:
  - Fire and smoke dampers must be accessible for **inspection and regular testing**, typically every 12 months under BS 9999 and HTM 03-01 (for healthcare).

8.7.4 Common Failures in Practice

- Using **combustible pipe insulation** in escape routes.
- Dampers omitted or incorrectly installed in ductwork penetrations.
- Service penetrations sealed with **non-rated foams or sealants.**
- Lack of maintenance access, making inspection of dampers impossible.

Summary

Pipe insulation and ductwork present significant risks if not correctly specified and protected. In escape routes, only **non-combustible insulation** should be used. All ductwork must incorporate **fire and smoke dampers** at penetrations and, where necessary, be constructed of **fire-resisting materials.** Correct sealing, certification, and ongoing maintenance are vital to ensure these hidden systems do not compromise the integrity of compartmentation or the safety of occupants.

8.8 Floor Coverings

Floor finishes contribute directly to **fire growth, flame spread, and smoke production**, especially in escape routes where safe evacuation depends on tenable conditions. While structural elements and wall linings tend to dominate compartmentation design, inappropriate floor coverings or adhesives can compromise safety by producing **dense smoke and toxic gases**, even if flame spread is limited. A fire strategy must therefore clearly specify performance standards for floor coverings, adhesives, and their installation.

8.8.1 Regulatory Classification

Floor coverings in the UK are classified under **BS EN 13501-1** using the **flooring-specific "fl" classes**:

- **Euroclass A1fl / A2fl** – Non-combustible (stone, ceramic tiles, concrete).
- **Euroclass Bfl-s1** – Very limited contribution to fire, very low smoke (common requirement for escape routes).
- **Euroclass Cfl–Efl** – Increasing levels of combustibility and smoke production, generally restricted to low-risk or domestic areas.
- **s1 / s2** – Smoke emission (s1 = low smoke, s2 = higher smoke).
- **d0 / d1** – Flaming droplets/particles (d0 preferred in escape routes).

Older UK classifications (BS 5287) have largely been replaced by the Euroclass system but may still be referenced in legacy projects.

8.8.2 Escape Routes

- **Requirement**: Floor coverings must be **Class Bfl-s1 or better**, ensuring minimal smoke and flame contribution.
- **Typical Compliant Materials**:
    - Rubber flooring.
    - Fire-rated vinyl or linoleum.
    - Ceramic or stone tiles.
    - Concrete screeds or resin-based finishes with certification.
- **Adhesives**: Often overlooked, adhesives must also be **fire-rated** to prevent early ignition and hidden flame spread beneath coverings.

8.8.3 Other Building Areas

- **General Circulation or Workspaces**:
    - Materials with **Cfl-s1 or Dfl-s1** may be acceptable in non-escape areas, subject to fire risk assessment.
    - Carpet tiles are commonly used but should carry Euroclass certification.
- **High-Risk Areas**:
    - Kitchens, laboratories, or industrial workshops require non-combustible or **A2fl/Bfl** rated flooring due to ignition hazards.
- **Residential / Low Occupancy Areas**:
    - Domestic carpet or laminate flooring (Class Dfl–Efl) may be acceptable where the risk profile is lower, though careful attention must still be paid to escape corridors.

8.8.4 Vulnerable Occupants and Special Considerations

- In **healthcare, schools, and care homes**, additional caution is required:
    - Floors must not release toxic fumes that could harm vulnerable occupants.
    - Slips, trips, and falls risk must also be balanced against fire performance.

  - Rubber and vinyl finishes are often specified for both fire safety and accessibility.

8.8.5 Installation, Testing, and Maintenance

- **Installation**: Products must be installed in line with the tested specification, including adhesives, underlays, and backing materials. Substituting materials voids certification.
- **Testing**: Floor coverings should be tested to **BS EN ISO 9239-1 (radiant panel test)** for flooring products.
- **Maintenance**: Floor coverings can degrade over time — wear, cleaning products, or retrofitted finishes may alter their performance. Regular inspection and specification control are recommended, especially in escape routes.

Summary

Floor coverings form an integral part of passive fire safety. In escape routes, only **Class Bfl-s1 or better** should be specified, ensuring low smoke, low flame spread, and no flaming droplets. In other areas, performance standards may be adjusted according to risk, but adhesives and installation details must also be fire-rated. Proper specification, certification, and maintenance ensure that flooring materials support, rather than compromise, the wider fire strategy.

8.9 External Walls and Cladding

The **external envelope** of a building plays a vital role in fire safety, both in terms of **preventing external fire spread** and protecting escape routes from external flames and smoke. Inadequate design or specification of cladding systems has been linked to several catastrophic fires, most notably the **Grenfell Tower tragedy in 2017**, which highlighted the dangers of combustible materials used in external wall systems.

8.9.1 Objectives of External Wall Fire Performance

- **Limit vertical and horizontal fire spread** on the external face of the building.
- **Prevent fire re-entry** into upper floors through windows, vents, or openings.

- **Protect adjacent buildings** by limiting radiant heat and flame projection.
- **Support compartmentation** by including cavity barriers at key junctions.

8.9.2 Regulatory Framework

- **Building Regulations Part B (ADB)**:
    - Requires that the external walls of buildings adequately resist fire spread.
    - Sets specific restrictions on materials used above certain heights.
- **Regulation 7(2)** (introduced in 2018 after Grenfell):
    - In England, all **new high-rise residential buildings (≥18 m)**, student accommodation, hospitals, and care homes must use materials achieving **Euroclass A1 or A2-s1,d0** throughout the external wall system.
    - This effectively bans combustible insulation and cladding panels in these buildings.
- **BS 8414 Fire Tests & BR 135**:
    - Large-scale façade fire tests that assess the fire performance of external wall systems.
    - Compliance routes may include testing to BS 8414 with assessment against BR 135 criteria.

8.9.3 Performance Requirements by Building Type

- **High-Rise Residential (≥18 m, UK)**
    - Cladding systems must be **non-combustible (Euroclass A1 or A2-s1,d0)**.
    - Applies to insulation, panels, membranes, and fixings within the wall build-up.
- **Low-Rise Residential and Non-Residential**
    - Greater flexibility permitted, subject to **boundary conditions**:
        - Where buildings are close to site boundaries, materials must limit external spread to neighbouring structures.
        - Where separation is greater, lower-performance materials may be acceptable, though risk assessment is essential.

Special Occupancies (Healthcare, Schools, Assembly Buildings)

Even at lower heights, higher performance (A2 or better) is recommended due to vulnerability of occupants.

8.9.4 Cavity Barriers and Fire-Stopping

- **Hidden Cavities** within cladding systems can act as chimneys for flame and smoke spread.
- **Cavity barriers** are required:
  - At **compartment floors and walls.**
  - Around **window and door openings**.
  - At the **edges of cladding systems** (roofline, wall junctions).
- **Performance Standards**:
  - Cavity barriers must provide at least **30 minutes fire resistance (E30)**, often higher in high-rise applications.
  - Tested to **BS EN 1366-4** or relevant European standards.

8.9.5 Lessons from Grenfell and Other Fires

- The Grenfell Tower fire demonstrated how combustible cladding and insulation can cause **rapid, uncontrolled fire spread**, overwhelming compartmentation and evacuation strategies.
- Other international cases (e.g., Dubai high-rise façade fires) reinforce the importance of **non-combustible external envelopes** in tall buildings.
- Fire strategies must **explicitly document** the external wall system, including:
  - Full material specification.
  - Fire classification (Euroclass A1/A2).
  - Location and detailing of cavity barriers.
  - Testing evidence (BS 8414 / BR 135 or equivalent).

8.9.6 Integration with Fire Strategy

The fire strategy must demonstrate that:

- External walls support the **compartmentation strategy**, preventing vertical fire spread across floors.

- Escape routes adjacent to façades are **protected from external flame and smoke exposure**.
- **Inspections and maintenance** are included in the management plan, particularly for rainscreen or ventilated cladding systems.

Summary

The external envelope is not just an architectural feature — it is a **critical fire safety barrier**. For high-rise residential and sensitive occupancies, **non-combustible external wall systems (Euroclass A1 or A2-s1,d0)** are now mandated. Across all building types, cavity barriers and robust detailing must ensure fire does not bypass compartmentation. The Grenfell Tower tragedy remains a stark reminder of the consequences of inadequate cladding design, underscoring the need for **explicit, evidence-based documentation** in every fire strategy.

8.10 Maintenance and Inspection of Passive Fire Protection

Passive fire protection (PFP) measures are only effective if their **integrity is preserved throughout the life of the building**. Unlike active systems, which are routinely serviced, passive systems are frequently overlooked once construction is complete. In practice, fire doors are wedged open, cavity barriers are omitted during refurbishment, or service penetrations are poorly sealed — all of which can fatally undermine compartmentation.

A robust fire strategy must therefore set out clear requirements for **maintenance, inspection, and control of changes** to ensure that PFP measures continue to deliver their intended performance.

8.10.1 Routine Inspections

- **Fire Doors**:
  - Should be inspected at least every **6 months** (or more frequently in high-traffic areas such as schools, hospitals, and hotels).
  - Checks include gaps, seals, closers, glazing, hinges, and signage.
- **Dampers and Smoke Control Elements**:
  - Fire and smoke dampers must be tested at commissioning and then at regular intervals (typically **annually** according to HTM 03-01 and Section 8.3.7).

- **Cavity Barriers and Fire-Stopping**:
  - Difficult to inspect once concealed; therefore, **as-built documentation and photographic evidence** are vital.
  - Opportunistic inspections should be carried out during maintenance or refurbishments.
- **Seals, Joints, and Intumescent Products**:
  - Must be checked for damage, wear, or inappropriate replacement during routine maintenance cycles.

8.10.2 Third-Party Certification

- **Products**:
  - All fire-resisting elements (doors, dampers, cavity barriers, intumescent paints) must be installed and maintained in line with the **tested and certified specification**.
  - Substitutions or "equivalent products" not tested in the same configuration can invalidate fire performance.
- **Installers**:
  - Fire strategies should recommend that only **third-party accredited contractors** (e.g., FIRAS, IFC Certification, BRE/LPCB approved) carry out PFP works.
  - This provides assurance of competency and traceability.

8.10.3 Record-Keeping and the Golden Thread

- **Fire Safety Logbooks**:
  - Should record all inspections, maintenance, and remedial works carried out on passive systems.
- **As-Built Drawings and Schedules**:
  - Must show compartment walls, fire doors, penetrations, and dampers, linked to product certification.
- **Golden Thread of Information**:
  - Particularly relevant under the **Building Safety Act 2022**, which requires higher-risk residential buildings to maintain transparent, digital fire safety records throughout the building's life.

- **Auditable Evidence:**
    - Photographic records of installations and repairs should be kept to demonstrate compliance.

8.10.4 Refurbishment and Alterations

- **Key Risk:** Fit-out contractors frequently compromise compartmentation by cutting new service penetrations or removing barriers.
- **Control Measures:**
    - All works must be subject to a **Permit-to-Work system**.
    - Fire-stopping penetrations should only be carried out by **certified installers** using tested systems.
    - Post-works inspections are mandatory to verify reinstatement of compartmentation.

8.10.5 Common Failures in Practice

- Fire doors wedged open or closers removed.
- Unsealed service penetrations after IT or M&E upgrades.
- Missing cavity barriers in roof voids or façade systems.
- Intumescent coatings painted over or damaged.
- Lack of record-keeping, leading to uncertainty over installed systems.

Summary

Passive fire protection is only effective if **maintained and inspected throughout the building's lifecycle**. Fire strategies must establish clear management requirements: **routine inspections, third-party certification, detailed record-keeping, and refurbishment control**. Embedding these requirements into building management processes ensures that compartmentation and other PFP measures continue to safeguard occupants long after construction is complete.

8.11 Case Example (Illustrative)

To illustrate the application of compartmentation and passive fire protection, consider a **new-build hospital ward** designed with vulnerable occupants in mind. The fire strategy includes the following measures:

- **Compartmentation and Sub-Compartmentation**
    - The ward is divided into **60-minute fire compartments**, ensuring that fire and smoke cannot spread beyond each ward zone for at least one hour.
    - Within each ward, **progressive horizontal evacuation (PHE)** is enabled by **30-minute sub-compartments**, allowing patients to be moved horizontally into an adjacent safe area without using stairs.
    - This approach supports the **Regulatory Reform (Fire Safety) Order 2005** and HTM 05-02 (Healthcare Guidance), reflecting the challenges of evacuating immobile patients.
- **Doors and Internal Separation**
    - All patient bedrooms are separated from circulation spaces with **FD30S doorsets**, providing 30 minutes of fire and smoke protection.
    - Doors are fitted with **cold smoke seals and self-closing devices**, tested and certified under BS EN 1634.
    - Staff are trained in assisted evacuation and the use of evacuation aids, with provisions for Personal Emergency Evacuation Plans (PEEPs).
- **Service Risers and Penetrations**
    - Every vertical riser is enclosed in **60-minute fire-resisting construction** and **fire-stopped at each floor level** to prevent vertical smoke spread.
    - **Mineral wool insulation** is used for all pipework within escape corridors, ensuring non-combustibility and preventing hidden fire spread.
- **Escape Routes and Protected Stairs**
    - Escape stair cores are constructed with **120-minute fire resistance**, exceeding the minimum requirements of Approved Document B, to reflect the high-risk occupancy profile.
    - Stair pressurisation systems maintain smoke-free conditions during evacuation and firefighting.

- **Surface Finishes and Floor Coverings**
    - All wall and ceiling finishes are specified to **Class B-s1,d0 (EN 13501-1)**, providing low smoke production and no flaming droplets.
    - Corridor flooring is **fire-rated vinyl with a non-slip finish**, meeting **Class Bfl-s1**, combining fire performance with patient safety in terms of slips and falls.
- **Management and Inspection Regime**
    - The fire strategy sets out a detailed inspection programme:
        - **Quarterly inspections** of all fire doors for gaps, seals, and closer performance.
        - **Annual audits** of fire-stopping measures, including penetrations and cavity barriers.
        - Inspections logged in a **digital fire safety management system** to maintain the golden thread of information.

Commentary

This case example demonstrates how **passive fire protection measures are layered**:

- **Compartmentation** ensures fires are contained.
- **Sub-compartments and fire doors** enable phased evacuation.
- **Service penetration control** prevents hidden vertical spread.
- **Escape routes** are reinforced with higher fire resistance, guaranteeing firefighter access.
- **Surface finishes and flooring** balance fire safety with functional needs.
- **Ongoing inspection and record-keeping** ensure measures remain effective in practice.

By integrating these elements, the hospital's fire strategy provides a **defensible, risk-appropriate design** that protects vulnerable occupants and supports long-term resilience.

8.12 Summary

Compartmentation and passive fire protection (PFP) form the **foundation of fire safety design**. Unlike active systems, which depend on detection, activation, or external intervention, passive measures are **built into the very fabric of the building**, providing constant, reliable protection throughout its life.

This chapter has shown that PFP is achieved through a **layered approach**, including:

- **Principles of Compartmentation** – dividing buildings into fire-resisting sections that limit spread and protect escape routes.
- **Fire Resistance Ratings** – specifying performance in terms of structural stability, integrity, and insulation, tailored to building use, height, and risk.
- **Internal Fire Separation** – between occupancies, high-risk areas, floors, and service risers, with sub-compartmentation in healthcare and other high-dependency occupancies.
- **Fire-Stopping and Penetrations** – ensuring services, joints, and cavities do not undermine compartment barriers.
- **Surface Finishes and Linings** – controlling flame spread and smoke production, particularly in escape routes.
- **Pipe Insulation and Ductwork** – eliminating hidden hazards and maintaining compartment integrity through dampers and non-combustible insulation.
- **Floor Coverings** – specifying low-smoke, low-flame materials in escape routes.
- **External Walls and Cladding** – preventing catastrophic external fire spread, with lessons drawn from Grenfell.
- **Maintenance and Inspection** – embedding management, inspection, and record-keeping into building operation to preserve PFP effectiveness.
- **Case Applications** – illustrating how PFP is layered into real projects, such as hospitals, to protect vulnerable occupants.

The key message is that **passive fire protection is not a "fit and forget" measure**. Its success depends on:

- Rigorous design and specification.
- Competent installation and third-party certification.
- Ongoing inspection, maintenance, and management.

Looking Ahead

The next chapter will build on these foundations by examining **active fire protection systems** — sprinklers, alarms, smoke control, and suppression technologies — and how they work in combination with passive measures to deliver a comprehensive fire safety strategy.

Chapter 9

Active Fire Protection Systems

9.1 Introduction

Active fire protection (AFP) systems are engineered measures designed to **detect, warn, control, or extinguish fire and smoke once an incident occurs.** Unlike **passive fire protection (PFP)**, which is integrated into the building fabric and provides constant protection, AFP systems require some form of **activation** — whether automatic (e.g., detectors, sprinklers) or manual (e.g., extinguishers, alarms triggered by occupants).

These systems often rely on **mechanical, electrical, and electronic components**, meaning their effectiveness is dependent on correct design, installation, commissioning, and ongoing maintenance. When integrated into a fire strategy, AFP systems work in tandem with passive measures, evacuation planning, and management controls to provide a **layered defence against fire risk.**

9.1.1 Critical Roles of Active Fire Protection

Active systems are central to a fire strategy for several reasons:

- **Early Detection and Warning**
    - Automatic fire detection (smoke, heat, flame detectors) alerts occupants quickly, even at night or in unstaffed areas.
    - Audible, visual, or voice alarms ensure all occupants — including those with sensory impairments — receive timely warning.
- **Fire Suppression and Control**
    - Sprinklers, water mist, gaseous, or foam systems suppress or control fire growth before it becomes uncontrollable.
    - In some occupancies (e.g., warehouses, data centres, high-rise residential), suppression is critical to limiting property loss and maintaining resilience.

- **Smoke Management**
  - Ventilation and smoke control systems prevent escape routes from becoming smoke-logged.
  - Atrium smoke extract, stair pressurisation, and automatic opening vents (AOVs) preserve tenability for evacuation and firefighting.
- **Supporting Firefighting Operations**
  - By slowing fire development, AFP systems **"buy time"** for the fire and rescue service to arrive and intervene.
  - Active systems may also integrate directly with firefighting facilities, e.g., sprinklers linked to rising mains or alarms connected to fire service response centres.

9.1.2 Integration with Fire Strategies

AFP systems cannot be designed in isolation. Their selection and specification must be aligned with:

- The **building's use and occupancy profile** (e.g., residential vs. industrial).
- The **compartmentation and PFP strategy** (active systems support but do not replace passive measures).
- **Evacuation strategy** (e.g., phased evacuation in high-rise offices often depends on voice alarms and smoke control).
- **Fire service access and facilities** (systems may extend survivability until crews arrive).

9.1.3 Scope of This Chapter

This chapter explores the **principal categories of active fire protection** and how they are addressed in fire strategies:

- **Detection and Alarm Systems** (automatic fire detectors, manual call points, voice alarms).
- **Automatic Suppression Systems** (sprinklers, water mist, gaseous, foam, and hybrid systems).
- **Smoke Control and Ventilation** (natural and mechanical systems, AOVs, pressurisation, extract).

- **Portable Firefighting Equipment** (extinguishers, hose reels).
- **Integration with Fire Service Operations** (system monitoring, alarm signalling, interfaces).

Each section will outline the **regulatory framework, design considerations, case studies, and common challenges**, showing how AFP systems provide critical protection when properly integrated into a comprehensive fire strategy.

Summary

Active fire protection systems are **dynamic safeguards** that work alongside passive protection to deliver life safety, property protection, and resilience. They ensure fires are **detected early, controlled rapidly, and managed effectively**, giving occupants time to escape and firefighters time to intervene. A fire strategy must therefore describe AFP provisions clearly, justify their selection, and ensure they are maintained as part of the building's safety management plan.

9.2 Fire Detection and Alarm Systems

The purpose of **fire detection and alarm systems** is to provide **timely warning** of fire, initiate evacuation, and trigger the activation of associated fire safety systems (e.g., door releases, smoke control, suppression). Early detection is often the difference between a controlled incident and a catastrophic loss of life or property.

A fire strategy must describe the **type, category, and extent of detection and alarm systems**, referencing relevant standards and justifying design decisions.

9.2.1 Automatic Fire Detection (AFD)

Automatic fire detection is used where fires need to be identified quickly or where reliance on human observation is insufficient.

- **Smoke Detectors**
    - **Optical (photoelectric)**: Detect visible smoke particles; effective for smouldering fires (e.g., upholstery, plastics).
    - **Ionisation**: Detect invisible combustion particles; faster response to flaming fires. Now less common due to environmental restrictions on radioactive sources.

   - **Aspirating Systems (VESDA)**: Continuously sample air via pipes; highly sensitive, ideal for critical environments such as data centres, museums, and cleanrooms.
- **Heat Detectors**
   - **Fixed Temperature**: Operate when a set threshold is reached (commonly 58°C or 72°C).
   - **Rate-of-Rise**: Detect abnormally rapid increases in temperature. Useful in kitchens or plant rooms where smoke detectors would produce false alarms.
- **Flame Detectors**
   - Detect infrared, ultraviolet, or combined IR/UV radiation.
   - Used in high-risk industrial areas (e.g., petrochemical plants, aircraft hangars) where fires develop rapidly with high radiant energy.

9.2.2 Manual Call Points (MCPs)

- Provide a means for occupants to raise the alarm manually.
- Located on **escape routes**, at building exits, and at every change of level.
- **Maximum travel distance** to an MCP is typically **45 m** (per ADB and BS 5839-1).
- Must be positioned at **1.4 m height**, clearly visible, and accessible to disabled occupants.
- Protective covers may be used to avoid false alarms in vulnerable settings (schools, public buildings).

9.2.3 Alarm and Warning Systems

Once activated, alarms must be **clearly perceivable** by all occupants.

- **Audible Sounders**
   - Bells, sirens, or electronic sounders providing a distinct fire alarm tone.
   - Sound pressure levels must exceed background noise by at least **5 dB(A)** (or **10 dB (A) in sleeping areas**).

- **Visual Alarms (Beacons)**
  - Essential for occupants with hearing impairments or in noisy environments (factories, clubs).
  - Typically use high-intensity flashing LEDs.
- **Voice Alarm / Public Address Systems**
  - Provide clear, intelligible voice messages instead of simple tones.
  - Reduce confusion and panic by delivering **specific instructions** ("Use stairwell A", "Do not use lifts").
  - Often used in **complex, high-occupancy, or public assembly buildings** (airports, shopping centres, stadiums).

9.2.4 Integration with Other Systems

Modern alarm systems often **interface with other fire safety measures**, including:

- **Magnetic Door Releases** – closing fire doors automatically.
- **Smoke Control Systems** – activating AOVs, pressurisation fans, or extract systems.
- **Lift Control** – sending passenger lifts to ground level and disabling normal operation.
- **Suppression Systems** – signalling sprinklers, water mist, or gas suppression panels.
- **Building Management Systems (BMS)** – transmitting alarms and fault signals to central monitoring.

9.2.5 Categories of System (BS 5839-1)

Fire detection and alarm systems for **non-domestic premises** are categorised by BS 5839-1:

- **Category L (Life Protection)**
  - **L1** – Maximum life protection: detectors throughout all areas.
  - **L2** – Detectors in escape routes and high-risk rooms.
  - **L3** – Detectors in escape routes and adjoining rooms.
  - **L4** – Detectors in escape routes only.
  - **L5** – Customised design based on specific risks.

- **Category M**
  - **No automatic fire detection** (no smoke or heat detectors)
  - Fire alarm is raised **manually**, typically by:
    - Operating a **manual call point (MCP)**, often called a *break-glass unit.*
    - Or, in smaller premises, simply shouting "fire" may suffice (where justified by risk assessment).
  - **Sounders** or **alarm bells** are installed to warn occupants once the alarm is raised.
  - Primarily intended to **alert occupants to evacuate**, not necessarily to summon the Fire and Rescue Service (though connection to an ARC may be added if required).
- **Category P (Property Protection)**
  - **P1** – Detectors throughout all areas for maximum protection.
  - **P2** – Detectors only in defined high-risk areas.

For **domestic premises**, systems are defined under **BS 5839-6** (Grades A–F, Category LD1–3).

9.2.6 Standards and Guidance

- **BS 5839-1:2017** – Non-domestic premises (design, installation, commissioning, maintenance).
- **BS 5839-6:2019** – Domestic dwellings.
- **EN 54 series** – European standards for system components.
- **Approved Document B** – Prescriptive requirements for detection and alarm in specific building types.

Summary

Fire detection and alarm systems are the **front line of active fire protection**, ensuring fires are detected rapidly, alarms are communicated effectively, and associated fire safety systems are activated. The fire strategy must state the **category of system provided, its coverage, integration with other systems, and compliance with BS 5839-1/6**.

9.3 Fire Suppression Systems

While passive fire protection and detection systems are essential, many buildings also require **active suppression** to control or extinguish fires at an early stage. Suppression systems are intended to **limit fire growth**, reduce smoke and heat release, and **protect occupants and property** until the fire and rescue service arrives. Their inclusion in a fire strategy depends on building type, occupancy, and risk profile.

9.3.1 Sprinkler Systems

- **Purpose**: Automatically suppress or control a fire, preventing it from reaching flashover.
- **Operation**: Each sprinkler head is independently heat-activated (commonly 68°C–74°C). Only heads directly exposed to fire discharge water.
- **Standards**:
  - **BS EN 12845** – Fixed firefighting systems – Automatic sprinkler systems (commercial/industrial).
  - **BS 9251** – Sprinkler systems for residential and domestic occupancies.
- **Applications**:
  - **Residential**: Increasingly mandated in **new high-rise residential buildings (>11 m in England, >18 m in Scotland and all residential buildings in Wales)**,
  - **Commercial**: Widely used in warehouses, shopping centres, and large public buildings.
- **Benefits**:
  - Proven to reduce fatalities and property damage by up to 80%.
  - Reliable (success rate >95% when correctly maintained).
- **Limitations**:
  - Risk of water damage.
  - Requires robust water supply (mains-fed or storage tanks with pumps).

9.3.2 Water Mist Systems

- **Operation**: Discharge ultra-fine water droplets (<1 mm diameter) that absorb heat and displace oxygen around the fire.
- **Advantages**:
  - High cooling efficiency with minimal water consumption.
  - Reduces smoke and toxic gases.
  - Limits water damage — ideal for **heritage buildings, museums, archives, and hospitals**.
- **Standards**:
  - **BS 8458** – Fixed water mist systems for residential/domestic use.
  - **BS 8489** – Water mist systems for commercial and industrial use.
- **Limitations**:
  - Performance depends on droplet size, nozzle design, and room geometry.
  - Requires extensive fire testing for approval in each application.

9.3.3 Gaseous Suppression Systems

- **Purpose**: Used where water is impractical or damaging. Flood the protected volume with an inert or chemical gas to extinguish fire.
- **Types**:
  - **Inert Gases** (e.g., Inergen, Argonite): Reduce oxygen concentration to below combustion threshold but above human survivability limits.
  - **Chemical Gases** (e.g., FM-200, Novec 1230): Interrupt the chemical chain reaction of fire.
- **Applications**:
  - IT/data centres.
  - Control rooms.
  - Archives, museums, and sensitive equipment areas.
- **Standards**:
  - **BS EN 15004** – Gaseous fire-extinguishing systems.

- **Considerations**:
  - Requires gas-tight enclosures and integrity testing (door fan tests).
  - Evacuation time must be factored in before discharge.

9.3.4 Foam Suppression Systems

- **Purpose**: Applied where flammable liquids or hydrocarbon hazards are present. Foam forms a blanket, suppressing vapour release and excluding oxygen.
- **Applications**:
  - Petrochemical plants.
  - Aircraft hangars.
  - Fuel storage depots.
  - High-hazard industrial facilities.
- **Types**:
  - **Low-Expansion Foam** – large-scale fuel fire control.
  - **Medium/High Expansion Foam** – filling volumes such as warehouses or tunnels.
- **Standards**:
  - **NFPA 11** and **BS EN 13565**.

9.3.5 Portable Fire Extinguishers

Even with advanced suppression systems, **portable extinguishers remain an essential first line of defence.**

- **Types and Applications**:
  - **Water** – Class A fires (wood, paper, textiles).
  - **Foam** – Class A & B fires (solids and flammable liquids).
  - **CO_2** – Electrical fires and small Class B fires.
  - **Dry Powder** – Versatile, but high residue (Class A, B, C fires).
  - **Wet Chemical** – Specifically for Class F fires (deep fat fryers, catering kitchens).
- **Requirements**:
  - Provision based on **risk assessment** (PAS 79, BS 5306-8).

    - Typically one 13A-rated extinguisher per 200 m^2 floor area.
    - Additional specialist extinguishers for high-risk areas (e.g., CO_2 in server rooms).
- **Training**:
    - Staff must be trained to recognise fire classes and extinguisher use.
    - Extinguishers should not be relied upon where fire growth is rapid — evacuation remains the priority.

9.3.6 Integration into Fire Strategies

A comprehensive fire strategy should:

- Identify where **automatic suppression** is required by law or risk assessment.
- Specify **standards** (e.g., BS 9251, BS EN 12845, BS 8458, BS EN 15004).
- Document water supply arrangements, system reliability, and maintenance regimes.
- Ensure **interfacing with detection, alarms, and smoke control** for coordinated response.
- Provide for **ongoing inspection and certification** under the Regulatory Reform (Fire Safety) Order 2005.

Summary

Fire suppression systems are vital tools in preventing small fires from escalating into major incidents. From **sprinklers in high-rise residential buildings** to **gaseous systems in data centres** and **foam suppression in petrochemical facilities**, their role in fire strategies is to **control fire growth, protect life, safeguard property, and enhance resilience**. Even the simplest provision — **portable extinguishers with trained staff** — can be decisive in stopping fire spread and protecting escape routes.

9.4 Smoke Control and Ventilation

Smoke, rather than flame, is the primary cause of fire-related injuries and fatalities. It reduces visibility, disorients occupants, and carries toxic gases such as carbon monoxide and hydrogen cyanide. For firefighters, smoke obscures the seat of the fire and creates untenable conditions in escape routes.

Effective **smoke control and ventilation systems** are therefore a cornerstone of active fire protection. Their role is not to extinguish the fire, but to:

- **Maintain tenable conditions** in escape routes during evacuation.
- **Assist firefighting operations** by improving visibility and reducing heat.
- **Limit smoke spread** to unaffected areas.
- **Protect property** by reducing secondary damage from heat and soot.

9.4.1 Natural Smoke Ventilation

- Relies on the **natural buoyancy** of hot smoke to rise and escape through **high-level openings**.
- Typical devices include:
    - **Automatically Opening Vents (AOVs)** in stairwells, lobbies, and atriums.
    - **Operable windows or rooflights** triggered by smoke detection.
    - **Smoke shafts** that channel smoke to the exterior.
- **Advantages**:
    - Simple, reliable (fewer moving parts).
- **Limitations**:
    - Performance depends on wind direction and pressure.
    - Less effective in tall buildings or complex geometries.

9.4.2 Mechanical Smoke Ventilation

- Uses **powered fans** to actively extract smoke and maintain a clear layer above escape routes.
- Common applications:
    - **Atriums and shopping centres**.
    - **Underground car parks**.
    - **Large assembly buildings** (stadia, airports).
- **Design Objectives**:
    - Maintain a **smoke-free layer of at least 2.5 m above floor level** to allow evacuation.

  - Ensure system capacity matches predicted smoke production (based on fire size and growth rate).
- **Integration**: Mechanical systems often work with **make-up air inlets** to replace the extracted volume.

9.4.3 Pressurisation Systems

- Supply **clean air under pressure** into escape routes (e.g., stairwells, firefighting lobbies).
- Prevent smoke ingress by maintaining a positive pressure difference between protected spaces and adjacent fire compartments.
- **Design Challenges**:
  - Pressure must be high enough to resist smoke entry, but not so high that doors become difficult to open (typically ≤100 N opening force).
  - Systems must allow controlled air relief to avoid overpressurisation.
- **Applications**:
  - High-rise residential and office buildings.
  - Healthcare and transport hubs where evacuation takes longer.

9.4.4 Make-Up Air Systems

- Provide **replacement air** for the volume extracted by smoke control systems.
- Prevents negative pressure that could collapse ducts or disrupt smoke layers.
- Must be carefully balanced:
  - Too much make-up air can **dilute smoke and spread it further**.
  - Too little can cause system failure.
- Typically achieved via **low-level openings, automatic vents, or mechanical supply fans**.

9.4.5 Standards and Guidance

- **BS EN 12101 series** – Smoke and heat control systems:
  - Part 2: Natural smoke and heat exhaust ventilators.

    - Part 3: Powered smoke and heat exhaust ventilators.
    - Part 6: Pressure differential systems.
    - Part 7: Smoke control ducts.
    - Part 8: Smoke control dampers.
- **BS 7346** – National guidance on smoke control systems.
- **Approved Document B** – prescriptive requirements for smoke ventilation in certain building types (e.g., common stairs in residential flats).
- **BS 9999** – Performance-based smoke control design guidance.

9.4.6 Common Issues in Practice

- Poor **maintenance** (fans not tested, vents painted shut).
- **Inadequate commissioning** leading to imbalanced flows.
- **Lack of integration** with alarms, fire doors, or building management systems.
- Occupants propping open **fire doors**, compromising pressurisation systems.

Summary

Smoke control and ventilation systems are critical to the safety of both occupants and firefighters. A fire strategy must:

- Define whether natural, mechanical, or pressurisation systems are appropriate.
- Ensure systems comply with **BS EN 12101** and relevant guidance.
- Integrate operation with alarms, door releases, and firefighting procedures.
- Provide for **regular inspection and testing** to maintain reliability.

In essence, while sprinklers control fire growth, smoke control systems ensure people can still **see, breathe, and escape**.

9.5 Emergency Lighting and Power Supplies

In the event of a fire, mains power is often one of the first systems to fail — either through direct damage, deliberate isolation by the fire service, or cascading effects of smoke and heat. For this reason, **emergency lighting and backup**

power supplies are essential to maintain life safety systems during evacuation and firefighting.

A fire strategy must clearly specify how **emergency illumination, exit signage, and power redundancy** are provided, ensuring critical systems remain operational until the building has been evacuated and fire service operations are complete.

9.5.1 Emergency Lighting

- **Purpose**: To provide safe illumination of escape routes, staircases, refuges, and assembly points if mains power fails.
- **Performance Requirements** (BS 5266-1:2016, BS EN 1838):
    - **Minimum lighting levels**:
        - 1 lux along the centre line of escape routes.
        - 5 lux in high-risk task areas (e.g., plant rooms, control centres).
    - **Uniformity**: Avoid sudden changes in brightness that could disorient occupants.
    - **Duration**: Typically 1 hour minimum, 3 hours for places of public assembly or sleeping accommodation.
- **Types of Systems**:
    - **Self-contained luminaires**: Each fitting has its own battery pack.
    - **Central battery systems**: A dedicated central battery or UPS supplies multiple fittings.
- **Maintenance**: Monthly functional checks and annual full-duration tests (per BS 5266).

9.5.2 Exit Signs

- **Purpose**: To provide clear, unambiguous guidance to the nearest safe exit.
- **Requirements**:
    - Signs must be **illuminated or internally lit** and visible in all conditions.

    - Symbols and pictograms must comply with **BS EN ISO 7010** (green running man pictogram with directional arrow).
    - Signs must remain visible during smoke conditions — often achieved with higher brightness or supplementary floor-level signage.
- **Best Practice**: Use **photo-luminescent (glow-in-the-dark) materials** as an additional layer of redundancy in case both mains and emergency systems fail.

9.5.3 Backup Power Supplies

- **Purpose**: To ensure all fire safety systems remain operational even during power loss.
- **Systems Supported**:
    - Fire detection and alarm systems.
    - Emergency lighting and exit signs.
    - Smoke control fans and AOVs.
    - Firefighting lifts and rising mains pumps.
    - Communication systems (e.g., fire telephones, radios and emergency voice communication systems).
- **Options**:
    - **Local battery backup**: Integrated into emergency luminaires, alarm panels, and AOV actuators.
    - **Central batteries / UPS**: Larger systems for critical facilities (e.g., control rooms, hospitals).
    - **Standby generators**: For large or high-risk buildings, capable of running for several hours.
- **Design Standards**:
    - BS EN 50171 (central power supply systems).
    - BS 9999 and BS 9991 (design guidance for backup supply durations).
    - BS EN 81-72 (firefighting lifts backup requirements).

9.5.4 Integration and Reliability

- Emergency lighting and power must be **automatically activated** within seconds of power failure.
- Fire strategy documents must specify:
  - The **system type** (self-contained vs central).
  - The **duration of supply** (minimum 1–3 hours depending on occupancy).
  - **Maintenance arrangements** (monthly checks, annual full-discharge testing).
- Redundancy should be considered in high-risk occupancies — e.g., **dual generators** or **duplicate feeds** for critical infrastructure.

9.5.5 Common Failures in Practice

- Inadequate testing of emergency lighting (batteries failing before rated duration).
- Exit signage obscured by later fit-outs or decorative finishes.
- Generators not maintained, failing to start during outages.
- Overloading of UPS systems beyond their rated capacity.

Summary

Emergency lighting and power supplies are **life-critical systems** that ensure occupants can evacuate safely and firefighters can operate effectively even when mains power is lost. The fire strategy must reference **BS 5266, BS EN 1838, BS EN 50171**, and other applicable standards, specify the **type and duration of emergency supply**, and embed **testing and maintenance requirements** into the building's management plan.

Together, these provisions ensure that **darkness and power loss never compromise life safety**.

9.6 Integration of Active Systems

Modern fire safety design is built on the principle of a **layered and coordinated response**. Active systems such as detection, alarms, smoke ventilation, suppression, and lifts are no longer standalone technologies. Instead, they are integrated to function as part of a **cohesive fire safety strategy**, ensuring that fire events trigger the right sequence of actions at the right time.

However, integration introduces complexity. If not properly designed and managed, systems can conflict, fail to activate, or even compromise safety. A fire strategy must therefore describe **how active systems interact, how control is prioritised, and how independence is preserved where required**.

9.6.1 Typical System Interactions

- **Fire Detection and Alarm → Other Systems**
  - Triggers **smoke control** systems (AOVs, fans, pressurisation).
  - Initiates **door release mechanisms**, ensuring escape routes are clear.
  - Sends a signal to **ground passenger lifts** (returning them to a safe floor and disabling normal operation).
  - Provides input to **firefighting lifts**, enabling fire service use.
- **Sprinkler Systems → Fire Alarm**

Flow switches in sprinkler risers signal the **fire alarm panel** when water is discharged.

Can trigger evacuation messages and fire service notification.

- **Voice Alarm Systems → Public Address**
  - In assembly buildings (e.g., airports, stadia, shopping centres), the **voice alarm system overrides background music or PA announcements**, broadcasting clear evacuation instructions.
- **Firefighting Facilities → Alarm and Control Systems**
  - Fire service control panels provide a **central interface**, showing system status (alarms, smoke control zones, suppression discharge).
  - Allows firefighters to **manually override automated sequences** if operationally required.

9.6.2 Building Management Systems (BMS)

- In many modern developments, the BMS acts as the **hub for system coordination**, integrating HVAC, lighting, security, and fire safety.
- While useful for monitoring and diagnostics, there are **critical safeguards**:

  - Fire safety systems must operate **independently of BMS software** to ensure fail-safe performance.
  - Priority must always be given to fire safety functions — e.g., in the event of conflict, fire alarm signals override all other controls.
  - Interfaces must be **hard-wired or fail-safe** (e.g., door release relays) rather than dependent on network communication alone.

9.6.3 Principles of Effective Integration

- **Fail-Safe Design** – Systems must default to a safe state in the event of power failure or communication loss (e.g., fire doors close, smoke vents open).
- **Zoning and Cause-and-Effect Programming** – Fire alarm cause-and-effect matrices define exactly what each detection event triggers. These should be documented in the fire strategy and tested during commissioning.
- **Redundancy** – Critical functions (e.g., alarm signals, sprinkler flow monitoring) should not rely on a single point of failure.
- **Testing and Commissioning** – Integrated systems must undergo **end-to-end scenario testing**, simulating real fire conditions to ensure sequences operate as intended.
- **Documentation** – Clear schematics and cause-and-effect tables should be included in the fire strategy and building logbook for future reference.

9.6.4 Common Challenges in Integration

- **Over-Reliance on BMS** – Fire functions should never be dependent on non-fire software that could fail or be overridden.
- **Incompatible Interfaces** – Different suppliers' equipment may not communicate effectively without proper protocol integration (e.g., BACnet, Modbus).
- **Human Factors** – Building managers and occupants may not understand system interactions, leading to misinterpretation during incidents.
- **Lack of Updates** – Refurbishments often alter alarm zoning or door layouts, but the integration matrix is not updated, leaving the system ineffective.

Summary

Integration of active fire protection systems ensures that alarms, suppression, smoke control, and evacuation support operate as a **coordinated safety net** rather than isolated measures. The fire strategy must define:

- **Which systems interact and how** (documented cause-and-effect).
- **Where fail-safe independence is required** (e.g., fire alarm overrides).
- **Testing and maintenance protocols** to ensure reliability throughout the building's life.

Done correctly, integration provides clarity, speed, and robustness in fire emergencies, ensuring that occupants are warned, escape routes are protected, and firefighters have the information and control they need.

9.7 Maintenance and Testing

Active fire protection systems are only effective if they remain in full working order throughout the life of the building. Unlike passive measures, they rely on **mechanical, electrical, and electronic components** that can degrade, fail, or be disabled if not properly managed.

A fire strategy must therefore set out a **clear maintenance and testing regime**, aligned with British Standards and regulatory requirements, to ensure that systems are always ready to operate in an emergency.

9.7.1 Routine Checks

- **Weekly Tests** (as per BS 5839-1 for fire alarms):
    - Test at least one manual call point per week, rotating through different zones.
    - Confirm alarm sounders operate correctly.
    - Record all tests in the fire logbook.
- **Monthly Checks**:
    - Visual inspection of fire alarm panels, extinguishers, sprinklers, and emergency lighting.
    - Test emergency lighting briefly to confirm function.
    - Operate at least one smoke vent or AOV to confirm response.

9.7.2 Scheduled Servicing

- **Quarterly Servicing**:
    - Alarms and detection systems should be inspected by a **competent contractor**.
    - Smoke control systems, dampers, and AOVs tested for correct operation.
- **Annual Servicing**:
    - Comprehensive inspection of all active systems.
    - Full 3-hour discharge test of emergency lighting (BS 5266-1).
    - Sprinkler system checks: water supply, pumps, flow switches.
    - Gas suppression integrity testing (door fan test).
    - Firefighting lifts tested to **BS EN 81-72**.

9.7.3 Competence and Certification

- Only **competent, third-party certified organisations** should service life safety systems. Recognised schemes include:
    - **LPCB (Loss Prevention Certification Board)**.
    - **BAFE (British Approvals for Fire Equipment)**.
    - **IFC Certification**.

Certification ensures that maintenance is undertaken in accordance with the relevant standards, and gives regulators, insurers, and building managers confidence in the system's reliability.

9.7.4 Record-Keeping and the Golden Thread

- All inspections, tests, and remedial actions must be recorded in a **fire safety logbook**, which may be paper-based or digital.
- Under the **Building Safety Act 2022**, high-risk residential buildings must maintain a **"golden thread" of information**, including digital fire safety records accessible throughout the building's life.
- Records should include:
    - Test dates, methods, and results.
    - Faults identified and corrective actions.
    - Names of contractors and certification details.

9.7.5 Consequences of Poor Maintenance

- **Case Example – Lakanal House (2009)**: Poorly maintained fire doors and inadequate fire safety management contributed to multiple fatalities.
- **Case Example – Grenfell Tower (2017)**: Failures in smoke ventilation maintenance and unclear inspection records were cited in the Inquiry.
- Many serious incidents demonstrate that even well-designed active systems are ineffective without continuous, documented maintenance.

Summary

A fire strategy must embed **maintenance and testing requirements** into the building's operational framework. This includes:

- **Routine checks** by building staff.
- **Regular servicing** by competent, certified contractors.
- **Comprehensive record-keeping** as part of the golden thread.
- **Ongoing compliance monitoring** by the Responsible Person under the RRO.

Ultimately, **an unmaintained active system is a failed system**. The fire strategy must therefore make clear that ongoing testing, servicing, and certification are not optional, but central to the building's fire safety plan.

9.8 Human Factors and Active Systems

No matter how advanced or automated, **active fire protection systems ultimately rely on human interaction**. From raising the initial alarm to responding to evacuation messages and operating firefighting equipment, people form the last link in the fire safety chain. A fire strategy must therefore address the **human interface with active systems** and ensure that training, communication, and management are designed to support effective response.

9.8.1 Raising the Alarm

- **Manual Call Points (MCPs)**:
    - Depend on occupants recognising signs of fire and activating them promptly.
    - Delays in raising the alarm are a common cause of extended evacuation times.

  - Training and awareness campaigns are essential to encourage **early use without hesitation**.
- **False Alarms**:
  - Excessive false alarms lead to complacency, where occupants may hesitate to respond.
  - Fire strategies should incorporate **false alarm reduction measures** (e.g., multi-sensor detectors, staff investigation procedures).

9.8.2 Portable Fire Extinguishers

- Extinguishers are effective only if staff:
  - Can identify the **right type of extinguisher** for the fire (e.g., water for Class A, CO_2 for electrical, wet chemical for deep fat fryers).
  - Are trained to **use them safely** without putting themselves at risk.
- Fire strategies should:
  - Provide **basic extinguisher training** for staff.
  - Make clear that extinguishers are for **incipient-stage fires only** — evacuation remains the priority.

9.8.3 Responding to Alarm and Voice Messages

- **Audible Alarms**: Occupants may delay action, assuming the signal is a drill or false alarm.
- **Voice Alarm/Public Address Systems**:
  - More effective than bells or tones in prompting orderly evacuation.
  - Messages must be **clear, concise, and language-appropriate** for the expected occupants.
  - In international hubs (e.g., airports, hotels), multilingual or pictorial aids may be required.
- **Behavioural Factors**:
  - People often seek confirmation before acting — known as **"confirmation bias"**.

- Well-designed voice alarms help overcome hesitation by providing **specific instructions** ("Leave by Stair A now").

9.8.4 Operating Firefighting Equipment

- Fire service equipment (e.g., hose reels, dry risers, firefighting lifts) may also require trained staff to make them usable before the fire service arrives.
- **Trained security or facilities teams** may act as first responders, tackling small fires or guiding evacuation until firefighters take over.
- Staff must understand the **limits of their role**: life safety comes first, not firefighting beyond safe capacity.

9.8.5 Training and Awareness

- **Staff Induction and Refreshers**: All staff should receive fire safety training on induction, with refresher sessions at least annually.
- **Fire Drills**: Should be carried out regularly to test human interaction with alarms, evacuation procedures, and firefighting equipment.
- **Personal Emergency Evacuation Plans (PEEPs)**: Special provisions for disabled or vulnerable occupants must be clearly understood by staff.
- **Signage and Instructions**: Simple, visible instructions on how to activate alarms, use extinguishers, and follow evacuation routes are critical.

9.8.6 Human Reliability and System Design

- Systems should be designed to **minimise reliance on perfect human behaviour**.
- Examples:
    - Automatic door closers reduce dependence on staff remembering to shut fire doors.
    - Clear signage reduces reliance on familiarity with the building.
    - Pre-recorded voice messages ensure consistency during emergencies.
- Where human action is required, the system should **support correct behaviour** through intuitive design (e.g., MCPs at exits, extinguishers at eye level).

Summary

Active fire protection systems succeed or fail depending on the **people who use them**. A fire strategy must therefore:

- Ensure training equips occupants to raise alarms and respond effectively.
- Provide clear instructions through **intelligible alarms, signage, and drills**.
- Recognise human limitations and design systems that are **simple, intuitive, and fail-safe**.

In this way, technology and people work together to provide a robust, realistic fire safety framework.

9.9 Case Example (Illustrative)

A large **underground car park** beneath a shopping mall provides an example of how multiple active fire protection systems are integrated into a single fire strategy.

- **Automatic Fire Detection**
    - The car park is fitted with **heat and smoke detectors** designed to operate even in dusty and vehicle-polluted environments.
    - Aspirating detectors are used in some zones to provide **early warning of smouldering fires**, such as from electrical faults in parked vehicles.
- **Alarm System and Communication**
    - On detection, the fire alarm panel initiates:
        - **Evacuation signals in the car park** to warn staff and any occupants.
        - **Voice alarm messages in the shopping mall above**, instructing visitors to evacuate safely and avoid using lifts.
    - The alarm also signals directly to the **fire service control room**, providing details of the incident zone.
- **Sprinkler Suppression**
    - A **BS EN 12845 sprinkler system** is installed throughout the car park.
    - In the event of fire, sprinklers suppress flames and prevent flashover, limiting fire growth until the fire service arrives.

  - Sprinkler flow switches send signals back to the alarm system for logging and response.

- **Mechanical Smoke Extraction**
  - The car park is equipped with **high-capacity extract fans** designed to clear smoke and maintain a **smoke-free layer at head height** along escape routes.
  - Extract is supported by **make-up air inlets** positioned at low level, ensuring that smoke removal is balanced and effective.
  - Fans are zoned so that only the affected area is activated, reducing unnecessary smoke spread.
- **Backup Power Supply**
  - A **standby generator** provides one hour of backup power for smoke extraction, sprinklers, and alarm systems, ensuring that life safety systems remain operational even if mains power fails.
  - Automatic transfer switches (ATS) ensure a seamless changeover within seconds of power loss.
- **Integration and Interfaces**
  - Fire alarm cause-and-effect programming ensures that:
    - **Detectors trigger the smoke extract system** and release fire doors on escape routes.
    - **Sprinkler activation signals the alarm system** to escalate response.
    - **Voice alarm overrides public address/music systems** in the shopping mall.
  - A **fire service control panel** at the mall entrance provides firefighters with real-time system status, zone plans, and manual override controls.

Commentary

This case illustrates the **layered nature of active fire protection**:

- **Detection and alarms** provide rapid warning to occupants and the fire service.
- **Sprinklers** control fire growth at source, reducing heat and smoke production.

- **Smoke extraction** maintains tenability in escape routes and supports firefighting.
- **Backup power** ensures system reliability even under severe conditions.
- **Integration** through the alarm panel ensures all systems work together rather than in isolation.

Summary

By combining **detection, alarms, sprinklers, smoke ventilation, power resilience, and integrated control**, the fire strategy for this underground car park provides a **robust, multi-layered defence**. Occupants can evacuate safely, the shopping mall above is protected, and firefighters are supported with clear information and effective facilities.

9.10 Summary

Active fire protection systems represent the **dynamic, responsive layer** of a building's fire safety strategy. Whereas passive measures are fixed into the structure, active systems operate in real time, **detecting, warning, suppressing, and managing smoke** to protect both occupants and the building.

This chapter has demonstrated that a robust fire strategy must:

- **Provide Early Warning** – through well-designed fire detection and alarm systems, ensuring all occupants are alerted quickly and evacuation can begin without delay.
- **Control Fire Growth** – via suppression systems such as sprinklers, water mist, gaseous agents, and foam, which contain or extinguish fires before they escalate.
- **Manage Smoke Movement** – using natural, mechanical, and pressurisation systems to maintain tenability in escape routes and support firefighting efforts.
- **Ensure Resilience of Systems** – through emergency lighting, illuminated signage, and backup power supplies, so life safety functions remain operational during power failure.
- **Integrate Responses** – by linking alarms, suppression, smoke control, lifts, and voice communication systems into a coordinated cause-and-effect sequence that is both effective and fail-safe.

- **Embed Maintenance and Testing** – ensuring systems remain reliable throughout the building's life, with clear responsibilities, certification, and record-keeping forming part of the golden thread.
- **Account for Human Factors** – recognising that technology alone cannot save lives; training, drills, and clear communication are essential to ensure occupants and staff respond appropriately.

The **case examples** provided — from the hospital ward to the underground car park — illustrate how active systems are applied in real-world contexts, reinforcing that no single system is sufficient in isolation. Instead, **it is their combination and integration with passive protection that creates a robust, layered defence.**

Looking Ahead

The next chapter will shift focus from automated and mechanical systems to the role of the **fire and rescue service**. Even with the best active and passive protections in place, firefighter intervention remains a critical factor in life safety and property protection. Chapter 10 will explore how fire strategies must provide **access, facilities, and information** that allow emergency services to respond safely and effectively.

Chapter 10

Fire Service Facilities and Access

10.1 Introduction

A comprehensive fire strategy must not only ensure the **safe evacuation of occupants** but also provide the **fire and rescue service (FRS)** with the necessary facilities to carry out effective firefighting and rescue operations. Firefighter intervention is a critical component of life safety: inadequate access, poorly designed shafts, or insufficient water supplies can compromise their ability to respond, placing both occupants and responders at unnecessary risk.

Modern fire strategies therefore place equal weight on:

- **Occupant evacuation provisions**, and
- **Firefighter access and operational facilities.**

These two dimensions are interdependent: occupants rely on firefighters to control or extinguish fires when self-evacuation is not possible, and firefighters depend on safe, functional infrastructure to carry out rescues and prevent escalation.

The Importance of Fire Service Facilities

- **Life Safety**
    - Firefighter intervention may be the only means of rescue for mobility-impaired occupants, trapped residents, or unconscious casualties.
    - In high-rise buildings, effective fire service operations are critical where evacuation may be partial, delayed, or phased.
- **Property Protection and Resilience**
    - Well-designed firefighting facilities improve the chances of containing fires to their room or compartment of origin, protecting valuable assets and ensuring business continuity.
    - Insurers increasingly recognise the value of robust firefighting design in reducing loss severity.

- **Firefighter Safety**
    - Poorly designed or absent facilities expose responders to unnecessary risks, as seen in tragic incidents worldwide where crews became trapped or disoriented.

Core Facilities Considered in Fire Strategies

A fire strategy should document and justify all provisions for firefighter access and intervention, including:

- **Access for appliances** – suitable routes, turning circles, and hardstandings.
- **Firefighting shafts** – protected stairwells with associated lobbies.
- **Firefighting lifts** – allowing rapid transport of personnel and equipment to upper floors.
- **Rising mains (wet and dry)** – ensuring water delivery within tall or deep buildings.
- **Hydrant provision** – external water supplies, supplemented by tanks where required.
- **Smoke control systems** – maintaining tenable conditions for firefighting operations.
- **Pre-incident information** – building plans, signage, and digital systems to aid decision-making.

Global and Historical Lessons

The significance of fire service facilities has been repeatedly demonstrated in practice:

- **World Trade Center (2001, USA)** – lack of protected firefighting access routes and limited water supply severely hindered fire service operations in a catastrophic high-rise scenario.
- **Grenfell Tower (2017, UK)** – inadequate dry riser provision, lack of compartment integrity, and restricted access for appliances contributed to operational challenges.
- **Shanghai High-Rise Fire (2010, China)** – difficulty accessing upper floors highlighted the importance of firefighting lifts and shafts.

These incidents illustrate that **failure to provide adequate facilities not only compromises firefighting effectiveness but can also exacerbate loss of life.**

Purpose of This Chapter

This chapter sets out the **requirements, design considerations, and best practices** for providing fire service facilities in buildings. It will explore:

1. **Appliance access and site planning** – ensuring fire crews can reach and deploy effectively.
2. **Internal firefighting facilities** – shafts, lifts, rising mains, and water supplies.
3. **Smoke control for firefighters** – how design supports safe entry and operations.
4. **Information provision** – signage, plans, and digital tools.
5. **Case studies and emerging approaches** – learning from incidents and new technologies.

By embedding these elements into a fire strategy, designers ensure that firefighters are **not fighting the building as well as the fire** — but are instead supported by an environment that maximises the chances of safe, effective intervention.

10.2 Legal and Standards Framework

The provision of facilities for the fire and rescue service is not optional — it is a statutory requirement embedded in the **Building Regulations** and further supported by **British Standards, European Standards, and local fire service guidance**. A credible fire strategy must demonstrate how the design complies with these frameworks, while also recognising that in complex buildings, **enhanced or engineered provisions** may be necessary to ensure resilience.

10.2.1 Building Regulations – Approved Document B (ADB)

- **Part B functional requirements** mandate that:
    - *"The building shall be designed and constructed so that there are appropriate provisions for the fire service."*
- **ADB Volume 1 (Dwellings)**
    - Covers access to small residential buildings, including access for appliances and hydrant locations.
- **ADB Volume 2 (Buildings other than dwellings)**
    - Provides detailed guidance on:

- **Vehicle access routes** – widths, turning circles, hardstanding's.
- **Appliance access to facades** – maximum distances for hose reach.
- **Hydrant provision** – proximity and flow capacity.
- **Internal firefighting shafts** – requirements triggered by building height, depth, and floor plate size.

ADB therefore forms the **minimum compliance baseline** for firefighting facilities.

10.2.2 BS 9999 – Code of Practice for Fire Safety in the Design, Management and Use of Buildings

- Offers a **risk-based, flexible approach** beyond the prescriptive ADB.
- Provides detailed design guidance for:
    - **Firefighting shafts** – with allowances for management levels and risk profiles.
    - **Firefighting staircases and lobbies** – performance requirements, dimensions, and pressure differentials.
    - **Access strategies** – integration of fire service access with evacuation planning.
- Particularly useful for **complex, high-rise, or mixed-use buildings** where prescriptive rules may be too restrictive.

10.2.3 BS 9991 – Fire Safety in the Design, Management and Use of Residential Buildings

- Residential-specific guidance, complementing ADB Volume 1.
- Key provisions include:
    - **Firefighting shafts in tall residential buildings**.
    - **Sprinkler interaction with firefighting facilities**.
    - **Provisions for stay-put vs simultaneous evacuation strategies**.
- Increasingly relevant for high-rise housing following the **Grenfell Tower Inquiry recommendations**.

10.2.4 BS EN 81-72 – Firefighting Lifts

- European standard providing requirements for the **design and installation of firefighting lifts**.
- Covers:
    - Lift car dimensions and capacity (to accommodate equipment and stretchers).
    - Water protection measures for lift shafts and machinery.
    - Priority controls for firefighter use.
    - Fire-resisting lobbies serving lifts.
- Critical for **high-rise and deep basements**, where manual transport of equipment would be impractical.

10.2.5 Local Fire and Rescue Service Guidance

National standards are supplemented by **local fire authority requirements**, which often influence planning approval:

- **Hydrants** – specific flow/pressure requirements, location preferences, or need for private hydrants.
- **Appliance access** – acceptable road gradients, bridge weight limits, or turning facilities.
- **Pre-incident planning** – provision of building information (drawings, digital systems, fire service boxes).
- **Site-specific requirements** – e.g., additional access routes for prisons, hospitals, or industrial sites.

Early consultation with the local fire and rescue service is therefore essential to ensure compliance and avoid costly redesign later.

10.2.6 Other Relevant Standards and Guidance

- **BS EN 12101 series** – smoke and heat control systems, critical for supporting firefighter access.
- **NFPA 14 (International context)** – Standard for the Installation of Standpipe and Hose Systems, often referenced in global projects.
- **Building Safety Act 2022** (UK) – strengthens the "golden thread" of accountability, ensuring that firefighting facilities are not only provided but maintained throughout the building lifecycle.

Summary

The legal and standards framework for firefighting facilities is multi-layered:

- **Building Regulations (Part B / ADB)** – statutory baseline requirements.
- **BS 9999 and BS 9991** – enhanced and occupancy-specific guidance.
- **BS EN 81-72** – technical requirements for firefighting lifts.
- **Local fire service guidance** – site-specific operational needs.

A robust fire strategy must reference these standards explicitly, showing where compliance is prescriptive, where performance-based design is applied, and how **local operational input has been incorporated**.

10.3 Vehicle Access

The ability of fire appliances to reach and position near a building is fundamental to effective firefighting operations. Without suitable access roads, turning circles, and hardstanding's, fire crews may be unable to deploy hoses, ladders, or aerial platforms, delaying intervention and jeopardising life safety.

A fire strategy must therefore clearly demonstrate how vehicle access has been considered, referencing **Approved Document B (ADB), BS 9999, and local fire service requirements**.

10.3.1 Access Roads

Fire appliances require dedicated access roads designed for their **width, weight, and manoeuvrability**:

- **Minimum widths**:
    - 3.7 m minimum carriageway width (ADB/BS 9999).
    - Reduced to 3.1 m where the road serves only one building and no aerial access is required.
- **Vertical clearance**: 4.0 m minimum height clearance for high reach appliances.
- **Turning circles**:
    - ADB Table 15.2 states, Minimum inside radius 7.7 m; outside radius 12.3 m.
    - Hammerhead or turning facilities required where reversing distances exceed 20 m.

- **Load-bearing capacity**:
    - Roads and surfaces must support 12.5 tonnes (standard pumping appliances).
    - Where aerial appliances are anticipated, local services may require up to 17–24 tonnes capacity.

Best practice: Use **swept-path analysis software** to demonstrate that vehicles can manoeuvre throughout the site without obstruction.

10.3.2 Proximity to Entrances

- Fire appliances must be able to position within **18 m of the principal building entrance** or of a **dry/wet fire main inlet**.
- For multi-access sites, multiple entrance points may be necessary to ensure hose runs do not exceed practical lengths (typically 45–60 m per hose line).
- Where buildings are deep or irregular in form, **strategic access points** should be identified in the fire strategy to align with internal firefighting shafts.

10.3.3 Aerial Appliance Access

For high-rise or tall buildings, aerial ladders and platforms are essential for both firefighting and external rescues:

- At least **one accessible façade** must allow aerial appliance positioning.
- **Hardstandings** must be:
    - Minimum 3.7 m wide.
    - Located within 2.0–3.0 m of the façade.
    - Capable of supporting **24 tonnes gross vehicle load**.
- **Vertical reach**: Local fire services should be consulted, as appliance reach varies (commonly 30–32 m, some extend to 64 m).
- Obstructions such as **balconies, trees, or overhead cables** must be avoided.

10.3.4 Obstructions and Constraints

Access is frequently compromised by poor planning or later modifications:

- **Landscaping** – trees, planting beds, and bollards reducing usable width.

- **Street furniture** – benches, lamp posts, or bike racks obstructing hose routes.
- **Security features** – rising bollards, gates, or hostile vehicle mitigation (HVM) measures restricting appliance entry.
- **Car parking** – unauthorised parking in access roads can render facilities unusable.

Best practice: Provide **controlled access measures** (e.g., lockable or automatic bollards linked to fire alarms) and clearly marked "Keep Clear – Fire Access" zones.

10.3.5 Special Considerations

- **Large sites and campuses**
  - Hospitals, universities, and airports may require **multiple access routes** with internal road networks.
- **Basements and underground structures**
  - May need separate appliance approach points to reach dry riser inlets and smoke extract systems.
- **Mixed-use developments**
  - Shared loading bays or service yards should not conflict with fire appliance access.

10.3.6 Case Examples

- **Grenfell Tower (2017, UK)**
  - Appliance access was limited to one side of the building; aerial appliances could not reach upper floors.
  - Highlights the consequences of restricted access on operational effectiveness.
- **Urban high-rise developments**
  - In some London sites, retrospective provision of access routes has been required after planning, due to inadequate initial consultation with the fire service.

Summary

Vehicle access is a **critical enabler** of firefighting effectiveness. A fire strategy must demonstrate that:

- **Access roads** are suitable for width, load, and turning radii.
- **Entrances and fire mains** are within 18 m of appliance positioning points.
- **Aerial appliances** can reach at least one façade where required.
- **Obstructions** such as landscaping and barriers are avoided or controlled.
- **Swept-path analysis** confirms manoeuvrability across the site.

By embedding these requirements into the design from the outset, fire strategies ensure that the fire service can intervene **quickly, safely, and effectively**.

10.4 Firefighting Shafts

In tall and complex buildings, dedicated **firefighting shafts** are essential to enable the fire and rescue service to access upper floors quickly and safely. These shafts provide a **protected route** for personnel and equipment, a **safe staging area** near the fire floor, and a **smoke-protected environment** that allows sustained firefighting operations.

10.4.1 Components of a Firefighting Shaft

A compliant shaft typically comprises three integrated elements:

1. **Firefighting Stair**
   - Provides protected vertical access to all floors, enclosed within **120-minute fire-resisting construction**.
   - Sized to allow safe two-way movement of firefighters ascending and occupants descending (if evacuation is ongoing).
2. **Firefighting Lift**
   - Designed and installed to **BS EN 81-72**.
   - Allows rapid vertical movement of crews and equipment, particularly in high-rise buildings where stair ascent would be impractical.
   - Must remain operational during fire conditions with protected power supply and water ingress protection.

3. **Firefighting Lobby**
    - Provides a **smoke-protected buffer zone** between the shaft and fire floor.
    - Enclosed with **FD60S fire doors** (self-closing, smoke-sealed).
    - Serves as a staging area for laying out hoses, establishing communication, and controlling entry to the fire floor.

10.4.2 Functions of Firefighting Shafts

Firefighting shafts serve three critical functions:

- **Protected Vertical Access**
    - Crews can ascend to the fire floor without being exposed to smoke or flames.
- **Smoke Control**
    - Lobbies and **pressurisation systems** prevent smoke infiltration into the shaft, maintaining tenable conditions.
- **Operational Base**
    - Lobbies act as a controlled point from which firefighting hoses are advanced, minimising risk of smoke spread into shafts.

10.4.3 Design Requirements

Firefighting shafts must comply with the following key provisions (as per **ADB, BS 9999, BS 9991, and BS EN 81-72**):

- **Fire Resistance**
    - Shaft enclosure: **120 minutes fire resistance**.
    - Lobby and stair doors: **FD60S self-closing**.
- **Pressurisation**
    - Shafts must be mechanically pressurised (or naturally ventilated where permitted) to maintain positive pressure and prevent smoke ingress.
    - Typical design pressure differential: **50 Pa** between shaft and accommodation areas.

- **Size and Provision**
  - Required where:
    - The building has a floor more than **18 m above fire service access level**.
    - The building has a basement more than **10 m deep**.
  - In very tall buildings, multiple shafts may be required to reduce hose runs and support redundancy.
- **Layout Considerations**
  - Lobbies must be directly accessible from corridors, not through other rooms.
  - Travel distance from lobby to the furthest point of a floor should not exceed **60 m** (ADB).
  - Firefighting lifts must open only into protected lobbies, never directly into accommodation.

10.4.4 Enhanced and Engineered Solutions

- **BS 9999** allows flexibility in shaft provision, for example reducing the number of shafts if additional risk mitigation measures (sprinklers, enhanced management) are provided.
- **Performance-based designs** may include extended hose reach lengths or alternative pressurisation strategies but must be validated through CFD modelling and operational consultation with the fire service.

10.4.5 Case Examples

- **Grenfell Tower (2017, UK)**
  - Only one firefighting shaft was available; stair congestion and smoke spread hampered operations.
  - Demonstrates the risks of **insufficient shafts** and the importance of maintaining **compartment integrity**.
- **High-rise office towers (London, Middle East, Asia)**
  - Modern designs typically include **multiple shafts**, separated for redundancy, ensuring continued operation even if one is compromised.

Summary

Firefighting shafts are the **backbone of high-rise firefighting strategy**, comprising:

- **Firefighting stairs** for protected access.
- **Firefighting lifts** for rapid personnel and equipment transport.
- **Lobbies** for smoke control and operational staging.

With 120-minute enclosure, FD60S doors, and effective pressurisation, firefighting shafts ensure that fire crews can operate **close to the fire floor** without losing their protected egress route — safeguarding both firefighter and occupant safety.

10.5 Firefighting Lifts

In tall and deep buildings, stair access alone is insufficient for rapid fire service intervention. Where the fire floor is located more than **18 m above** or more than **10 m below** the fire service access level, **firefighting lifts** are required. These lifts allow crews to reach upper floors or deep basements quickly, carrying heavy equipment, and in some cases evacuating mobility-impaired occupants under controlled conditions.

Firefighting lifts are therefore not ordinary passenger lifts adapted for fire use, but **purpose-built installations** designed to remain operational and safe in the hostile environment of a fire.

10.5.1 Regulatory Basis

- **Approved Document B (ADB)** requires firefighting lifts in buildings where fire service access by stairs alone would be insufficient (above 18 m or below 10 m).
- **BS EN 81-72:2020** is the definitive standard for design and installation, covering everything from shaft construction to lift controls.
- **BS 9999 and BS 9991** expand on provisions, offering risk-based guidance for complex and residential buildings.

10.5.2 Design Considerations

Firefighting lifts must meet strict technical requirements:

- **Capacity**
    - Minimum **630 kg capacity** (8 persons), though larger lifts are often specified (1,000–1,275 kg) to accommodate firefighters with breathing apparatus, stretchers, and equipment.
- **Access**
    - Lifts must be accessible **only from protected firefighting lobbies** (never directly from accommodation).
    - Lobbies must be enclosed in 60-minute fire-resisting construction and provided with smoke control (ventilation or pressurisation).
- **Power Supply**
    - Must have a **secondary power source** (e.g., generator, dual incoming supply) to ensure operation if the primary source fails.
    - Power supplies must be routed separately and protected from fire.
- **Water Protection**
    - Lift shafts and machinery must be designed to prevent water ingress from sprinklers or firefighting hoses.
    - Drainage systems should be incorporated.
- **Controls**
    - Firefighting lifts are fitted with **dedicated firefighter controls** (EN 81-72 compliant).
    - In firefighting mode, the lift is under the **exclusive control of the fire service**, overriding passenger use.
    - Recall functions bring the lift to the access floor upon alarm activation.
- **Speed and travel**
    - Travel speed must allow rapid access to the top floor within **60 seconds (1.0m/s)** where practicable.
    - Doors must remain open long enough for firefighters to deploy with equipment.

10.5.3 Operational Role

Firefighting lifts serve several operational functions:

- **Rapid deployment** – allow fire crews to reach upper levels or basements within minutes, carrying hoses, BA sets, and heavy gear.
- **Staging** – deliver equipment to the firefighting lobby, where crews can assemble and plan operations.
- **Assisted evacuation** – may be used for controlled evacuation of mobility-impaired occupants under firefighter supervision.

10.5.4 Integration with Firefighting Shafts

Firefighting lifts are always part of a **firefighting shaft**, located alongside the protected stair and lobby. This ensures redundancy:

- If the lift fails, the firefighting stair remains available.
- If the stair becomes smoke-logged, the lift may still provide protected vertical movement.

10.5.5 Case Examples

- **Shanghai High-Rise Fire (2010)**
  - Lack of firefighting lifts hindered intervention above the 20th floor. Crews were forced to climb stairs carrying equipment, delaying operations.
- **Modern London high-rises**
  - Firefighting lifts are standard, often oversized to accommodate stretchers, recognising their dual use for firefighter deployment and assisted evacuation.

Summary

Firefighting lifts are mandatory in buildings above 18 m or below 10 m relative to fire service access. They must:

- Comply with **BS EN 81-72**.
- Provide a minimum capacity of **630 kg**, ideally larger.
- Be accessible only via **protected firefighting lobbies**.
- Have **backup power and water ingress protection**.
- Operate under **firefighter-exclusive control** in emergencies.

A fire strategy must clearly specify the **location, specification, and operation** of all firefighting lifts, ensuring regulators and fire services can confirm that the provision is robust, compliant, and reliable.

10.6 Rising Mains

Rising mains are a critical firefighting facility in tall buildings, allowing crews to deploy water directly to upper floors without the impracticality of dragging hoses up staircases. Correctly designed and located, they provide a rapid means of water delivery and form an integral part of the firefighting shaft system.

A fire strategy must clearly document the provision of rising mains, including **type (dry or wet), locations, inlets, outlets, and maintenance regime**, as these factors directly affect operational effectiveness.

10.6.1 Types of Rising Mains

- **Dry Risers**
    - Installed where the highest floor is between **18 m and 60 m above fire service access level**.
    - Normally empty but fitted with inlet connections at ground level for fire appliances to charge with water.
    - Outlet landing valves are provided at each floor, typically within firefighting lobbies or stairs.
    - Advantage: simple, low-maintenance system.
    - Limitation: dependent on fire appliance pumping capacity and access to inlets.

Wet Risers

- Required where the highest floor exceeds **60 m above access level**.
- Permanently charged with water from **on-site tanks and pumps**, ensuring immediate availability.
- Outlets are similar to dry risers, but pressure is maintained by booster pumps and monitored via control systems.
- Advantage: immediate and reliable water supply at height.
- Limitation: higher installation and maintenance costs; reliance on power supplies.

10.6.2 Design Requirements

- **Location of Inlets**
    - Must be positioned within **18 m of appliance parking points** for rapid connection.
    - Clearly marked, accessible, and free from obstructions.
- **Landing Valves (Outlets)**
    - Located at every floor level (and in basements) within **protected lobbies or stair enclosures**.
    - Positioned to allow firefighters to connect hoses before entering the fire floor.
    - Should include isolating valves to allow flexibility of use.
- **Pipework and Pressure**
    - Pipework must be protected to at least **60 minutes fire resistance**.
    - For wet risers, pumps must be capable of delivering flow and pressure to the highest outlet simultaneously with system demand.
    - Pressure-regulating valves may be required to prevent excessive outlet pressures at lower floors.
- **Testing and Maintenance**
    - *Dry risers*: visual inspection every 6 months; full wet pressure test annually.
    - *Wet risers*: functional testing of pumps, tanks, and valves at regular intervals (monthly checks for pumps; annual full flow tests).
    - Records must be maintained as part of the fire safety management system.

10.6.3 Integration with Firefighting Shafts

Rising mains are typically located within **firefighting shafts** alongside stairs, lifts, and lobbies. This ensures that:

- Firefighters can charge hoses safely from within a protected environment.
- Hoses can be deployed directly onto the fire floor from a **smoke-controlled lobby**.

- Vertical water delivery is integrated with vertical access, reducing time lost during intervention.

10.6.4 Case Examples

- **Grenfell Tower (2017, UK)**
  - A dry riser was present but became inadequate due to rapid external fire spread and loss of compartmentation. Fire crews were forced into difficult conditions to maintain water supply at height.
- **Modern supertall buildings (Middle East, Asia)**
  - Wet risers are standard, often supported by **dual water tanks and pumps** to provide redundancy. In some cases, intermediate tanks at mid-level floors reduce pressure demands.

10.6.5 Best Practice in Fire Strategies

When documenting rising mains, strategies should include:

- **Plans and elevations** showing riser runs, inlet positions, and outlet locations.
- **Specification of dry vs wet systems**, including pumps, tanks, and redundancy.
- **Maintenance schedules** as part of the fire safety management plan.
- **Operational assumptions** (e.g., that appliances can reach inlets within 18 m, or that generators will maintain pump supply).

Summary

Rising mains are essential for **efficient firefighting in tall and complex buildings**.

- **Dry risers**: required for 18–60 m buildings, charged by appliances.
- **Wet risers**: required above 60 m, permanently charged by pumps and tanks.
- Both must be integrated into **firefighting shafts**, with accessible inlets, protected outlet locations, and robust maintenance regimes.

By clearly documenting these systems, fire strategies provide assurance that water can be delivered to any floor reliably, forming a cornerstone of operational firefighting capability.

10.7 Water Supplies

An **adequate and reliable water supply** is the foundation of all firefighting operations. Whether used to charge rising mains, feed sprinklers, or support external appliance operations, water provision must be assured under a range of conditions. A fire strategy must therefore demonstrate that hydrants, tanks, and pumps are provided to meet both **regulatory requirements** and **operational needs of the fire and rescue service (FRS)**.

10.7.1 Hydrants

- **Regulatory requirement**
    - Approved Document B (ADB) requires a hydrant to be located within **90 m of every building entrance**.
- **Design considerations**
    - Hydrants must be accessible, clearly marked, and kept free of obstructions such as parked vehicles or landscaping.
    - Flow rates should typically achieve **1,500 litres/min** at 1 bar pressure, though higher rates may be required for industrial or high-rise premises.
- **Private vs public hydrants**
    - In some developments (e.g., campuses, airports, remote sites), **private hydrants** connected to site mains may be necessary to supplement municipal supply.
- **Best practice**
    - Hydrant positions should be confirmed with the local water authority and FRS during design.
    - Fire strategies should map hydrant locations and flow-test data to verify performance.

10.7.2 On-Site Storage Tanks and Pumps

For high-rise, remote, or high-hazard buildings, reliance on external hydrants may not be sufficient. In such cases, **dedicated on-site storage and pumping systems** are required:

- **Storage tanks**
  - Capacity typically sized to provide **60–90 minutes of firefighting water supply**, depending on building use and risk profile.
  - May be sectional steel, GRP, or concrete tanks, often located in basements or service yards.
  - Must be protected from contamination and accessible for maintenance.
- **Pumping systems**
  - Fire pumps must deliver adequate flow and pressure to feed wet risers, hydrants, and suppression systems.
  - Pumps must be designed to **BS EN 12845** (sprinklers) or relevant system standards.
  - Automatic start with manual override is required, and control panels must be in fire-protected enclosures.
- **High-rise practice**
  - In very tall buildings, intermediate storage tanks may be placed at mid-height levels to reduce pressure demands on risers and avoid excessive pump sizes.

10.7.3 Redundancy and Resilience

Given the critical role of water supply, strategies must ensure **resilience against single-point failures**:

- **Dual pumps** – one duty and one standby, each capable of full capacity.
- **Independent power supplies** – diesel-driven backup pumps or generators to ensure continued operation during mains failure.
- **Duplicate tanks** – in some cases, two tanks are provided (or one tank subdivided) to allow maintenance without compromising protection.
- **Pipework protection** – water mains routed through protected shafts to prevent damage during fire.

10.7.4 Case Examples

- **Atherstone-on-Stour Warehouse Fire (2007, UK)**
  - Limited water supply hampered firefighting efforts; reliance on distant hydrants delayed operations.

- **Tall building fires in the Middle East**
    - Demonstrated the importance of **redundant on-site tanks and dual pump systems** to sustain operations when public water networks proved unreliable.
- **Grenfell Tower (2017, UK)**
    - Firefighting operations were hindered by difficulties maintaining water supply at upper levels, underlining the importance of reliable pumping and riser design.

10.7.5 Best Practice in Fire Strategies

A robust fire strategy should explicitly set out:

- Location and capacity of **hydrants** (with flow-test data).
- Details of **storage tanks, pumps, and redundancy measures**.
- **Hydraulic calculations** demonstrating system adequacy.
- Maintenance and testing regime to ensure long-term reliability.

Summary

Reliable water supplies are indispensable for firefighting.

- **Hydrants**: must be within 90 m of building entrances and capable of delivering required flow/pressure.
- **On-site tanks and pumps**: required in high-rise, remote, or high-hazard buildings to ensure sustained supply.
- **Redundancy**: dual pumps, backup power, and duplicate tanks enhance resilience.

By explicitly documenting these provisions, a fire strategy ensures that firefighters can depend on an **uninterrupted and adequate water supply**, even under adverse conditions.

10.8 Fire Control Rooms and Panels

In large or complex buildings, particularly those with extensive fire safety systems, a **fire control room** (sometimes called a *fire command centre*) may be required. These rooms act as the **operational hub for incident management**, enabling the fire and rescue service (FRS) to access building information, monitor alarms, and take control of life-safety systems during a fire.

A well-designed control room improves coordination, reduces response time, and provides a secure base of operations in the early stages of an incident.

10.8.1 Functions of Fire Control Rooms

A fire control room should allow the fire service to:

- **Monitor alarms and system status** – understanding where alarms are triggered, what zones are affected, and whether faults exist.
- **Operate firefighting systems** – activating or overriding smoke control, fans, damper release, or door-hold devices.
- **Access building information** – floor plans, fire strategy extracts, and emergency contact numbers.
- **Establish communications** – with both fire crews in the building and external command units.

10.8.2 Fire Alarm Panels

- **Role**: The fire alarm control and indicating equipment (CIE) provides a real-time display of alarm conditions, device activations, and system faults.
- **Design**:
    - Zoned and addressable displays allow precise identification of affected areas.
    - Panels must be visible, intuitive, and accompanied by zone charts.
    - Must comply with **BS 5839-1** for non-domestic premises.
- **Best practice**:
    - Duplicate repeater panels at key entrances can support early intervention.
    - Firefighters should be trained in interpreting the display and using panel controls.

10.8.3 Control Interfaces

Large or complex buildings often include **integrated fire system controls**, which must be accessible through the control room:

- **Smoke control systems** – activation of extract fans, stair pressurisation, or atrium venting.
- **Lifts** – recall and control of firefighting lifts as per **BS EN 81-72**.

- **Doors and shutters** – override controls for magnetic door releases, fire curtains, and shutters.
- **Suppression systems** – monitoring status of sprinklers, wet risers, or gaseous suppression.

Integration ensures that firefighters can directly control life-safety measures without relying on dispersed local switches.

10.8.4 Communication Systems

- **Firefighter telephones** – hard-wired communication points within firefighting shafts, connected back to the control room.
- **Radios** – provision for FRS radio repeaters or leaky-feeder systems to ensure communications deep within basements or high-rise cores.
- **Public address / voice alarm** – ability to broadcast evacuation instructions (BS 5839-8).
- **Best practice**: The control room should be acoustically treated to ensure radio clarity even during noisy incidents.
- **Emergency Voice Communication System**: enables clear, two-way communication between building occupants, fire wardens, and emergency services during an incident to coordinate safe evacuation and firefighting operations.

10.8.5 Building Information

The control room should house all **critical building information**, including:

- Up-to-date **floor plans and fire drawings** laminated or in digital display format.
- A copy of the **fire strategy**, highlighting escape routes, firefighting shafts, and system design assumptions.
- **Emergency contacts** for building management, keyholders, and system maintainers.
- **Pre-incident plans** prepared in consultation with the fire service.

10.8.6 Location and Design Requirements

- **Accessibility**: Should be directly accessible from a fire service entrance, ideally at ground level.

- **Protection**: Enclosed in **120-minute fire-resisting construction**, with FD60S fire doors.
- **Space and layout**: Sufficient space for multiple firefighters to operate, with clear labelling and logical arrangement of panels and controls.
- **Security**: Restricted access to authorised personnel, with systems safeguarded against tampering.
- **Environment**: Ventilation, lighting, and temperature control to maintain reliable system operation.

10.8.7 Case Examples

- **King's Cross Underground Fire (1987, UK)**
  - Lack of integrated information systems led to delayed understanding of fire spread.
- **Modern airport terminals**
  - Typically include sophisticated fire control centres with digital building management system (BMS) integration, enabling coordinated smoke control, evacuation, and suppression monitoring.

Summary

Fire control rooms and panels are **vital for incident management** in large or complex buildings. They provide:

- **Fire alarm monitoring** (BS 5839-compliant).
- **Control interfaces** for smoke control, lifts, doors, and suppression.
- **Communication systems** for firefighters and occupants.
- **Building information** to guide operations.

Clear, secure, and well-labelled control rooms give the fire and rescue service the tools they need to **manage an incident effectively, safely, and quickly**.

10.9 Smoke Control for Firefighting

Smoke is the **greatest hazard** faced by firefighters. It contains toxic gases, reduces visibility, and raises temperatures, creating conditions in which both occupants and fire crews cannot survive for long without protection. Even when breathing apparatus (BA) is available, firefighters require **smoke-free access routes** to advance hoses and perform rescues effectively.

Fire strategies must therefore set out how smoke control systems will support **fire service intervention**, not just occupant evacuation.

10.9.1 Stair Pressurisation

- **Purpose**: Keeps escape stairs and firefighting shafts clear of smoke, ensuring firefighters can ascend safely.
- **Design criteria**:
  - Pressurisation systems must provide a pressure differential (typically **50 Pa**) between stair and accommodation areas.
  - Air supply fans must deliver sufficient airflow even under door-open conditions (e.g., when firefighters move through).
  - Relief vents are required to avoid excessive door opening forces.
- **Standards**:
  - Designed to **BS EN 12101-6** (smoke control systems).
- **Strategy requirement**: Confirm which stairs are pressurised, their fan locations, power supplies, and redundancy provisions.

10.9.2 Smoke Vents and Extraction Systems

- **Lobby and corridor smoke vents**
  - Automatically open to release smoke into shafts, reducing spread to firefighting stairs.
- **Atrium smoke control**
  - Extract fans and roof vents remove smoke from large spaces, ensuring visibility and tenability for crews.
- **Basement smoke extraction**
  - Critical where natural venting is impossible. Fans and ducts must maintain smoke clearance until fire service operations are complete.
- **Design principle**: Systems should maintain **tenable conditions (visibility >10 m, temperature <60 °C at 2 m height)** to enable firefighting entry.

10.9.3 Compartmentation Integrity

- **Importance**: Even well-designed smoke control systems fail if compartmentation is breached.
- **Requirements**:
    - All penetrations (services, ducts, cabling) must be sealed with tested fire-stopping.
    - Fire doors must be **self-closing, smoke-sealed (FD30S or FD60S)**, and properly maintained.
- **Common failures**: Poor workmanship, damaged seals, or uncontrolled openings can lead to **smoke migration** into firefighting routes.

10.9.4 Reliability of Smoke Control Systems

- **Power supply**
    - Fans and controls must have **secondary power sources** (diesel generator or dual supplies).
- **Testing and maintenance**
    - Systems must be tested weekly (functional), quarterly (service checks), and annually (full performance test).
- **Monitoring**
    - Control rooms must display system status, faults, and override controls (see Section 10.8).

10.9.5 Case Examples

- **King's Cross Underground Fire (1987, UK)**
    - Lack of effective smoke control and delayed evacuation led to extensive smoke spread, causing 31 fatalities.
- **Channel Tunnel Fire (1996, UK/France)**
    - Demonstrated the importance of robust ventilation to allow firefighters to enter and extinguish deep underground fires.
- **High-rise fires in Dubai (2010s)**
    - External cladding fires caused massive smoke infiltration, overwhelming stairs not properly pressurised.

10.9.6 Best Practice in Fire Strategies

A fire strategy must confirm:

- Which stairs and shafts are **pressurised**, and to what design standard.
- Locations and design capacities of **smoke vents** and extract fans.
- Assumptions about **compartmentation integrity** and how this is maintained.
- Redundancy in **power supply and fan operation**.
- Maintenance schedules to ensure long-term reliability.

Summary

Smoke control for firefighting is as critical as water supply or risers. Key provisions include:

- **Stair pressurisation** – protecting firefighter access routes.
- **Smoke vents and extract systems** – clearing lobbies, corridors, atria, and basements.
- **Compartmentation integrity** – preventing uncontrolled smoke spread.
- **Reliability and redundancy** – ensuring systems function when needed most.

By documenting the **design, capacity, and maintenance regime** for smoke control, a fire strategy assures that firefighters can operate effectively, maximising the chances of both rescue and fire containment.

10.10 Coordination with Fire Services

The effectiveness of firefighting facilities depends not only on technical design but also on **alignment with the operational capabilities of the local Fire and Rescue Service (FRS)**. A fire strategy should therefore document how the design team has engaged with the relevant authority at key project stages — from concept through to handover and occupation.

Early and ongoing consultation reduces delays during approval, prevents costly retrofits, and ensures that firefighting provisions are **practical, usable, and familiar** to the crews who will rely on them.

10.10.1 Pre-Planning Engagement

- **Purpose**: Establishes expectations before formal submissions.
- **Key discussion points**:

- **Appliance access** – confirmation that roads, turning areas, and hardstandings are suitable for local fleet dimensions and weight limits.
- **Water supplies** – agreement on hydrant provision, storage tanks, and flow rates.
- **Rising mains and shafts** – number, location, and hose reach relative to the FRS's operational preferences.
- **Special risks** – hazardous processes, fuel storage, or unusual occupancies requiring additional provisions.

- **Outcome**: A preliminary understanding that the proposed strategy aligns with operational requirements.

10.10.2 Familiarisation Visits

- **Timing**: Should occur **before building handover** and ideally at key construction stages.
- **Content**:
  - Walkthroughs of firefighting shafts, lobbies, and control rooms.
  - Demonstrations of fire alarm panels, smoke control interfaces, and firefighting lifts.
  - Orientation around hydrant locations, inlet connections, and generator rooms.
- **Benefits**: Familiarity significantly improves response efficiency in an actual incident.

10.10.3 Information Sharing

- **Pre-incident information** should be provided to the FRS, both in hard copy and digital formats where possible. This may include:
  - Building layouts and firefighting shaft locations.
  - Details of suppression systems, tanks, and pumps.
  - Known hazards (chemicals, fuel storage, lithium-ion batteries, etc.).
  - Evacuation strategies and refuge locations.
- **On-site provision**: Fire control rooms or secure information boxes at entrances should hold:

    - Laminated plans, fire strategy extracts, and contact details.
    - Instructions for unusual systems (e.g., smoke extraction controls).

10.10.4 Ongoing Coordination

Engagement should not end at handover:

- **Periodic liaison** – particularly where occupancies change, or where alterations to systems are made.
- **Joint training exercises** – beneficial for complex sites such as hospitals, airports, and shopping centres.
- **Feedback loop** – incidents, false alarms, or near misses should prompt review and updates to the fire strategy.

10.10.5 Case Examples

- **Successful Example**: In many UK airports, early fire service involvement in access design, hydrant positioning, and smoke control led to smooth approvals and facilities tailored to specific operational needs.
- **Failure Example**: In several urban high-rises, inadequate consultation resulted in appliances being unable to access façades or hydrant flows being insufficient, requiring costly post-construction modifications.

Summary

Coordination with the fire service is not just good practice — it is a **cornerstone of defensible fire strategy development**. It should include:

- **Pre-planning discussions** to align on access, water, and facilities.
- **Familiarisation visits** before handover to ensure crews know the building.
- **Information sharing** through plans, control rooms, and pre-incident packs.
- **Ongoing engagement** throughout the building lifecycle.

Early and continuous consultation ensures that firefighting provisions are **realistic, effective, and trusted** by the people who will depend on them in an emergency.

10.11 Case Example (Illustrative)

To demonstrate how firefighting access and facilities are brought together, consider a **25-storey office tower in a dense urban environment**. The building is designed to accommodate approximately 2,500 occupants and includes retail units at ground level, open-plan offices on typical floors, and plant space at roof and basement levels.

Firefighting Shafts and Lifts

- Two **firefighting shafts**, each comprising a firefighting stair, firefighting lift (BS EN 81-72 compliant), and protected lobby.
- Shafts are positioned at opposite ends of the floor plate, ensuring redundancy and hose coverage to all areas.
- Lifts have a **1,000 kg capacity** to accommodate crews with breathing apparatus and stretchers.
- Shafts are enclosed in **120-minute fire-resisting construction** with lobby doors rated FD60S and supported by pressurisation fans to maintain tenable access.

Rising Mains

- **Dry risers** serve from ground to the 18th floor.
    - Each riser has a ground-level inlet located within 18 m of appliance hardstandings.
- **Wet risers** serve floors 19–25.
    - Fed by dual electric/diesel pumps drawing from on-site storage.
    - Equipped with pressure-regulating valves to control excessive outlet pressures on lower floors.

Water Supplies

- A **150,000-litre on-site tank** provides dedicated firefighting water storage.
- **Dual pumps** (one duty, one standby), each sized for full system demand.
- Pumps powered from **separate electrical supplies**, with a diesel-driven generator backup for redundancy.
- External hydrants provided at 90 m intervals along the perimeter road, tested for adequate flow and pressure.

Vehicle and Appliance Access

- Fire appliances can access **three sides of the building** via 6 m-wide service roads.
- **Aerial appliance hardstanding** provided on the south façade, capable of supporting 24-tonne loads.
- Appliance parking positions are located within **18 m of rising main inlets** and entrances to firefighting shafts.

Fire Control Room

- Located adjacent to the main entrance for immediate fire service access.
- Enclosed in **120-minute fire-resisting construction** with secure access.
- Facilities include:
    - **Fire alarm panel (BS 5839-1)** with zone chart and repeater functions.
    - **Smoke control panel** for pressurisation fans, atrium extract, and basement extract.
    - **Communication systems** – fire service telephones linked to lobbies, and integration with leaky-feeder radio system.
    - **Building information** – laminated floor plans, refuge locations, and system schematics.

Smoke Control Systems

- **Stair pressurisation** maintains tenable conditions in both firefighting shafts.
- **Atrium smoke extract** sized to clear 1,500 m^3/min via high-level roof vents.
- **Basement extract fans** provide ten air changes per hour in firefighting mode.
- All systems connected to emergency generator backup.

Fire Service Coordination

- The local fire and rescue service engaged during design to agree access, riser strategy, and hydrant provision.
- **Familiarisation visits** conducted during commissioning to demonstrate operation of the control room, risers, and lifts.

- A pre-incident information pack lodged with the fire service, including plans, contacts, and hazard registers.

Integration and Operational Effectiveness

This case illustrates how multiple firefighting provisions combine into a coherent strategy:

- **Redundant shafts and lifts** ensure access even if one is compromised.
- **Hybrid riser provision** balances cost and operational effectiveness.
- **Dedicated tank and dual pumps** guarantee water supply above 60 m.
- **Access on three sides with aerial appliance hardstanding** provides robust external intervention.
- **Fire control room** centralises information and system control, enabling efficient incident command.

Summary

The 25-storey office tower demonstrates how fire strategies integrate **access, risers, water, control rooms, and smoke systems** into a defendable, practical, and operationally effective design. By embedding these elements from the outset and consulting with the fire service, the strategy provides confidence that firefighting operations can be carried out safely and efficiently in even the most challenging scenarios.

10.12 Summary

Fire service facilities form one of the **core pillars of a robust fire strategy**. They are not simply technical additions but essential systems that enable firefighters to intervene safely, protect occupants, and contain fires before they escalate into catastrophic events.

Throughout this chapter, we have seen how facilities must be designed, documented, and maintained to support **rapid, safe, and effective intervention**:

- **Vehicle Access** – ensuring appliances can approach, position, and deploy equipment without obstruction.
- **Firefighting Shafts** – protected vertical routes providing safe staging areas and close access to fire floors.
- **Firefighting Lifts** – critical in tall and deep buildings, allowing rapid deployment of crews and equipment.

- **Rising Mains (Dry and Wet)** – enabling fast water delivery to upper floors, linked directly to firefighting shafts.
- **Water Supplies** – hydrants, tanks, and dual pumps ensuring a reliable and redundant flow for firefighting operations.
- **Fire Control Rooms and Panels** – central hubs providing information, system control, and communication during incidents.
- **Smoke Control Systems** – maintaining tenable conditions in stairs, lobbies, basements, and atria, safeguarding crews as well as occupants.
- **Coordination with Fire Services** – ensuring facilities are aligned with local operational capability, validated through consultation and familiarisation.

These facilities are only effective when considered **holistically**. An access road that does not connect to a riser inlet, a firefighting lift without backup power, or a control room that is difficult to locate all compromise operational effectiveness. The fire strategy must therefore present these elements as an **integrated package** designed to work together under fire conditions.

Finally, these facilities must not be viewed as one-off construction features. They form part of the **"golden thread"** of building safety, requiring **regular inspection, testing, and management oversight** throughout the building's life.

Looking Ahead

The next chapter will focus specifically on **High-Rise Buildings**, where the importance of firefighting facilities becomes even more pronounced.

Longer travel distances, reliance on lifts, and extended evacuation times mean that **firefighter access and system reliability are paramount**.

Lessons from incidents such as Grenfell Tower demonstrate the consequences of inadequate facilities in tall residential blocks.

High-rise strategies require enhanced coordination, redundancy, and resilience across every firefighting provision.

By understanding the principles covered in this chapter, we can now explore how they are adapted and strengthened in the **unique context of vertical evacuation and high-rise firefighting**.

Chapter 11
High-Rise Buildings

11.1 Introduction

High-rise buildings — generally defined in the UK as those with a floor level more than **18 metres above ground** (with further thresholds at 30 m and 60 m for certain regulatory provisions) — represent some of the most complex and high-risk challenges in fire safety design. Their scale, vertical form, and reliance on mechanical systems mean that strategies which may be adequate in low-rise structures often require significant adaptation or enhancement in a tall building context.

The **unique challenges of high-rise fire safety** include:

- **Vertical Evacuation** – Large numbers of occupants may need to descend many flights of stairs, placing pressure on stair capacity, refuge areas, and evacuation strategies (simultaneous, phased, or stay-put).
- **Extended Firefighting Distances** – Fire service personnel may be required to operate at considerable heights or depths, carrying equipment through long vertical shafts and along corridors. This increases reliance on **firefighting lifts, shafts, and rising mains.**
- **Façade Fire Spread** – The external envelope of high-rise buildings has come under intense scrutiny following Grenfell Tower (2017). Combustible cladding systems have demonstrated how external fire spread can compromise multiple floors in minutes.
- **Reliance on Lifts and Mechanical Systems** – Evacuation lifts, pressurisation fans, smoke control systems, and water pumps are often essential for life safety. Their reliability during fire conditions is therefore a central concern.
- **Compartmentation and Stay-Put Strategies** – High-rise residential design often relies on the principle that fires can be contained within a single flat for a defined period. The integrity of compartmentation, fire doors, and service penetrations is therefore critical.

11.1.1 Why High-Rise Fire Safety Matters

- **Life Safety** – The high occupant density of towers (residential, commercial, or mixed-use) means that even small system failures can put hundreds of lives at risk.
- **Resilience and Business Continuity** – High-rise offices, hotels, and hospitals often house critical services or thousands of workers. Fire disruption can have major economic impacts.
- **Firefighter Safety** – Fire crews must operate in extreme conditions, often at height, with long hose runs and delayed water delivery. Designing for firefighter access and safety is as vital as designing for occupant protection.
- **Public Confidence** – High-rise fires attract intense public and media scrutiny, and the Grenfell disaster has reshaped both regulation and public trust in high-rise safety. Fire strategies for such buildings must therefore be especially robust, transparent, and defensible.

11.1.2 Scope of This Chapter

This chapter sets out how fire strategies address the **inherent risks of high-rise buildings**, covering:

- The **regulatory framework** for tall buildings in the UK and internationally.
- The **role of evacuation strategies** (stay-put, phased, simultaneous, use of evacuation lifts).
- The design and integration of **firefighting facilities** (firefighting shafts, lifts, rising mains, and control rooms).
- The challenges of **façade fire spread** and external wall design.
- The importance of **management, maintenance, and resident engagement** in sustaining long-term safety.
- **Case studies** and lessons learned from historical high-rise fires.

Summary

High-rise buildings demand fire strategies that go beyond prescriptive compliance, integrating **engineering analysis, active and passive measures, and management controls** to deliver a comprehensive safety framework. This introduction highlights why tall structures require a different mindset and sets the stage for the technical discussions that follow.

11.2 Why High-Rise Buildings Are Different

High-rise buildings are not simply "taller versions" of their low-rise counterparts. Their height, density, and reliance on engineered systems fundamentally alter how fire develops, how occupants evacuate, and how firefighters respond. These differences mean that **standard prescriptive guidance is often insufficient**, and fire strategies must carefully consider the unique challenges posed by tall structures.

11.2.1 Extended Evacuation Times

- Occupants in high-rise buildings may need to descend **dozens of floors** to reach ultimate safety. This increases evacuation time significantly compared with low-rise buildings.
- Evacuation times are influenced by:
  - **Occupant density** (e.g., crowded office floors vs. low-density residential).
  - **Stair width and number of stairs.**
  - **Evacuation strategy** – whether simultaneous, phased, or stay-put.
- Fire strategies must ensure that **escape routes remain tenable for longer durations**, often through enhanced compartmentation, smoke control, and pressurisation.

11.2.2 Reliance on Firefighting Lifts and Facilities

- Beyond **18 m above ground level**, carrying hoses and heavy equipment up stairs becomes **impractical and dangerous**.
- Fire strategies must therefore provide:
  - **Firefighting lifts** (BS EN 81-72 compliant) within protected shafts.
  - **Rising mains** (dry or wet, depending on height) to deliver water at upper levels.
  - **Firefighting lobbies** and pressurised stairs to give crews safe staging points.
- These facilities are not optional; they are essential for **fire service intervention** in tall buildings.

11.2.3 Wind and Stack Effects

- **Stack effect** occurs when heated smoke rises through tall shafts, stairwells, and lift cores, drawing in cooler air from lower levels.
- This can:
  - Pull smoke into escape routes.
  - Cause smoke to spread **vertically between floors** even without direct fire penetration.
  - Stall smoke flow in atriums when cooler air is entrained, reducing buoyancy.
- **Wind effects** at height can also disrupt smoke ventilation, particularly in buildings with large façade openings.
- Fire strategies must use **smoke modelling and pressure control systems** to account for these complex air movement dynamics.

11.2.4 Façade Fire Spread

- External fire spread poses a heightened risk in high-rise buildings because:
  - Fire can bypass internal compartmentation by **spreading up combustible cladding** or through poorly detailed cavities.
  - Flames can break out of windows and re-enter upper floors, trapping occupants above.
- Post-Grenfell regulations now require external walls of high-rise residential buildings to achieve **non-combustible classifications (Euroclass A1 or A2)**.
- Fire strategies must explicitly assess and justify the **performance of cladding, insulation, and cavity barriers**, with clear evidence of compliance.

11.2.5 Complex Occupancy Loads

- Many high-rise towers are **mixed-use developments**, combining residential, office, retail, hospitality, and leisure spaces.
- Each use has **different evacuation characteristics**:
  - **Residential**: Stay-put or phased evacuation.
  - **Offices**: Phased evacuation.

  - **Retail**: Simultaneous evacuation with voice alarm systems.
  - **Hotels and Hospitals**: Progressive horizontal evacuation, with provisions for vulnerable persons.
- Strategies must manage **interaction between occupancies**, ensuring, for example, that a fire in a shopping podium does not compromise evacuation routes for residential apartments above.

11.2.6 Summary

High-rise buildings demand a fire safety strategy that addresses:

- Longer evacuation times.
- Dependence on firefighting lifts and rising mains.
- Complex smoke movement due to wind and stack effects.
- Increased façade fire spread risks.
- Mixed-use occupancies with competing evacuation needs.

These factors make high-rise design **fundamentally different from low- and mid-rise structures**. A "one-size-fits-all" approach is inadequate; instead, strategies must adopt a **performance-based, holistic framework** that integrates engineering, management, and resilience.

11.3 Regulatory Framework

High-rise buildings are subject to some of the most stringent fire safety requirements in the UK regulatory system. This reflects the **heightened risks of vertical evacuation, façade fire spread, and firefighter access**. A fire strategy for a tall building must demonstrate clear compliance with statutory requirements, guidance documents, and post-Grenfell reforms.

11.3.1 Approved Document B (ADB)

Approved Document B remains the **primary statutory guidance** accompanying Part B of the Building Regulations. For high-rise buildings, it sets out specific provisions on:

- **Stair Cores and Escape Provisions**
  - Minimum number of stairs increases with building height and occupancy.

    - Protected stair enclosures with **minimum 60–120 minutes fire resistance**, often supported by smoke ventilation or pressurisation systems.
- **Firefighting Shafts**
    - Required where floor levels exceed **18 m above** or **10 m below** fire service access level.
    - Incorporate **firefighting stairs, lobbies, and lifts**, enclosed in fire-resisting construction.
- **Cladding and External Fire Spread**
    - Restrictions on combustible materials in external walls of buildings above 18 m.
    - Requirement for **cavity barriers** at compartment floors and around window openings.
- **Basements and Atriums**
    - Additional smoke ventilation and firefighting shaft requirements apply.

ADB provides the **baseline prescriptive requirements** but allows deviations where equivalent performance can be demonstrated through fire engineering.

11.3.2 BS 9999 – Code of Practice for Fire Safety in the Design, Management and Use of Buildings

- A **risk-based, performance-oriented standard** offering more flexibility than ADB.
- For high-rise buildings, BS 9999 enables:
    - Extended **travel distances** if supported by enhanced fire precautions.
    - Risk-adjusted **stair and exit sizing**, taking account of occupancy type and management controls.
    - Tailored **smoke control and fire service facilities** where justified by engineering analysis.
- Widely used for **complex or innovative high-rise designs** that fall outside ADB's prescriptive limits.

11.3.3 BS 9991 – Residential Guidance

- Provides detailed fire safety guidance for **blocks of flats, maisonettes, and high-rise residential towers.**
- Key aspects include:
    - **Stay-put strategies** for compartmented residential blocks, supported by 60+ minutes fire resistance between flats.
    - Guidance on **automatic fire detection and alarm systems** in common parts and dwellings.
    - Provisions for **sprinklers** and **smoke ventilation systems** in corridors and stairs.
- Increasingly relevant for **mixed-use towers** where residential components require tailored strategies.

11.3.4 Post-Grenfell Legislation and Reform

The **Grenfell Tower tragedy (2017)** exposed serious failings in the regulation of high-rise safety. Major reforms since include:

- **Building (Amendment) Regulations 2018**
    - Banned the use of **combustible materials** in the external walls of residential buildings, hospitals, student accommodation, and care homes over 18 m in height.
    - All external wall systems must now be **Euroclass A1 or A2-s1,d0.**
- **Building Safety Act 2022**
    - Introduced a new regime for **"higher-risk buildings"** (HRBs), covering residential buildings over 18 m or 7 storeys, and hospitals and care homes of the same height.
    - Established the **Building Safety Regulator (BSR).**
    - Requires the **"golden thread"** of safety information to be maintained throughout the building lifecycle.
- **Fire Safety Act 2021** and **Fire Safety (England) Regulations 2022**
    - Clarified that external walls and flat entrance doors must be considered in fire risk assessments.
    - Introduced new duties for building owners/managers, including **monthly checks of lifts and firefighting equipment** and provision of up-to-date floor plans to fire services.

11.3.5 Summary

High-rise buildings sit at the intersection of **prescriptive regulation (ADB), risk-based codes (BS 9999, BS 9991), and post-Grenfell reforms**. A defensible fire strategy must:

- Clearly reference the chosen route to compliance.
- Demonstrate conformity with statutory bans on combustible cladding.
- Document how ongoing management requirements (e.g., golden thread, monthly checks) will be met.

In short, **regulatory compliance for high-rise buildings is both more detailed and more demanding** than for low-rise structures, reflecting the life safety challenges they pose.

11.4 Evacuation Strategies for High-Rise Buildings

The choice of **evacuation strategy** is one of the most critical decisions in a high-rise fire strategy. Unlike low-rise buildings, where **simultaneous evacuation** is often assumed, tall buildings require tailored approaches that reflect their height, compartmentation, occupant characteristics, and management arrangements.

11.4.1 Simultaneous Evacuation

- **Description**: All occupants evacuate the building at the same time once the alarm is raised.
- **Application**:
  - Feasible in **smaller high-rise towers** with limited occupant numbers.
  - More common in mid-rise buildings (below 18–20 storeys).
- **Advantages**:
  - Simple, easy to communicate and manage.
  - No reliance on phased systems or advanced communication.
- **Limitations**:
  - Can overwhelm stairs and exit routes in large towers.
  - Prolonged evacuation times in tall buildings may result in untenable conditions before everyone is clear.

11.4.2 Phased Evacuation

- **Description**: Occupants leave in stages, beginning with the fire floor and the floor above, followed by adjacent levels, then progressively more distant levels.
- **Application**:
  - Standard approach in **high-rise office buildings**.
  - Often combined with voice alarm/public address systems and fire wardens.
- **Advantages**:
  - Reduces congestion in stairs by limiting the number of evacuees at any one time.
  - Buys time for controlled evacuation while maintaining business continuity during false alarms.
- **Requirements**:
  - **Robust alarm and voice communication systems** (BS 5839-8).
  - **High integrity compartmentation** to keep occupants safe while waiting.
  - Firefighting lobbies and smoke control systems to protect escape routes.
- **Limitations**:
  - Requires **high levels of management control** and trained staff.
  - Vulnerable to occupant non-compliance (people may self-evacuate early).

11.4.3 Stay Put / Defend-in-Place

- **Description**: Occupants remain in their flats or units unless directly affected by fire or instructed to leave.
- **Application**:
  - Common in **residential high-rises** in the UK.
  - Assumes each dwelling is a **120-minute fire-resisting compartment** with FD30S/FD60S entrance doors.

- **Advantages**:
  - Avoids mass stair congestion, allowing firefighters to operate unimpeded.
  - Practical for vulnerable populations who may struggle with immediate evacuation.
- **Requirements**:
  - **High-quality compartmentation** between flats and between flats and common areas.
  - Reliable fire doors, cavity barriers, and minimal breaches in service penetrations.
- **Limitations**:
  - **Compromised compartmentation** (e.g., poor maintenance, unauthorised alterations) can render this strategy unsafe, as seen in Grenfell Tower.
  - Requires robust resident communication and trust in the system.

11.4.4 Progressive Horizontal Evacuation (PHE)

- **Description**: Occupants are moved horizontally into adjacent fire-resisting compartments, then vertically if necessary.
- **Application**:
  - Healthcare occupancies (hospitals, care homes) where patients cannot self-evacuate.
- **Advantages**:
  - Avoids immediate vertical evacuation of vulnerable persons.
  - Allows staff to relocate patients in manageable stages.
- **Requirements**:
  - **Sub-compartmentation** of floors with 30–60 minutes fire resistance between zones.
  - Sufficient staff trained in patient movement and Personal Emergency Evacuation Plans (PEEPs).

- **Limitations**:
  - Labour-intensive; depends on staff availability.
  - Requires strict management of fire doors and compartment integrity.

11.4.5 Selecting the Strategy

- The chosen evacuation strategy must be:
  - **Explicitly stated in the fire strategy**.
  - Supported by design features (compartmentation, alarms, smoke control).
  - Linked to **management procedures and staff training**.
- Hybrid approaches are common, e.g.:
  - Residential towers adopting stay-put for dwellings but simultaneous evacuation for communal areas.
  - Mixed-use towers using phased evacuation for offices and defend-in-place for apartments.

Summary

Evacuation in high-rise buildings cannot rely on a single prescriptive model. Instead, the fire strategy must consider:

- The **occupancy type** (residential, office, healthcare, mixed-use).
- The **robustness of compartmentation** and active systems.
- The **human factors** influencing evacuation behaviour.

By tailoring the approach to the building, fire strategies can balance safety, practicality, and resilience, ensuring that all occupants — from office workers to hospital patients — can reach safety during a fire.

11.5 Compartmentation and Structure

Compartmentation is the **cornerstone of fire safety in high-rise buildings**, ensuring that fire and smoke are contained to their area of origin and do not rapidly compromise multiple storeys or flats. In tall structures where full evacuation may not be immediate, **compartmentation underpins stay-put and phased evacuation strategies**, buying essential time for occupants and firefighters.

11.5.1 Dwellings and Units

- In high-rise residential buildings, **each flat or maisonette must be a self-contained fire-resisting compartment**.
- Minimum fire resistance is typically:
    - **60 minutes** for walls and floors separating flats (BS 9991 baseline).
    - **90–120 minutes** in buildings exceeding 30 m, or where higher risks are present.
- **Flat entrance doors** must be **FD30S or FD60S** (fire- and smoke-sealing), self-closing, and fitted with intumescent strips and cold smoke seals.

11.5.2 Compartment Floors and Walls

- **Storey separation** between flats, offices, or other uses is generally required to be **60–120 minutes fire resisting**.
- For mixed-use towers, compartment floors between retail podiums and residential towers are typically **120 minutes** to account for increased risk.
- Structural floors may also act as **horizontal fire barriers**, supporting evacuation strategies and firefighting operations.

11.5.3 Vertical Shafts and Risers

- Vertical elements such as **lift shafts, risers, service ducts, and stair cores** are critical weak points in compartmentation.
- Requirements include:
    - Shafts enclosed in **minimum 120 minutes fire-resisting construction**.
    - **Service penetrations** sealed with tested fire-stopping systems.
    - **Fire dampers** in ducts at compartment lines, tested to **BS EN 1366-2**.
    - **Pressurisation systems** for firefighting stairs and lifts to prevent smoke ingress.

11.5.4 Façade Systems and Cavity Barriers

- **Façade fires** can bypass compartment floors and walls if cavities are left unprotected.
- Cavity barriers are therefore mandatory:
  - At **compartment floors and walls.**
  - Around **window openings.**
  - At the **perimeter of cladding systems.**
- The **Grenfell Tower fire (2017)** exposed the catastrophic consequences of inadequate cavity barriers and combustible façades. Modern regulations (Building (Amendment) Regulations 2018) now require **non-combustible materials** (Euroclass A1/A2) in external wall systems for relevant high-rise buildings.

11.5.5 Structural Fire Resistance

- High-rise structures must remain **stable for the full evacuation and firefighting duration.**
- Structural frame fire resistance is typically:
  - **60 minutes** for buildings under 18 m.
  - **90–120 minutes** for buildings between 18–60 m.
  - **120–180 minutes** for super-tall structures (>60 m).
- Fire protection may be achieved by:
  - **Concrete construction** (inherently fire-resisting).
  - **Fire-protected steel** (encasement, board, or spray).
  - **Hybrid systems** with fire-rated composites.

11.5.6 Interaction with Evacuation Strategies

- **Stay-put strategies** rely entirely on the compartmentation of flats, lobbies, and stairs.
- **Phased evacuation** requires compartments to resist fire long enough for later phases to evacuate safely.
- Fire strategy documents must **explicitly state the fire resistance ratings**, supported by calculations, product certificates, or performance-based assessments.

Summary

Compartmentation and structural fire resistance are what make high-rise fire strategies **credible and defensible**. They provide the containment necessary to prevent fire spread, protect escape routes, and sustain the building long enough for safe evacuation and firefighting.

A fire strategy must therefore:

- Identify each compartment boundary.
- Specify resistance levels (60–180 minutes).
- Confirm compliance of doors, shafts, risers, and façade cavity barriers.
- Link compartmentation directly to the chosen evacuation policy.

Without robust compartmentation, **high-rise strategies collapse**, as demonstrated by past failures in the UK and internationally.

11.6 Firefighting Facilities

High-rise buildings present unique challenges for fire service intervention due to **height, access distance, and reliance on mechanical systems**. Without dedicated firefighting facilities, crews would be unable to transport equipment, deploy water, or maintain safe conditions within the timeframes necessary to save lives.

A fire strategy must therefore set out **what facilities are provided, where they are located, and how they are maintained**. These provisions are not optional enhancements — they are **core life safety requirements**.

11.6.1 Firefighting Shafts

- **Purpose**: To provide firefighters with safe, protected vertical access and staging areas.
- **Components**:
    - **Firefighting Stair**: Enclosed, fire-resisting stair (usually 120 minutes).
    - **Firefighting Lift**: Providing vertical transport of equipment and personnel.
    - **Firefighting Lobby**: A protected lobby separating the stair/lift from accommodation, with smoke ventilation or pressurisation.

- **Requirements** (ADB, BS 9999):
  - Required in buildings where the **top floor is more than 18 m above** or more than 10 m below fire service access level.
  - Shafts must be located to give **coverage within maximum hose lengths** (typically 45 m from outlet to furthest point).
- **Design Considerations**:
  - Pressurisation or smoke ventilation of lobbies to prevent smoke ingress.
  - Fire-resisting construction with self-closing doors (FD60S or better).

11.6.2 Firefighting Lifts

- **Purpose**: To allow rapid transport of crews and equipment to upper levels, reducing time lost in stair climbing.
- **Requirements**:
  - Mandatory where floors are more than **18 m above** or **10 m below** access level.
  - Must comply with **BS EN 81-72**, including:
    - **Minimum 630 kg capacity (8 persons)**.
    - **Access only via protected lobbies**.
    - **Dual power supply** (mains plus emergency backup).
    - **Fire service controls** with override priority.
    - **Water ingress protection** (drainage and sealed electrics).
- **Best Practice**: In super-high-rise towers, **multiple firefighting lifts per shaft** are recommended to ensure redundancy.

11.6.3 Rising Mains

- **Purpose**: To allow firefighters to connect hoses at upper floors without dragging lines vertically.
- **Types**:
  - **Dry Risers**:
    - Installed where floors are between **18–60 m above access level**.

- Consist of empty pipework connected to an inlet at ground level.
- Fire service pumps supply water on demand.

  - **Wet Risers**:
    - Required where floors exceed **60 m**.
    - Permanently charged with water via pumps and tanks.
    - Must have **redundant pumps** and emergency power supplies.

- **Design Notes**:
  - Outlets (landing valves) must be provided within 45 m of any part of the floor.
  - Inlets and valves must be accessible, protected, and clearly labelled.
  - Regular testing and maintenance are essential to ensure reliability.

11.6.4 Fire Control Rooms

- **Purpose**: To provide a **single point of control and information** for fire services.
- **Contents**:
  - Fire alarm control panels (BS 5839).
  - Smoke control and ventilation controls.
  - Sprinkler and riser status panels.
  - Communication systems (firefighter telephones, radios).
  - Building plans and pre-incident information.
- **Design**:
  - Located on the fire service access level.
  - Fire-resisting construction (minimum 120 minutes).
  - Clearly signed, with secure but immediate access for fire crews.

11.6.5 Integration and Testing

- Facilities must be integrated with the building's fire strategy, ensuring:
  - **Smoke control works in tandem** with firefighting shafts and lobbies.

  - **Rising mains and water supplies** are tested and maintained to BS requirements.
  - **Control room information** reflects real-time building status.
- Fire strategies should include:
  - Commissioning test records.
  - Ongoing inspection and servicing schedules.
  - Procedures for liaison with the local Fire and Rescue Service.

Summary

Firefighting facilities are fundamental to high-rise fire strategies. They provide the access, water, and control necessary for firefighters to operate safely and effectively. Without protected shafts, firefighting lifts, rising mains, and a dedicated control room, **fire service intervention at height becomes unsafe or impossible.**

By clearly documenting these provisions, a fire strategy ensures that firefighters are not disadvantaged by building design, but instead supported to carry out life-saving operations.

11.7 Smoke Control

In high-rise buildings, **smoke presents a greater hazard than flames.** It spreads rapidly through shafts, lobbies, and stairwells, reducing visibility, impairing breathing, and rendering escape routes unusable. The **stack effect** (air movement caused by temperature and pressure differences in tall structures) can draw smoke upwards, while wind pressures can spread it laterally through façades.

Because evacuation and firefighting in high-rises often take **much longer** than in low-rise buildings, smoke control is **not optional** but fundamental to the fire strategy. The system must be **robust, fail-safe, and maintainable for extended durations**, typically **30–60 minutes or more**, depending on the evacuation profile.

11.7.1 Stair Pressurisation Systems

- **Purpose**: To keep firefighting and escape stairs clear of smoke.
- **How It Works**:
  - Fans supply clean air into stairwells at positive pressure.

    - Pressure differential prevents smoke ingress when doors are opened.
- **Design Considerations**:
    - Must balance between too little pressure (smoke leakage) and too much pressure (doors too heavy to open).
    - Designed to **BS EN 12101-6**.
    - Typically require **backup power supplies** for resilience.
- **Application**: Standard in most high-rise office and residential towers.

11.7.2 Mechanical Smoke Extract Systems

- **Purpose**: To remove smoke from lobbies, corridors, or entire floors.
- **How It Works**:
    - High-capacity fans extract smoke through ducts to external discharge points.
    - Often paired with **make-up air systems** to balance airflow.
- **Applications**:
    - Office towers with deep floorplates.
    - Underground car parks beneath high-rise buildings.
- **Design Standards**:
    - **BS EN 12101-3 and 12101-7**.
    - Must achieve designed extract rates to maintain a **smoke-free layer above head height**.
- **Challenges**: Ducts and fans must be **fire-rated** to maintain integrity under fire exposure.

11.7.3 Natural Smoke Shafts

- **Purpose**: Low-energy solution for removing smoke from common lobbies and corridors in high-rise residential towers.
- **How It Works**:
    - Smoke vents or automatic opening vents (AOVs) open to a vertical shaft discharging smoke outside.
    - Relies on buoyancy and wind to drive smoke flow.

- **Applications**:
  - Common in residential blocks designed to **ADB and BS 9991**.
  - Simpler and less maintenance-intensive than mechanical systems.
- **Limitations**:
  - Less reliable in adverse wind conditions.
  - May not cope with large smoke volumes compared to mechanical systems.

11.7.4 Atrium Smoke Control

- **Purpose**: To prevent smoke filling large, open spaces such as atriums that connect multiple storeys.
- **Methods**:
  - **Smoke reservoirs** with high-level extraction.
  - **Curtains or baffles** to channel smoke into controlled zones.
  - **Mechanical extract fans** designed using CFD modelling.
- **Applications**:
  - Mixed-use towers with retail podiums and hotel/residential atriums.
- **Design Standards**:
  - **BS EN 12101-5 and 12101-6**.
  - Often justified through **BS 7974 performance-based analysis**.

11.7.5 Integration with Fire Strategy

- Smoke control systems must be **coordinated with alarms, sprinklers, and fire service overrides**.
- Strategies should include:
  - **Cause-and-effect logic** (e.g., "Detector activation in Lobby A → open shaft vent → start fan → release doors").
  - **Manual overrides** for firefighters.
  - Clear diagrams of zoning and system control points.
- **Testing and Maintenance**:
  - Routine functional testing (monthly/annual).
  - Certification of fans, dampers, and vents to BS EN standards.

11.7.6 Summary

Smoke control in high-rise buildings is a **life-critical system**. It ensures escape routes remain clear, protects firefighting shafts, and prevents smoke spread into occupied zones. Fire strategies must:

- Identify the chosen smoke control method (pressurisation, mechanical extract, natural shafts, or atrium systems).
- Demonstrate compliance with BS EN 12101 and BS 9991/9999.
- Document cause-and-effect programming and fire service controls.
- Embed maintenance and testing regimes to ensure reliability throughout the building's life.

11.8 External Wall Systems

The performance of external wall systems is one of the most **critical and scrutinised elements** of high-rise fire safety. The **Grenfell Tower fire (2017)** demonstrated how combustible cladding and insulation materials, combined with inadequate cavity barriers, can enable fire to bypass internal compartmentation and spread across an entire façade within minutes. This tragedy reshaped UK regulation and public trust, placing external walls at the centre of high-rise fire strategy.

11.8.1 Regulatory Requirements

- **Combustible materials are no longer permitted** in external wall systems of:
    - Residential buildings over **18 m** in height.
    - Hospitals, care homes, and student accommodation above the same threshold.
- Under the **Building (Amendment) Regulations 2018**, all materials in these systems must achieve **Euroclass A2-s1,d0** or better (non-combustible).
- The **Building Safety Act 2022** and subsequent **Fire Safety (England) Regulations 2022** strengthened these requirements by:
    - Making external walls a mandatory element of **fire risk assessments**.
    - Introducing duties for **ongoing inspection and remediation** of unsafe cladding.

11.8.2 Cavity Barriers and Fire-Stopping

- Even non-combustible cladding systems require **robust detailing** to prevent smoke and flame spread.
- **Cavity barriers** are mandatory:
    - At **compartment floors and walls.**
    - Around **window and door openings**.
    - At **roof junctions** and edges of façade systems.
- Barriers must:
    - Be tested and certified to **BS EN 1366 and BS 476-20/22**.
    - Be installed continuously and sealed to prevent gaps.
- The failure of cavity barriers was a **key factor at Grenfell**, where fire exploited gaps to spread unchecked.

11.8.3 Inspection, Maintenance, and Remediation

- Fire strategies must recognise that façade performance can deteriorate over time due to:
    - **Weathering**, corrosion, or delamination of panels.
    - **Building alterations**, including window replacement or services penetration.
    - Poorly executed repairs that compromise barrier integrity.
- Best practice includes:
    - **Regular façade inspections**, recorded as part of the "golden thread" of information.
    - **Thermal imaging or intrusive surveys** to confirm cavity barrier presence and continuity.
    - Remediation plans where unsafe materials or poor installations are identified.

11.8.4 Broader Risk Considerations

- External wall systems must also consider:
    - **Fire spread between buildings** where boundary distances are small.
    - **Wind-driven fire behaviour**, which can accelerate flame propagation on tall façades.

    - **Compatibility with active systems**, e.g., ensuring sprinklers discharge effectively near glazed façades.
- Strategies should also assess **balconies**, which can introduce combustible decking or furniture, creating pathways for vertical fire spread.

11.8.5 Summary

External wall systems in high-rise buildings are now subject to **strict non-combustibility requirements** and rigorous inspection regimes. A fire strategy must:

- Confirm materials achieve **Euroclass A2-s1,d0 or better**.
- Demonstrate compliant detailing with **cavity barriers at all critical junctions**.
- Include a plan for **inspection, maintenance, and record-keeping** throughout the building's life.

The lessons of Grenfell underline that **façade safety is not simply a design issue, but a long-term management responsibility**.

11.9 Fire Service Considerations

The fire and rescue service plays a **critical role in the safety of high-rise buildings**, yet their ability to operate effectively is heavily influenced by building design. Unlike low-rise structures, where external firefighting access is often possible, tall buildings depend on **internal firefighting facilities**. A fire strategy must therefore explicitly consider how the fire service will access, navigate, and operate within the building under emergency conditions.

11.9.1 Access and Hardstanding

- **External Access**:
    - Fire appliances must be able to reach within **18 m of dry riser inlets** or principal entrances (ADB).
    - Hardstanding areas must be provided with adequate width, load-bearing capacity, and turning circles.

- **Aerial Appliance Access**:
    - Where possible, at least one façade should be accessible to aerial appliances.
    - This is often limited in very tall buildings, where heights exceed the reach of aerial platforms.
    - Still useful for access to podium levels, lower storeys, and rescue at mid-rise levels.

11.9.2 Internal Operations

- Firefighters in high-rises rely almost exclusively on **firefighting shafts** for vertical access.
- Shafts provide:
    - Protected stair access.
    - Firefighting lifts for rapid transport of personnel and equipment.
    - Lobbies acting as staging areas and smoke buffers.
- **Hose deployment** is via dry or wet risers, with outlets provided on each floor.
- Travel distances within floors must be minimised to reduce firefighter fatigue and ensure effective hose reach.

11.9.3 Duration and Intensity of Attack

- Fires in tall buildings may require **prolonged firefighting operations**, sometimes lasting hours.
- Extended duration requires:
    - **Resilient water supplies**, often via dual pumps and on-site tanks.
    - **Backup power** for smoke control, lifts, and alarms.
    - Staging areas for crew rotation and equipment storage, often within lobbies or designated refuge areas.
- Fire strategies should consider **crew fatigue, hose management, and thermal stress**, recognising that sustained vertical operations are physically demanding.

11.9.4 Information and Communication

- Accurate, accessible information is essential for effective firefighting:
    - **Fire control rooms** must house alarm panels, smoke control interfaces, riser and sprinkler status, and building plans.
    - **Up-to-date floor plans** are a legal requirement under the **Fire Safety (England) Regulations 2022** for high-rise residential buildings.
    - **Signage and wayfinding** within shafts and lobbies must support quick navigation, even in low visibility.
- Communication systems (e.g., fire telephones, leaky feeder radio systems) are vital, as conventional radios may fail in dense concrete and steel structures.

11.9.5 Liaison and Pre-Planning

- A strong fire strategy involves **consultation with the local Fire and Rescue Service (FRS)** at design stage.
- This should cover:
    - **Access arrangements**.
    - **Hydrant locations and water supplies**.
    - **Special hazards** (e.g., fuel storage, lithium-ion batteries, unusual structural forms).
- Pre-planning includes **familiarisation visits**, during which firefighters can learn the building layout and practice emergency procedures.

Summary

Fire service considerations are **central to high-rise fire strategies**. Access, firefighting shafts, water supplies, duration of attack, and communication systems must all be carefully designed and documented. Without these, firefighters may be unable to mount safe and effective operations, endangering both occupants and crews.

A robust strategy therefore ensures that:

- The fire service can **reach the building quickly**.
- They can **operate internally via protected shafts and risers**.

- They are supported with **reliable water, power, and communication systems**.
- They have **clear, accessible information** at the point of entry.

11.10 Human Behaviour in High-Rise Fires

While regulations and engineering focus on **systems and structures**, the outcome of a high-rise fire often depends on **human response**. Occupant behaviour in tall buildings is shaped by psychology, familiarity, trust in the building's systems, and communication during the incident. Fire strategies must therefore incorporate **realistic behavioural assumptions** rather than relying on idealised models of immediate evacuation.

11.10.1 Delay in Evacuation ("Risk Normalisation")

- Studies show that many residents in high-rise buildings **delay evacuation** after an alarm, often to:
    - Gather belongings.
    - Seek confirmation that the alarm is genuine.
    - Await instructions from management or fire services.
- This behaviour is common in **residential towers**, where alarms are sometimes perceived as false or routine.
- Delays increase **Required Safe Egress Time (RSET)**, meaning escape routes must remain tenable for longer periods.

11.10.2 Visitor and Transient Occupant Behaviour

- In mixed-use or commercial towers, **visitors and transient occupants** often default to the **entry route** (the familiar path they used to arrive), rather than the nearest escape stair.
- This can cause congestion in lobbies and delay evacuation.
- Clear **signage, wayfinding, and voice instructions** are therefore critical to guide unfamiliar users.

11.10.3 The Role of Voice Alarm Systems

- Research demonstrates that **voice alarm systems** (BS 5839-8 compliant) improve evacuation compliance compared with bells or sirens.

- Benefits include:
    - Providing **clear, unambiguous instructions**.
    - Reducing panic and confusion by explaining the nature of the emergency.
    - Supporting **phased evacuation strategies**, where some occupants are instructed to remain in place.
- Example: "Fire has been detected on the 10th floor. Occupants of floors 9, 10, and 11 please evacuate immediately. Other floors, please await further instruction."

11.10.4 Trust in Compartmentation and Evacuation Strategy

- **Phased and stay-put strategies** rely on occupant trust in the building's **compartmentation**.
- If residents or office workers believe the building fabric will not hold, they may self-evacuate en masse, overwhelming stairs and delaying firefighting access.
- Conversely, if people trust the system, they are more likely to comply with phased or defend-in-place instructions.
- This underlines the need for:
    - **Resident engagement and education** in residential towers.
    - **Regular fire drills** in offices and mixed-use buildings.
    - Transparent communication about **building safety measures**.

11.10.5 Special Considerations

- **Vulnerable Occupants**: Elderly, disabled, or mobility-impaired persons may require staff assistance or evacuation aids.
- **Children**: In schools within high-rise blocks, behaviour is shaped by staff leadership and drills.
- **Cultural Factors**: In international projects, cultural attitudes toward authority and risk can influence evacuation speed and compliance.

11.10.6 Summary

Human behaviour is one of the **least predictable but most decisive factors** in high-rise fire safety. A fire strategy must account for:

- Likely delays in response (risk normalisation).

- Misjudged exit choices by visitors.
- The importance of **voice alarms** in guiding safe and orderly evacuation.
- The reliance of phased and stay-put strategies on **occupant trust and engagement**.

Ultimately, **buildings must be designed not for ideal occupants, but for real people**, who may hesitate, make errors, or require reassurance during evacuation.

11.11 Case Study – Grenfell Tower (2017)

The **Grenfell Tower fire** in West London on **14 June 2017** remains one of the most tragic and transformative events in modern fire safety history. A kitchen fire in a fourth-floor flat spread externally via combustible cladding and insulation, rapidly engulfing the 24-storey residential tower. Seventy-two people lost their lives, hundreds were displaced, and the event triggered the most far-reaching reform of UK building safety regulation in decades.

11.11.1 The Façade and External Fire Spread

- The original 1970s tower was refurbished between 2012–2016, including the installation of **aluminium composite material (ACM) cladding with polyethylene cores** and combustible insulation.
- Once flames escaped the flat of origin, they ignited the cladding and insulation, allowing **rapid vertical and lateral fire spread** across the façade.
- This bypassed internal compartmentation and led to multiple flats being simultaneously compromised — a failure that **directly contravened the intended stay-put policy**.
- Lesson: **Non-combustible façades are essential** in high-rise residential and institutional buildings.

11.11.2 Compartmentation and the Stay-Put Strategy

- Grenfell Tower operated under a **stay-put policy**, consistent with UK high-rise residential practice, assuming each flat was a 60-minute fire-resisting compartment.

- However, façade fire spread and failures in fire doors and compartment lines led to **fire and smoke breaching multiple flats and common areas simultaneously**.
- Occupants, recognising the breakdown of compartmentation, began to **self-evacuate en masse**, overwhelming the stairs and complicating firefighter operations.
- Lesson: **Stay-put strategies are only viable when robust compartmentation is maintained throughout the building's life.**

11.11.3 Communication and Evacuation Challenges

- Residents reported **confusion and delays in communication** during the fire.
- No effective voice alarm system was present, and official instructions conflicted with what occupants could see (rapid external fire spread).
- Many residents ignored "stay-put" advice and attempted evacuation via the sole stairwell, which soon became congested and smoke-filled.
- Lesson: **Clear, adaptable communication systems (e.g., voice alarms, resident engagement, live updates)** are vital in high-rise buildings.

11.11.4 Regulatory and Management Failures

- Investigations revealed systemic failures, including:
    - **Inadequate testing and certification** of cladding systems.
    - Lack of proper **regulatory oversight** during refurbishment.
    - Weak enforcement of **maintenance and resident safety concerns**.
- The tragedy exposed flaws not just in technical design, but in the **regulatory framework, governance, and resident engagement processes**.
- *Lesson*: **Fire strategies must be reviewed and updated whenever refurbishments alter risk, and regulators must enforce compliance rigorously.**

11.11.5 Legacy and Reform

Grenfell catalysed a series of reforms that continue to reshape high-rise fire safety:

- **Building (Amendment) Regulations 2018** – ban on combustible materials in external walls of high-rise residential and institutional buildings.
- **Fire Safety Act 2021** – clarified that external walls and flat entrance doors fall within the scope of fire risk assessments.
- **Fire Safety (England) Regulations 2022** – mandated monthly checks of lifts, quarterly checks of firefighting equipment, and provision of building information to fire services.
- **Building Safety Act 2022** – established the Building Safety Regulator, the "golden thread" of building information, and a new duty-holder regime for higher-risk buildings.

11.11.6 Key Lessons for Fire Strategies

1. **Façade Systems** – must achieve Euroclass A2-s1,d0 or better, with cavity barriers at all compartment lines.
2. **Compartmentation Integrity** – must be preserved and verified throughout the life of the building.
3. **Evacuation Policies** – must be realistic, flexible, and supported by voice alarms and live communication tools.
4. **Regulatory Oversight** – must be robust, with third-party certification and rigorous enforcement of standards.
5. **Resident Engagement** – fire safety is not purely technical; residents must be informed, consulted, and prepared.

Summary

Grenfell Tower demonstrated the **devastating consequences** of weak façade performance, failed compartmentation, poor communication, and regulatory shortcomings. For modern fire strategies, Grenfell is both **a warning and a turning point**, underscoring that high-rise fire safety demands:

- **Non-combustible materials**,
- **Robust compartmentation**,
- **Clear communication systems**,

- **Regular review of strategies post-refurbishment**, and
- **Strong regulatory enforcement**.

11.12 International Perspectives

High-rise fire safety is a global challenge, and while the UK has developed its own regulatory framework, **international codes and practices offer valuable insight**. In particular, the USA, Middle East, and Asia have established approaches that differ in emphasis but share common objectives of **life safety, resilience, and fire service access**.

11.12.1 United States (NFPA 101 and International Building Code – IBC)

- **Evacuation Strategies**:
  - **Phased evacuation** is mandatory for most high-rise office buildings.
  - **Defend-in-place** is permitted only where compartmentation is extremely robust (e.g., hospitals).
  - Voice alarm systems are standard in tall commercial occupancies.
- **Fire Service Access**:
  - IBC requires **fire command centres** with full control of alarms, smoke systems, and communication.
  - Firefighter lifts are mandatory in taller buildings, often supported by multiple firefighting shafts.
- **Sprinkler Protection**:
  - Nearly all new high-rise buildings must be fully sprinklered to NFPA 13 standards, a more **universal requirement than in the UK**.
- **Lesson for the UK**: The USA's **mandatory active protection systems** and **command centres** emphasise operational control and redundancy.

11.12.2 Middle East and Asia (Adoption of IBC/NFPA Standards)

- Many Middle Eastern and Asian cities have rapidly urbanised skylines, adopting **IBC and NFPA codes** either directly or as the foundation for national building codes.

- **Key Characteristics**:
  - **More stringent fire service access requirements** than the UK, including wider firefighting shafts and enhanced lobby protection.
  - **Water supply redundancy** is a major focus, with multiple tanks and high-capacity pumps required for super-tall towers.
  - **Higher reliance on performance-based design**, particularly in countries such as the UAE and Saudi Arabia, where iconic architecture demands flexibility.
- **Lesson for the UK**: These regions illustrate the benefits of **resilience-driven requirements** (redundant water, shafts, and smoke systems) in extreme high-rise contexts.

11.12.3 Hong Kong and Singapore – Refuge Floors in Super-Tall Towers

- Both jurisdictions mandate **refuge floors** at prescribed intervals (often every 20–25 storeys) in super-tall towers.
- **Purpose of Refuge Floors**:
  - Provide protected spaces where occupants can gather before continuing evacuation.
  - Equipped with fire-resisting construction, ventilation, lighting, and sometimes communication points.
- **Why Required**:
  - In towers over 200 m, full evacuation can take **well over an hour**.
  - Refuge floors reduce evacuation fatigue, congestion, and smoke exposure.
- **Lesson for the UK**: Refuge floors are not currently a UK requirement but could be **considered for very tall towers**, particularly those over 50–60 storeys.

11.12.4 Comparative Observations

- ***UK vs USA***: UK relies more heavily on **compartmentation and stay-put strategies**, while the USA emphasises **universal sprinklers and phased evacuation**.

- ***UK vs Middle East***: UK has less focus on **system redundancy**, whereas Middle East codes mandate **dual supplies and enhanced resilience**.
- ***UK vs Asia***: UK does not yet require **refuge floors**, a common feature in Asia's tallest towers.

Summary

International perspectives demonstrate that while the UK approach is grounded in **compartmentation and prescriptive codes**, other regions place more emphasis on:

- **Universal sprinklers (USA)**.
- **System redundancy and resilience (Middle East)**.
- **Refuge floors and staged safety zones (Asia)**.

For high-rise strategies, this comparison highlights that the UK may benefit from **blending international best practice** into its evolving regulatory regime, particularly as super-tall and mixed-use towers become more common.

11.13 Summary

High-rise buildings represent the **greatest challenge in fire safety design**. Their height, occupant loads, and reliance on engineered systems mean that **traditional low-rise solutions are insufficient**. A robust fire strategy for high-rise buildings must therefore be **comprehensive, multi-layered, and resilient**, balancing prescriptive codes with performance-based analysis.

Key requirements include:

- **Compartmentation and Structural Integrity** – Each flat or unit must be a robust fire-resisting compartment, with floors, shafts, and façades designed to contain fire and smoke for extended durations.
- **Evacuation Strategies** – Whether phased, stay-put, or progressive horizontal evacuation, strategies must be explicitly stated, realistic, and supported by communication systems such as voice alarms.
- **Firefighting Facilities** – Firefighting shafts, lifts, risers, hydrants, and fire control rooms must be provided, ensuring fire crews can operate safely and effectively at height.
- **Smoke Control** – Pressurisation, natural shafts, or mechanical extract systems are critical to maintaining tenable escape routes and supporting fire service intervention.

- **External Wall Safety** – Façade performance has become a defining issue post-Grenfell. Non-combustible materials and properly installed cavity barriers are now mandatory in residential and institutional high-rises.
- **Human Behaviour** – Strategies must account for delayed evacuation, reliance on familiar routes, and the need for trust in compartmentation and management systems.

The **Grenfell Tower tragedy** has permanently altered expectations of high-rise safety in the UK and internationally. It highlighted that **design compliance alone is not enough** — continuous inspection, resident engagement, and regulatory enforcement are equally vital to sustaining safety throughout the building's life.

Ultimately, high-rise fire strategies must deliver **confidence and resilience**: confidence for occupants that they are safe in their homes or workplaces, and resilience in the systems that protect them, even under worst-case fire conditions.

The next chapter will turn to **Large Floor Area Buildings** (such as shopping centres, exhibition halls, and warehouses), where the challenges are not vertical evacuation but instead **horizontal evacuation, large fire loads, smoke spread, and the need for suppression systems** capable of controlling extensive fires.

Chapter 12

Large Floor Area Buildings (Shopping Centres, Warehouses)

12.1 Introduction

Large floor area buildings present some of the **most complex fire safety challenges** faced by designers, regulators, and fire services. Unlike high-rise towers, where vertical evacuation dominates the strategy, these buildings demand a focus on **horizontal movement of large numbers of people**, often across wide open-plan spaces with few internal subdivisions.

Examples include:

- **Retail complexes and shopping centres** – with extremely high occupant densities, varied tenancies, and transient visitor populations unfamiliar with layouts.
- **Exhibition halls and arenas** – which may hold tens of thousands of people for short periods, requiring rapid and controlled mass evacuation.
- **Warehouses and logistics facilities** – which store vast quantities of combustible goods, leading to very high fire loads and rapid potential fire growth.
- **Industrial sheds or aircraft hangars** – where specialist hazards such as fuel, chemicals, or vehicles amplify risks.

Key Challenges

1. **Scale and Occupancy** – A single fire compartment may be larger than an entire multi-storey building, holding thousands of people.
2. **Escape Distance** – With few internal walls, occupants may face long travel distances to exits, far exceeding those in prescriptive codes.
3. **Fire Growth and Loading** – Warehouses and retail malls contain high fire loads, and fires can grow quickly without suppression.
4. **Smoke Movement** – In open spaces, smoke stratification and spread can render escape routes untenable unless controlled.

5. **Management and Evacuation** – Occupant unfamiliarity, crowd behaviour, and reliance on staff direction are major factors.

Fire Strategy Objectives

A fire strategy for large floor area buildings must:

- **Ensure life safety** by designing escape routes that are wide, short, and supported by clear wayfinding.
- **Provide robust smoke control** through natural or mechanical extract systems, designed using computational modelling.
- **Integrate suppression systems** such as sprinklers, water mist, or foam for high-risk occupancies.
- **Maintain structural integrity** to prevent early collapse under high fire loads.
- **Support firefighting access** in often very deep or complex layouts.
- **Embed management procedures** including crowd evacuation planning, drills, and staff training.

Why These Strategies Matter

Incidents such as the **Bradford City stadium fire (1985)**, the **Düsseldorf Airport terminal fire (1996)**, and multiple warehouse fires worldwide highlight the devastating consequences when fire strategies for large open areas fail. These tragedies reinforce the need for **bespoke, performance-based design**, going beyond prescriptive standards to address the realities of scale, smoke, and occupancy.

Summary

Large floor area buildings demand **fire strategies that balance prescriptive rules with advanced engineering methods**. They are often too complex to rely on code-based guidance alone. Instead, strategies must integrate **smoke control, suppression, evacuation modelling, and strong management systems**, ensuring both life safety and resilience.

12.2 Key Characteristics of Large Floor Area Buildings

Large floor area buildings are among the most challenging environments to design safe and effective fire strategies for. Unlike conventional multi-storey buildings, their **scale, complexity, and open layouts** introduce unique hazards that cannot always be addressed through prescriptive code requirements. These structures often fall into two broad categories — **publicly occupied spaces** (shopping centres, malls, exhibition halls, stadia) and **industrial/storage facilities** (warehouses, logistics hubs, distribution centres). Each type presents a different balance between **life safety**, **property protection**, and **business continuity**.

12.2.1 Shopping Centres and Malls

Shopping centres and multi-level retail complexes are characterised by **large occupant numbers, transient populations, and mixed fire risks.**

- **High Occupant Turnover**
    - Thousands of people may enter and leave within a single hour.
    - Occupant profiles are highly variable, ranging from families with children to elderly or mobility-impaired persons.
    - This creates a **peak load challenge** for evacuation planning, especially during sales periods or festive seasons.
- **Large Atrium Spaces**
    - Atriums often extend through multiple floors, acting as **focal circulation zones** and providing visual orientation.
    - While architecturally attractive, they introduce significant **vertical smoke movement risks**. Without robust smoke ventilation, an atrium can act as a chimney, rapidly rendering escape routes untenable.
- **Public Unfamiliarity with Layouts**
    - Unlike office staff who undergo regular fire drills, shoppers are usually unfamiliar with the building.
    - Research shows that in emergencies, most occupants will attempt to exit by the **entry route** rather than using alternative exits, even if those are closer.
    - Fire strategies must therefore provide **redundant, well-signposted exits** and use **voice alarm systems** to counteract instinctive behaviours.
- **Multiple Tenants and Fire Loads**
    - Each retail unit may present a different fire risk profile — clothing shops (high plastic and textile load), electronics retailers (high smoke toxicity), restaurants (kitchen fire hazards).

  - Strategies must account for the **worst-case unit fire load** and prevent fire spread between shops through compartmentation and fire shutters.

12.2.2 Warehouses and Logistics Hubs

Warehouses and logistics facilities typically have **lower life safety risks but much higher property and business continuity concerns.**

- **High Fire Loads**
  - Large quantities of palletised goods, often plastics and packaging, create **very high fire load densities** (measured in MJ/m^2).
  - Fires can grow rapidly to flashover conditions, overwhelming unsprinklered warehouses in minutes.
- **Large Open Compartments**
  - Warehouses often exceed **20,000 m^2**, beyond the prescriptive limits of ADB.
  - Subdividing such spaces with compartment walls is rarely practical; reliance shifts instead to **suppression systems and engineered smoke ventilation.**
- **Automated Storage and Retrieval Systems (ASRS)**
  - Increasingly common in modern logistics hubs, these robotic systems stack goods in **high racks up to 30 m tall.**
  - Firefighting access is limited, sprinklers may be obstructed by racking, and fires can spread vertically if in-rack sprinklers are not provided.
  - This makes **specialist sprinkler design Early Suppression Fast Response (ESFR), in-rack))** and **CFD smoke modelling** critical.
- **Lower Occupant Numbers**
  - Typically occupied by trained staff who know the layout.
  - Life safety risks are lower than in retail, but evacuation routes must remain **clear of stock and vehicles**, and alarms must be **audible above machinery noise.**
  - The main concern is often **property loss and supply chain disruption**, as warehouse fires can destroy millions in goods and take years to rebuild (e.g., the Ocado Andover fire, 2019).

12.2.3 Exhibition Halls and Arenas

Although not always included in prescriptive "warehouse/retail" categories, exhibition halls, convention centres, and arenas share many characteristics with large floor area buildings.

- **High Density Occupancies** – Events may hold tens of thousands of people, with complex layouts of temporary stands and displays.
- **Temporary Fire Loads** – Exhibitions may involve timber staging, fabrics, electrical cabling, or combustible displays.
- **Variable Occupancy Patterns** – Occupants arrive and leave en masse, requiring crowd management and wide exits.
- **Open Roof Spaces** – High ceilings require engineered smoke ventilation, often with very large extract fans.

12.2.4 Key Risks Across Large Floor Area Buildings

Regardless of type, these buildings share common fire safety challenges:

- **Extended Travel Distances** – Occupants may have to travel 50–100 m or more to reach an exit, exceeding prescriptive limits and requiring performance-based justification.
- **Smoke Movement** – Smoke stratification in open spaces can quickly make escape routes untenable unless controlled with mechanical or natural venting.
- **Suppression Dependence** – Sprinklers or specialist suppression systems are often the only realistic means of controlling fire growth in large compartments.
- **Crowd Behaviour** – In retail and public spaces, evacuation relies heavily on clear communication and crowd management.
- **Fire Service Access** – Deep floor plates, high racks, and large volumes hinder fire service operations, demanding well-placed risers, hydrants, and fire control rooms.

12.2.5 Lessons from Past Incidents

- **Bradford City Stadium Fire (1985)** – showed how combustible seating and poor maintenance in an open-stand environment could lead to rapid fire spread, killing 56.

- **Düsseldorf Airport Fire (1996)** – smoke spread through large concourse areas killed 17 people, highlighting the importance of engineered smoke ventilation.
- **Ocado Warehouse Fire, Andover (2019)** – destroyed a state-of-the-art automated distribution centre, showing the vulnerability of ASRS facilities despite modern design.
- **Kuwait Mall Fire (2009)** – a major fire in a shopping centre demonstrated how **unfamiliarity and delayed evacuation** can lead to fatalities even with exits available.

Summary

Large floor area buildings present **diverse fire challenges** depending on their function:

- **Shopping centres** require strategies centred on **evacuation and smoke control for large, unfamiliar crowds.**
- **Warehouses** demand robust **suppression and structural resilience** to prevent catastrophic loss.
- **Exhibition halls and arenas** combine the high occupant densities of retail with the high fuel loads of warehouses, requiring bespoke fire strategies.

In all cases, **compartmentation alone is insufficient**; strategies must rely on **suppression, smoke control, and management measures** to mitigate the scale of fire risk.

12.3 Regulatory Framework

Large floor area buildings — whether retail, warehousing, or exhibition spaces — stretch the limits of traditional prescriptive codes. Their sheer size, occupant densities, and complex layouts often make strict adherence to Approved Document B (ADB) impractical. Instead, a **hybrid approach** combining prescriptive guidance with risk-based and performance-based methodologies is typically required.

12.3.1 Approved Document B (ADB)

- **Scope and Limitations**:
    - ADB sets functional requirements for **means of escape, travel distance, compartmentation, and fire service access.**

  - For large floor areas, ADB's maximum compartment sizes are restrictive: for example, warehouses are typically limited to **20,000 m²** without subdivision.
  - Travel distance restrictions (e.g., **18 m dead-end, 45 m two-way escape**) are often exceeded in open-plan malls or warehouse settings.
- **Sprinkler Triggers**:
  - ADB requires sprinklers in certain very large storage and retail premises. For example:
    - **Single-storey shop or self-service store >2,000 m²**.
    - **Warehouse >20,000 m²**.
  - This effectively recognises that suppression becomes essential where compartmentation cannot be applied.
- **Smoke Control in Atriums**:
  - ADB provides baseline prescriptive rules but generally defers to engineered analysis (often BS 7974) for complex multi-level atriums.

12.3.2 BS 9999 – Risk-Based Design

- **Flexibility**: BS 9999 adopts a **risk profile matrix**, which allows longer travel distances and reduced stair provision in exchange for management systems, sprinklers, and detection.
- **Travel Distance Adjustments**:
  - For low-risk occupancies with good management and suppression, escape distances can be extended beyond ADB values.
  - *Example*: ADB's 45 m limit for two-direction escape can be extended to **up to 60 m or more** under BS 9999 with compensatory measures.
- **Application to Retail**:
  - Shopping centres benefit significantly, as zoned evacuation strategies combined with sprinklers and smoke ventilation can justify extended distances.

- **Application to Warehousing**:

Supports engineered designs for very large compartments, especially where full subdivision is impractical.

12.3.3 BS EN 12845 – Sprinkler Systems

- The **European standard for automatic sprinkler systems**, widely adopted in the UK.
- **Hazard Classifications**:
    - *Ordinary Hazard (OH)* for standard retail occupancies.
    - *High Hazard (HH)* for warehousing and storage, with further classification by goods type (plastics, aerosols, flammable liquids).
- **Design Densities**:
    - Specifies water discharge densities, tank sizes, and pump capacities.
    - Example: A high-hazard warehouse storing plastics may require **12.5 mm/min design density** over 260 m^2.
- **Integration with Fire Strategy**:
    - Sprinklers in large compartments are not optional add-ons but central to achieving a defensible fire strategy.

12.3.4 BS 7974 – Performance-Based Fire Engineering

- Provides a **systematic framework for engineered analysis**, critical where prescriptive solutions fail.
- Applications in large floor area buildings include:
    - **CFD Smoke Modelling** – to prove smoke layer tenability in atriums or warehouses.
    - **Evacuation Modelling** – to demonstrate safe egress times for thousands of shoppers.
    - **Structural Fire Engineering** – to assess performance of long-span steel trusses in warehouses under fire load.
- Widely relied upon for **exhibition halls, stadiums, and retail atriums**, where prescriptive rules do not reflect real-world layouts.

12.3.5 European and International Standards

- **BS EN 13501** – Reaction-to-fire classification for linings, cladding, and finishes.
- **NFPA Standards (USA)** –
    - NFPA 13: Automatic Sprinkler Systems.
    - NFPA 92: Smoke Control Systems.
    - NFPA 101: Life Safety Code (widely applied to malls and assembly occupancies).
- **International Building Code (IBC)** – adopted in the Middle East and Asia, often requiring stricter measures such as refuge floors and multiple smoke control zones.
- These provide useful benchmarks where UK practice must be **aligned with multinational insurers or international clients.**

12.3.6 Local Authority and Insurer Requirements

- **Local Fire and Rescue Service** may impose conditions through planning, particularly for:
    - Access roads and appliance staging.
    - Firefighting water supplies.
- **Insurers** (e.g., FM Global) often demand standards beyond statutory minimums:
    - ESFR sprinklers in warehouses.
    - Enhanced compartmentation or smoke control.
    - Robust business continuity measures.

Summary

The regulatory framework for large floor area buildings is **layered and multi-source**:

- **ADB** provides the statutory baseline but often proves too restrictive.
- **BS 9999** offers flexibility through risk-based adjustments.
- **BS EN 12845 and BS 7974** underpin suppression and performance-based smoke/evacuation design.
- **International and insurer requirements** often exceed UK statutory provisions, particularly for warehouses and global logistics hubs.

Fire strategies for these buildings therefore require **explicit referencing and justification**, ensuring that both prescriptive and engineered elements are documented and defensible.

12.4 Means of Escape

Means of escape in large floor area buildings is one of the most challenging aspects of fire strategy design. Unlike high-rise buildings, where vertical movement dominates, here the challenge lies in **horizontal travel across wide open spaces**, often with thousands of occupants in unfamiliar environments. Prescriptive limits in ADB are frequently exceeded, meaning performance-based justification is almost always necessary.

12.4.1 Travel Distances

- **Prescriptive Limits**:
  - Approved Document B (ADB) sets maximum travel distances:
    - *18 m* in dead-end conditions.
    - *45 m* where two escape directions are available.
  - These distances are often unachievable in large malls, warehouses, or exhibition halls where floor plates may extend over 100 m.
- **Risk-Based Adjustments (BS 9999)**:
  - BS 9999 permits extensions where **sprinklers, smoke control, and good management** are provided.
  - For example, in a low-risk shopping centre with sprinklers and a voice alarm, travel distances may be extended up to *60 m* or more.
- **Performance-Based Justification (BS 7974)**:
  - CFD smoke modelling and ASET/RSET calculations can prove that conditions remain tenable for the full evacuation time, even where distances exceed prescriptive limits.
  - For warehouses, modelling may demonstrate that staff can safely evacuate within 2–3 minutes before smoke descends to dangerous levels.

12.4.2 Exit Numbers and Widths

- **Exit Sizing**:
  - ADB recommends **5 mm per person** for doorways when the door width is serving more than 220 persons.
  - Large occupancies (e.g., a shopping mall with 10,000 visitors) require multiple exits with wide dispersal to prevent congestion.
- **Distributed Exits**:
  - In shopping centres, exits must be located around the perimeter of the floor plate, ensuring no occupant has to return via the main entrance.
  - In warehouses, exits must be **clear of stored goods and vehicles**, with aisles maintained as escape paths.
- **Redundancy**:
  - Exits must be independent and remote, ensuring a single fire cannot block all available routes.
  - The *45° separation rule* is often applied to confirm adequate spacing.

12.4.3 Atrium Considerations

Atriums are a hallmark of modern retail design but present complex fire safety challenges.

- **Benefits**:
  - Provide visual orientation and natural wayfinding, reducing confusion during evacuation.
  - Can act as natural smoke reservoirs if equipped with roof vents or mechanical extraction.
- **Risks**:
  - Act as chimneys, allowing rapid vertical smoke spread.
  - Require **robust smoke ventilation design**, often involving smoke curtains, reservoirs, and extract fans sized via CFD.
- **Evacuation Implications**:
  - Tenants at upper levels may rely on **bridges and balconies** leading into the atrium, which must remain tenable.

- ○ Strategies must prove via modelling that smoke layers stay above head height during the evacuation period.

12.4.4 Case Example – Shopping Centre

A large regional shopping mall in the UK had maximum travel distances of **70 m**, exceeding ADB. A fire strategy based on BS 9999 and BS 7974 included:

- Full **sprinkler protection** throughout.
- **Mechanical smoke extraction** from the atrium, designed to maintain a **3 m smoke-free layer**.
- **Voice alarm system** for clear evacuation messages.
- Justification via **evacuation modelling**, showing safe clearance within 8 minutes.

Result: Building Control approved extended travel distances on the basis of **layered compensatory measures**.

12.4.5 Case Example – Warehouse

A 40,000 m^2 distribution centre had travel distances up to **120 m**. To justify this:

- The building was fully sprinklered to BS EN 12845 with ESFR heads.
- Smoke venting provided **1 m^2 of vent area per 200 m^2 of floor area**, supported by CFD modelling.
- Occupant load was limited to 50 staff, all trained in evacuation procedures.
- RSET calculations confirmed evacuation within 3 minutes, well below ASET (time before smoke layer descends to 2 m above floor level).

12.4.6 Key Lessons

- Prescriptive travel distances are rarely achievable in large floor areas — **suppression, smoke control, and modelling are essential compensatory measures**.
- Exits must be **wide, numerous, and distributed**; concentration at a single entrance/exit is unsafe.
- Atriums require particular care: while they assist orientation, they must be coupled with engineered smoke systems to prevent chimney effects.

- Performance-based justifications (BS 7974) provide the evidence base regulators require when prescriptive guidance cannot be met.

Summary

In large floor area buildings, means of escape cannot be addressed by prescriptive design alone. Strategies must combine **risk-based design (BS 9999), performance analysis (BS 7974), active systems (sprinklers, smoke control), and robust management procedures**. The goal is to ensure that despite long travel distances and high occupant numbers, **all occupants can reach safety before escape routes become untenable**.

12.5 Smoke Control

Smoke is the most dangerous product of fire. In large floor area buildings, the **scale of the space and the volume of combustible material** mean that smoke can accumulate rapidly, obscure exits, and render escape routes untenable long before flames reach occupants. Effective smoke control is therefore **the single most critical fire engineering measure** in these environments.

12.5.1 Objectives of Smoke Control

The design of smoke control systems in malls, warehouses, and exhibition halls must achieve the following:

- **Maintain Tenability** – ensure that escape routes and occupied zones remain free of smoke for the duration of evacuation.
- **Preserve Visibility** – maintain visibility of signage and exits (minimum 10–15 m visibility is generally required for safe movement).
- **Protect Escape Routes and Firefighting Access** – keep stair cores, lobbies, and firefighter entry points clear of smoke.
- **Support Suppression** – ensure smoke does not impede sprinkler performance or spread fire through stratification.
- **Enable Firefighting Operations** – create conditions where fire crews can safely locate and extinguish the fire.

12.5.2 Smoke Control in Shopping Centres and Malls

Retail malls are characterised by **multi-level atriums, high footfall, and mixed fire loads**.

- **Mechanical Extract Systems**
    - High-capacity fans extract smoke from atriums and malls.
    - Typically designed to maintain a **2.5–3 m smoke-free layer** above floor level for at least 30 minutes (BS 7974 guidance).
    - Extract volumes are calculated based on fire size, atrium geometry, and make-up air provision.
- **Smoke Reservoirs and Curtains**
    - Smoke curtains form **reservoirs** above shop fronts or atrium balconies.
    - This contains smoke in defined zones and prevents uncontrolled spread.
- **Natural Ventilation**
    - Roof vents can discharge smoke using buoyancy and wind effects.
    - Often combined with mechanical systems to improve resilience.
- **Make-Up Air Provision**
    - Replacement air is critical — without it, extraction systems create negative pressure, drawing smoke into escape routes.
    - In shopping centres, make-up air is often supplied through external doors, stair pressurisation, or dedicated fans.

12.5.3 Smoke Control in Warehouses and Logistics Hubs

Warehouses face different challenges, dominated by **high fire loads and large open compartments.**

- **Roof Vents**
    - Large-area natural vents discharge smoke and hot gases upward.
    - Typically provided at **1 m² vent area per 200–250 m² of floor area.**
    - Operation can be gravity-operated (fusible link) or powered (mechanical).
- **CFD Modelling**
    - Computational Fluid Dynamics (CFD) tools such as FDS are often used to verify that **smoke layers remain above 2 m** during evacuation.

- This is especially important in high-rack storage, where geometry affects smoke movement.

- **Integration with Sprinklers**
  - Smoke vents must not open prematurely, as hot gases rising through open vents can **disrupt sprinkler spray patterns**, reducing suppression effectiveness.
  - *Best practice*: vents open only after sprinklers have controlled the fire, or operate in phased sequence.

12.5.4 Special Challenges in Large Spaces

- **Atrium Stack Effect** – hot gases rise rapidly, pulling cooler air from lower floors and spreading smoke vertically.
- **Warehouse Stratification** – smoke can form **hot upper layers** that suddenly descend when cooled or disturbed, creating flashover or backlayering hazards.
- **Multi-Zone Control** – in exhibition halls or malls, smoke systems may need to create **separate control zones** to prevent smoke spread across large floor plates.
- **Power Supply Resilience** – mechanical smoke extract fans require **backup generators or UPS systems** to remain functional during a fire.

12.5.5 Standards and Guidance

- **BS EN 12101** – Smoke and heat control systems (design, testing, and installation).
- **BS 7346-4** – Code of practice for smoke control in shopping complexes.
- **BS 7974-2** – Fire engineering methodology for smoke control analysis.
- **NFPA 92 (USA)** – International benchmark for smoke control design, especially in atriums and covered malls.

12.5.6 Case Examples

Case 1: Retail Atrium Fire (Düsseldorf Airport Terminal, 1996)

- Smoke spread through a large concourse and adjacent retail units, killing 17 people.
- Lack of effective smoke control meant toxic gases filled escape routes before evacuation could be completed.

- *Lesson*: atrium smoke systems must be **engineered, tested, and maintained** to prevent catastrophic failures.

Case 2: Logistics Warehouse Fire (UK, 2013)

- A 30,000 m^2 unsprinklered warehouse was destroyed after smoke vents activated prematurely.
- Rising hot gases disrupted sprinkler patterns, preventing suppression.
- *Lesson*: smoke vent and sprinkler **coordination is essential** in warehouse fire strategies.

Case 3: Westfield Shopping Centre (London, 2008)

- Atrium smoke system designed using CFD analysis to maintain a **10 m smoke-free layer**.
- Integration with voice alarms ensured phased evacuation of shoppers.
- *Lesson*: performance-based smoke control can enable safe evacuation even in **extremely large enclosed malls**.

Summary

Smoke control is the **cornerstone of fire strategies** for large floor area buildings.

- In shopping centres, systems must protect atriums, shop units, and escape routes using **mechanical extract, smoke reservoirs, and make-up air**.
- In warehouses, **roof venting, CFD modelling, and careful integration with sprinklers** are key.
- Failures in smoke control have been the defining factor in multiple fire tragedies.
- Strategies must explicitly demonstrate — through calculations, modelling, and standards compliance — that smoke will remain controlled long enough for **safe evacuation and effective firefighting intervention**.

12.6 Fire Suppression

In large floor area buildings, fire suppression is often the **single most decisive factor** in preventing catastrophic outcomes. Unlike passive measures such as compartmentation (which are often impractical at scale), suppression systems actively limit fire growth, reduce smoke generation, and protect both life safety and property.

12.6.1 The Role of Suppression

- **Control Fire Growth** – automatic systems detect and attack fire during the incipient stage, before flashover occurs.
- **Reduce Heat and Smoke Release** – slowing fire growth keeps conditions tenable for longer, supporting safe evacuation.
- **Support Firefighting Operations** – by containing fire to a manageable size, suppression systems make manual intervention feasible.
- **Protect Business Continuity** – in warehouses and logistics hubs, suppression can prevent total building loss and multi-year disruption.

12.6.2 Sprinkler Systems

Sprinklers are the most widely used and proven form of suppression in large compartments.

- **Standards**:
  - *BS EN 12845*: Design and installation of commercial sprinkler systems.
  - *BS 9251*: Residential sprinklers, relevant for retail complexes with integrated housing.
- **Mandatory Applications**:
  - Warehouses over **20,000 m²** (ADB requirement).
  - Retail stores above certain thresholds (e.g., >2,000 m² in England & Wales).
- **Effectiveness**:
  - Statistically, sprinklers control or extinguish fires in **95% of cases** where installed and maintained correctly (BFRS data).
- **Design Considerations**:
  - Hazard classifications (Ordinary, High, or Storage).
  - Minimum water discharge density.
  - Storage height and arrangement (critical in warehouses).

12.6.3 Specialist Suppression Systems

ESFR (Early Suppression Fast Response) Sprinklers

- Specifically designed for **high-rack storage warehouses**.

- Larger orifice size discharges high volumes of water, rapidly suppressing fires at their base.
- Prevents vertical fire spread in racks without the need for in-rack sprinklers (up to certain storage heights).

In-Rack Sprinklers

- Installed within racking systems for **very tall or dense storage** (beyond ESFR capability).
- Target fires deep inside storage arrays that roof-level sprinklers cannot penetrate.
- Must be carefully coordinated with storage layouts and pallet configurations.

Foam Suppression Systems

- Required where flammable liquids or hydrocarbon fuels are stored (e.g., aircraft hangars, chemical warehouses).
- Foam blankets the surface of liquid fires, preventing vapour release and flame spread.
- May be used in conjunction with sprinklers for mixed-hazard warehouses.

Water Mist Systems

- Produce fine droplets (<1 mm), cooling the flame and displacing oxygen.
- Useful in heritage exhibition halls or data centres where water damage must be minimised.
- Increasingly used in airport terminals and enclosed atriums.

Gaseous Suppression Systems

- Use inert gases (e.g., Argonite, Inergen) or chemical agents (e.g., FM-200, Novec 1230).
- Most common in IT/data rooms or security-critical facilities embedded within malls or logistics centres.
- Not suitable for large open compartments but vital for **localised protection of critical assets**.

12.6.4 Integration with Other Systems

- **Fire Alarms**: Suppression must be integrated with detection and alarm systems to trigger early response and evacuation.

- **Smoke Vents**: Coordination required to avoid premature smoke vent opening, which can **disrupt sprinkler spray patterns.**
- **Backup Power and Water Supplies**: Large installations often require dual pumps, diesel backup engines, and multiple water tanks to ensure resilience.

12.6.5 Management and Maintenance

- **Testing**: Weekly alarm tests, quarterly pump tests, and annual full inspections.
- **Third-Party Certification**: LPCB, FM Global, or BAFE schemes provide assurance of performance.
- **Impairment Control**: Procedures must be in place when suppression is taken offline (e.g., during maintenance).

12.6.6 Case Examples

Case 1 – IKEA Warehouse Fire, Sweden (2008)

- A high-rack warehouse fire was controlled by ESFR sprinklers before fire crews arrived.
- Only limited smoke damage occurred, and operations resumed within days.
- *Lesson*: ESFR sprinklers can prevent catastrophic loss even in massive storage facilities.

Case 2 – Ocado Warehouse Fire, Andover, UK (2019)

- Despite advanced robotics, the warehouse was unsprinklered.
- A small electrical fire spread rapidly, destroying the building and disrupting operations for over two years.
- *Lesson*: absence of suppression in high-value logistics hubs creates unacceptable resilience risks.

Case 3 – Shopping Mall Atrium Fire, Asia (2012)

- Fire in a retail unit spread into the mall atrium.
- Sprinklers activated and limited fire growth, while smoke extraction fans maintained visibility.
- All occupants evacuated safely; structural damage was minimal.

Summary

Suppression is not optional in large floor area buildings — it is **fundamental**.

- **Sprinklers** remain the backbone of protection, with ESFR and in-rack systems tailored to storage risks.
- **Foam, water mist, and gaseous systems** provide specialist solutions for high-risk occupancies.
- Effective suppression requires **integration with alarms, smoke control, and firefighting facilities**, supported by robust maintenance.
- Case histories demonstrate the stark contrast between suppressed and unsuppressed fires: one leads to **controlled incidents**, the other to **catastrophic loss**.

12.7 Fire Load and Compartmentation

Fire load and compartmentation are central to understanding how fire will develop and spread in large floor area buildings. Unlike smaller buildings, where subdivision into compartments is relatively straightforward, the **scale and function** of malls, warehouses, and exhibition halls often make traditional compartmentation impractical. Fire strategies must therefore balance **structural fire resistance, fire load management, suppression systems, and smoke control** to achieve equivalent safety.

12.7.1 Fire Load – Definition and Relevance

- **Definition**: Fire load is the total potential energy released by the combustion of contents and construction materials, usually expressed in **MJ/m²** of floor area.
- **Why It Matters**:
  - Determines **fire growth rate and severity**.
  - Influences design of **suppression systems** (sprinkler density, water supply).
  - Informs **structural fire resistance** calculations.
- **Typical Fire Load Intensities**:
  - Offices: ~400–600 MJ/m².
  - Retail: ~800–1,200 MJ/m² (varies with merchandise).
  - Warehouses: >2,000 MJ/m² (especially plastics, paper, tyres).

In high-rack storage warehouses, **localised fire load densities** may be several times higher than average values, requiring tailored sprinkler and venting solutions.

12.7.2 Fire Load in Warehouses and Logistics Facilities

- **Nature of Stored Goods**: Packaging materials, plastics, aerosols, and chemicals dramatically increase fire load.
- **Stacking Height**: The higher the racks, the more intense and vertically driven the fire risk. Fires can propagate upward rapidly if not suppressed at source.
- **Moving Hazards**: Forklifts, charging stations for electric vehicles, and conveyor systems can act as ignition sources.
- **Design Implications**:
    - Hazard classification under **BS EN 12845** or **NFPA 13**.
    - ESFR or in-rack sprinklers tailored to storage type.
    - CFD modelling to determine tenability given anticipated fire growth rates.

12.7.3 Fire Load in Shopping Centres and Malls

- **Mixed Occupancies**: Each retail unit contributes different risks:
    - Clothing stores – high textile/plastic load.
    - Electronics – toxic smoke from plastics and batteries.
    - Restaurants/food courts – cooking oils and kitchen fires.
- **Anchor Tenants vs Small Units**: Large department stores have vast fire loads, but smaller kiosks can also introduce hazards (e.g., LPG for catering stalls).
- **Atrium Fire Loads**: Kiosks or display stands in atriums add central ignition risks and must be controlled through fire strategy and management (often prohibited or strictly regulated).
- **Design Implications**:
    - Fire separation between shops and mall concourse via fire shutters or compartment walls.
    - Smoke curtains to limit spread across atrium voids.

12.7.4 Compartmentation Principles in Large Floor Area Buildings

- **Prescriptive Limits (ADB)**:
    - Warehouses generally limited to **20,000 m² per compartment**.
    - Shopping centres must provide compartmentation between units, anchor stores, and service corridors.
- **Challenges**:
    - A single warehouse or distribution centre may exceed 50,000 m², making subdivision with compartment walls impractical for operational reasons.
    - Exhibition halls and arenas are designed as single open spaces, precluding internal subdivision.
- **Performance-Based Alternatives**:
    - Reliance on sprinklers, smoke vents, and structural fire engineering to replace physical compartmentation.
    - Robust separation between **high-risk and low-risk areas** (e.g., kitchens, plant rooms, and storage areas).

12.7.5 Structural Fire Resistance

- **Loadbearing Structure**: Long-span steel frames in warehouses and malls require 60–120 minutes fire resistance.
- **Separation Walls**: Compartment walls between tenants or risk areas must resist fire for 60 minutes minimum (often 120 minutes for anchor stores).
- **Roofs**: Fire resistance of roof structures is often overlooked but critical in warehouses, especially where collapse would endanger firefighters or adjoining properties.

12.7.6 Case Examples

Case 1 – Supermarket Warehouse Fire (UK, 2005)

- A 23,000 m² warehouse storing packaging materials exceeded ADB compartmentation limits but relied on sprinklers.
- A major fire overwhelmed suppression due to inadequate water supply; the building was destroyed.

- *Lesson*: large fire loads demand not just sprinklers, but **robust water infrastructure and system reliability**.

Case 2 – Shopping Centre Atrium Fire (Middle East, 2012)

- A kiosk fire spread into the atrium; poor separation allowed smoke to engulf multiple levels.
- 19 fatalities occurred due to blocked escape routes and uncontrolled smoke spread.
- Lesson: even small ignition sources can exploit weaknesses in **compartmentation and fire load management**.

Case 3 – Distribution Centre, Germany (2018)

- CFD modelling was used to justify a single 40,000 m^2 compartment with ESFR sprinklers.
- Fire tests and modelling proved tenability in escape routes, leading to regulatory approval.
- *Lesson*: **performance-based compartmentation strategies** can succeed with strong engineering evidence.

12.7.7 Key Lessons

- **Warehouses**: Compartmentation is often unachievable — suppression and smoke control become primary.
- **Shopping Centres**: Compartmentation between retail units is essential, especially around atriums and food courts.
- **Exhibition Spaces**: Must rely on suppression and engineered smoke control, as internal subdivision is impossible.
- **Overall**: Fire load assessments are the foundation — they dictate not only suppression and smoke control requirements, but also whether prescriptive compartmentation limits can be waived.

Summary

Large floor area buildings challenge the traditional model of **fire load management and compartmentation**.

- **Warehouses** rely heavily on suppression due to extremely high fire loads and impracticality of subdivision.
- **Shopping centres** require careful separation of units and risk areas, reinforced by smoke curtains and shutters.

- **Exhibition halls and arenas** are designed as single open compartments, requiring engineered smoke and suppression systems.

Ultimately, the fire strategy must demonstrate — through prescriptive compliance or performance-based justification — that despite massive fuel loads, conditions for **safe evacuation, firefighting, and structural stability** will be maintained.

12.8 Evacuation Strategies

Evacuation in large floor area buildings presents some of the most complex challenges in fire safety design. Unlike high-rise buildings, where evacuation is largely vertical, these occupancies require the safe movement of **thousands of people horizontally across vast spaces**. The strategy must consider not only **distance and capacity**, but also **human behaviour, communication, and management procedures**.

12.8.1 Evacuation in Shopping Centres

Shopping centres and malls are among the most complex public evacuation environments:

- **High and Variable Occupancies**
    - Daily occupant loads can range from a few hundred during quiet hours to tens of thousands at peak times.
    - Special events, holidays, or sales create **unpredictable surges**, complicating evacuation modelling.
- **Simultaneous vs Zoned Evacuation**
    - Full simultaneous evacuation of a large mall is rarely practical — it risks **panic, congestion, and bottlenecks**.
    - **Zoned evacuation** is often used instead, where only the fire-affected zone (e.g., a shop unit or retail level) is evacuated initially.
    - Remaining areas are held in place until confirmation of wider spread, reducing overcrowding at exits.
- **Voice Alarm Systems**
    - Audible alarms alone are insufficient in crowded retail environments.

  - **BS 5839-8 voice alarm systems** provide tailored messages ("Please leave the first floor shops by the south exit") to reduce confusion.
  - Clear communication is particularly important where occupants are **unfamiliar with layouts** and may default to the entry route.
- **Wayfinding and Signage**
  - Atriums and long malls can disorient shoppers.
  - Dynamic exit signage, directional lighting, and staff-led guidance help channel people to safe routes.

12.8.2 Evacuation in Warehouses and Logistics Facilities

Although warehouses typically have **low occupant numbers**, evacuation design remains critical due to the speed and severity of potential fires:

- **Occupant Characteristics**
  - Usually trained staff, familiar with layout and procedures.
  - Often dispersed across large floor areas or working in mezzanines and high racks.
- **Evacuation Routes**
  - Must remain **clear of goods, forklifts, and vehicles** at all times.
  - Staff must not have to travel excessive distances (>60–120 m) without protected exits.
- **Alarm Audibility**
  - High ambient noise from machinery requires **visual alarms** (flashing beacons) as well as sirens.
  - Voice alarms may be needed for multilingual workforces.
- **Evacuation Time Objectives**
  - Occupants must evacuate within **2–3 minutes**, before sprinkler-controlled smoke descends to untenable levels (ASET vs RSET analysis).

12.8.3 Human Behaviour Considerations

Large floor area evacuations are as much about **psychology and management** as about physical design:

- **Delay in Response**
  - Shoppers may ignore alarms, seek confirmation, or finish transactions before leaving.
  - Staff may delay to secure tills, stock, or equipment.
- **Preferred Exits**
  - Studies show that most people instinctively head towards the **entry route**, even when alternative exits are closer.
  - This creates congestion at main entrances if not managed.
- **Crowd Dynamics**
  - Large numbers can create "funnelling" effects at stairheads or narrow corridors.
  - Evacuation modelling tools (e.g., Pathfinder, STEPS, MassMotion) help predict congestion and optimise exit placement.
- **Staff Role**
  - Trained staff are crucial in directing crowds, calming uncertainty, and guiding vulnerable occupants.
  - Many strategies rely on **staff-led phased evacuation** rather than full simultaneous alarms.

12.8.4 Progressive Horizontal Evacuation

In certain large floor area occupancies — particularly **hospitals, care homes, and exhibition halls with vulnerable persons** — evacuation may not be immediate or vertical:

- **Sub-Compartmentation**
  - Patients or immobile occupants are moved to adjacent compartments on the same floor, protected for 30–60 minutes.
- **Fire-Resisting Corridors and Lobbies**
  - Allow staged movement of beds or wheelchairs.
- **Final Vertical Evacuation**
  - Once safe, occupants are transferred to stair cores or lifts (where evacuation lifts are provided).

This staged approach balances **life safety with operational realities**, particularly in healthcare and dependent environments.

12.8.5 Integration with Fire Engineering Tools

- **ASET vs RSET**
  - Available Safe Egress Time (ASET): time before conditions become untenable.
  - Required Safe Egress Time (RSET): time needed for full evacuation.
  - Fire strategies must prove **RSET < ASET**, using conservative behavioural assumptions.
- **Evacuation Modelling**
  - Software tools simulate human movement, bottlenecks, and response times.
  - For shopping centres, models account for **multi-level evacuation**, escalators, and atrium bridges.
  - For warehouses, models confirm that staff can reach exits within required limits despite long travel distances.

12.8.6 Case Examples

Case 1 – Shopping Centre Evacuation, UK (2014)

- A food court fire triggered a zoned evacuation on two floors.
- Voice alarms directed shoppers to alternative exits, avoiding main atrium congestion.
- Total clearance achieved in **under 10 minutes** with no injuries.
- Lesson: **zoned alarms and staff control** are more effective than full simultaneous evacuation.

Case 2 – Warehouse Evacuation, Europe (2017)

- Fire broke out in an automated storage system.
- Staff evacuated in 2.5 minutes; suppression contained the fire.
- Modelling had predicted 3-minute clearance — real-world performance confirmed validity.

Case 3 – Exhibition Hall Fire, Asia (2009)

- An electrical fire during an event led to smoke spread.
- Progressive horizontal evacuation moved thousands of attendees through fire-resisting corridors.

- No fatalities, despite very high occupancy.
- Lesson: **protected horizontal evacuation routes** are essential in high-density assembly halls.

Summary

Evacuation strategies in large floor area buildings must go far beyond prescriptive exit counts:

- **Shopping centres** need zoned evacuations, voice alarms, and robust crowd management.
- **Warehouses** depend on short evacuation times, clear routes, and audible/visible alarms.
- **Exhibition halls and healthcare occupancies** require progressive horizontal evacuation supported by strong compartmentation.
- Performance must be demonstrated through **ASET/RSET analysis, evacuation modelling, and realistic assumptions about human behaviour**.

12.9 Management and Operational Issues

Designing fire safety systems for large floor area buildings is only half the challenge; **long-term management and operational discipline** determine whether the strategy actually works in practice. The Grenfell Tower Inquiry and numerous warehouse and mall fires worldwide have shown that even well-designed systems can fail if **maintenance, coordination, and training are neglected**.

12.9.1 Shopping Centres and Malls

- **Shared Responsibility Across Tenants**
  - Each retail unit may have its own risk profile (clothing, electronics, food), but safety is only as strong as the weakest link.
  - A centralised fire strategy must clearly define **tenant vs landlord responsibilities**.
- **Centralised Fire Safety Management**
  - A **mall fire safety manager** (often the Responsible Person under the Regulatory Reform Order 2005) must oversee:
    - Alarm and detection testing.

        - Smoke control maintenance.
        - Fire drills and evacuation plans across multiple units.
    - Leasing agreements should require tenant compliance with the overall fire strategy.
- **Training and Communication**
    - Staff across different shops must be trained in **alarm procedures, evacuation routes, and assisting vulnerable persons**.
    - Security and housekeeping staff often form the "first response" until fire services arrive.
- **Public Interface**
    - Mall staff play a critical role in guiding unfamiliar occupants during evacuations.
    - Voice alarms and signage must be backed by **visible staff presence.**
- **Special Risks**
    - Seasonal decorations, kiosks, and promotional displays can add unexpected fire loads.
    - Central management must enforce strict policies to prevent unapproved or unsafe additions.

12.9.2 Warehouses and Logistics Hubs

- **Housekeeping and Storage Practices**
    - Good housekeeping is critical: blocked aisles, overstacked pallets, and damaged racks all increase risk.
    - Fire load must be controlled through **storage separation, hazardous goods management, and spacing between racks**.
- **Operational Controls**
    - "No smoking" policies must be rigorously enforced in high-load storage areas.
    - Hot works (welding, cutting, electrical repairs) must follow **permit-to-work systems** with fire watch procedures.
- **Training and Staff Responsibilities**
    - Forklift drivers, plant operators, and robotics supervisors must be trained in:

        - Not obstructing exits.
        - Recognising fire hazards (e.g., overheating batteries, damaged packaging).
        - Raising alarms promptly.
    - Staff should be familiar with emergency stop systems for conveyor belts or automated storage systems.
- **Integration with Insurers and Clients**
    - Many logistics hubs are insured by FM Global or similar organisations, requiring **audits and inspections** well beyond statutory minimums.
    - Fire strategies must therefore align not only with regulations but also with **business continuity demands**.

12.9.3 Common Issues Across Large Floor Area Buildings

- **Maintenance and Testing**
    - Sprinklers, smoke vents, alarms, and emergency lighting must be **routinely tested** and logged.
    - A breakdown in maintenance is a common theme in warehouse and mall fire investigations.
- **Emergency Planning**
    - Evacuation drills must be realistic and account for peak occupancies.
    - Shopping centres require **public-facing evacuation rehearsals**, while warehouses focus on **rapid staff clearance**.
- **Contractor Management**
    - Refurbishments, shop fit-outs, and warehouse alterations can compromise compartmentation and fire-stopping.
    - Strict contractor permits and post-work inspections are essential.
- **Cultural and Human Factors**
    - Multilingual or transient workforces (common in warehouses and retail) demand **clear, multilingual signage and training materials**.
    - A "safety culture" must be embedded, where all staff understand their role in maintaining fire safety.

12.9.4 Case Examples

Case 1 – Mall Fire Drill, UK (2016)

- A large shopping centre discovered that several tenants failed to participate in evacuation drills.
- Regulatory enforcement required the landlord to revise contracts to mandate tenant compliance.
- *Lesson*: **shared spaces require shared responsibility**.

Case 2 – Logistics Hub Fire, Europe (2019)

- A fire started at a battery charging station. Forklift operators attempted to extinguish it but delayed raising the alarm.
- By the time the alarm was triggered, the fire had grown beyond sprinkler control.
- *Lesson*: **staff training and prompt reporting are as vital as suppression systems**.

Case 3 – Warehouse Housekeeping Failures, USA (2020)

- Sprinkler systems failed to contain a fire due to racks being stacked too high and aisles obstructed.
- Insurance claim investigations found systematic breaches of fire load management policies.
- *Lesson*: **good housekeeping is as important as system design**.

Summary

Management and operational discipline underpin the success of every fire strategy:

- In **shopping centres**, coordination between multiple tenants, centralised management, and staff training are essential.
- In **warehouses**, strict housekeeping, hot works control, and staff training define resilience.
- Across all large floor area buildings, **testing, maintenance, and cultural buy-in** are non-negotiable.

Ultimately, even the most sophisticated suppression and smoke control systems cannot compensate for poor management. A fire strategy is therefore not just a technical document, but a **management framework for ongoing safety and resilience**.

12.10 Case Examples

Case 1: Shopping Centre Fire (Illustrative Example)

A large regional retail mall was designed with a central atrium spanning four levels. The fire strategy included:

- **Mechanical Smoke Extraction**: High-capacity extract fans serving the atrium, supported by smoke curtains forming reservoirs above shopfronts.
- **Sprinkler Protection**: Full coverage across retail units, anchor stores, and service corridors, designed to BS EN 12845.
- **Zoned Voice Alarm System**: Evacuation was planned on a phased basis — occupants in the affected zone were directed to exits while other zones were held in place to prevent congestion.
- **Tenant Responsibilities**: Each retail unit provided in-unit alarm audibility, extinguishers, and local staff training, coordinated under the central mall fire strategy.

Incident Outcome:

During a small fire in a food court kitchen, sprinklers controlled the blaze while smoke was extracted via the atrium system. The phased evacuation worked as intended, with security staff guiding shoppers to safe exits. Total evacuation time was under 8 minutes, with no injuries.

Lessons Learned:

- Zoned voice alarms reduce panic and congestion compared with simultaneous evacuation.
- Tenant participation and compliance with the central strategy are critical to success.
- Atrium smoke extraction must be sized to maintain at least a **2.5–3 m smoke-free layer** for effective evacuation.

Case 2: Warehouse Fire (Illustrative Example)

A logistics hub storing **plastic packaging materials** presented a very high fire load. The fire strategy included:

- **Suppression**: ESFR sprinklers tailored to high-rack storage, designed to suppress fires at the base of racks before vertical fire spread.

- **Smoke Control**: Roof vents supported by 10 m^3/s mechanical extract fans.
- **CFD Modelling**: Fire Dynamics Simulator (FDS) was used to confirm that, under a design fire scenario, a **3 m smoke layer** could be maintained above occupant head height for at least 20 minutes.
- **Staff Training**: Regular drills ensured that forklift operators and logistics staff could evacuate within 2–3 minutes of alarm activation.

Incident Outcome:

A pallet fire caused by a forklift battery fault activated the sprinklers. The system contained the fire to a single rack aisle, while mechanical extraction prevented smoke descent into escape routes. Staff evacuated within 3 minutes, and the fire service extinguished the blaze without structural damage.

Lessons Learned:

- ESFR sprinklers provide effective protection against high-rack fires when paired with adequate water supplies.
- CFD verification adds confidence that smoke systems will perform under realistic fire loads.
- Ongoing staff training is essential — without rapid evacuation, smoke could have compromised tenability despite suppression.

Case 3: International Example – Exhibition Hall Fire (Europe, 2009)

A major exhibition hall hosting over 5,000 visitors experienced an electrical fire during a trade show. The fire strategy included:

- **Sub-Compartmentation**: Fire-resisting partitions dividing the hall into zones.
- **Smoke Vents**: Natural vents in the roof automatically opened upon detection.
- **Voice Alarms**: Multilingual announcements directed visitors to the nearest exits.

Outcome: Despite heavy smoke production, the hall was evacuated safely within 12 minutes. Fire crews contained the fire to one section of the exhibition floor.

Lessons Learned:

- Multilingual communication is vital in international venues.
- Temporary displays and stands must not compromise smoke venting or exit paths.
- Progressive evacuation by sub-compartment can reduce congestion in very large open-plan halls.

Summary of Case Profiles

These case studies illustrate how **different large floor area occupancies demand tailored strategies**:

- *Shopping centres*: rely on **zoned evacuation, atrium smoke extraction, and coordinated tenant responsibilities**.
- *Warehouses*: depend on **suppression, CFD-supported smoke control, and staff training**.
- *Exhibition halls*: highlight the role of **sub-compartmentation, natural venting, and crowd communication**.

Each demonstrates the value of integrating **design, management, and operational practice** into a coherent fire strategy.

12.11 Summary

Large floor area buildings — whether shopping centres, exhibition halls, warehouses, or logistics hubs — present some of the **most demanding challenges in fire safety design**. Their vast internal volumes, high occupant loads, and often extreme fire loads mean that prescriptive codes alone are rarely sufficient. Instead, fire strategies must combine **prescriptive guidance, performance-based fire engineering, and strong management systems** to achieve acceptable levels of safety and resilience.

Key Lessons from Chapter 12

1. **Characteristics and Risks**
    - **Shopping centres:** high and fluctuating occupant densities, complex layouts, unfamiliar occupants, and fire load diversity across tenants.
    - **Warehouses:** extreme fire loads, high-rack storage, and automation — fewer people but greater risk to property and business continuity.

- **Exhibition halls and airports:** enormous open spaces with transient populations, requiring progressive evacuation strategies and robust smoke control.

2. **Regulatory and Standards Framework**
    - Approved Document B establishes baseline compartment limits, travel distances, and suppression triggers.
    - BS 9999 and BS 7974 provide **risk-based and performance-based routes** for designs that exceed prescriptive limits.
    - BS EN 12845 sprinklers and BS EN 12101 smoke systems form the backbone of engineered safety solutions.
3. **Means of Escape**
    - Prescriptive travel distance limits are routinely exceeded; evacuation modelling and smoke control are used to justify extensions.
    - Shopping centres often require **zoned evacuation supported by voice alarms**, while warehouses demand **rapid staff evacuation within minutes**.
4. **Smoke Control**
    - In malls, **mechanical extract and atrium reservoirs** prevent smoke descent into occupied zones.
    - In warehouses, **roof venting integrated with sprinkler systems** is essential to prevent catastrophic fire spread.
    - CFD modelling is widely adopted to prove smoke layer stability and evacuation tenability.
5. **Suppression Systems**
    - Sprinklers are indispensable — ESFR and in-rack sprinklers tailored for storage risks, with foam and water mist systems for specialist hazards.
    - Integration with smoke systems and robust water supplies is critical.
6. **Compartmentation and Fire Load**
    - Warehouses often exceed prescriptive compartment size limits, requiring performance-based alternatives.
    - Shopping centres demand **separation between units, atriums, and risk areas** to contain fires at source.

    - Fire load assessment informs not only suppression design but also evacuation modelling and structural fire resistance.

7. **Management and Operational Issues**
    - Shopping centres require **centralised coordination across multiple tenants** for drills, alarms, and fire strategy compliance.
    - Warehouses depend on **strict housekeeping, hot works control, and staff training**.
    - Poor management can undermine even the best-designed systems.
8. **Case Profiles**
    - Case studies show the stark differences between suppressed and unsuppressed fires: controlled incidents with minimal disruption versus catastrophic total losses.
    - They also highlight the importance of **integration, staff training, and coordinated evacuation systems**.

Final Reflection

The overarching lesson is that **large floor area buildings cannot rely on a single line of defence**. Their fire strategies must:

- Layer **passive protection, suppression, smoke control, and evacuation planning**.
- Be validated through **modelling, risk assessment, and robust engineering evidence**.
- Include strong **management systems and tenant/occupant engagement** to ensure that the design continues to function throughout the building's life.

Ultimately, the safety of these vast and complex environments depends on how effectively the **technical design, operational management, and human behaviour** are integrated into a **holistic fire strategy**.

Chapter 13

Mixed-Use Developments

13.1 Introduction

Mixed-use developments have become one of the defining features of modern urban environments. They bring together different building functions within a single development or complex — for example:

- **Retail at lower levels** with **residential or hotel accommodation above**.
- **Offices or co-working spaces** integrated with **shopping malls, restaurants, or leisure facilities**.
- **Transport hubs** (railway stations, bus interchanges, airports) that also incorporate **retail, entertainment, or hospitality spaces**.
- **Public sector "hub buildings"** combining civic offices, libraries, schools, or healthcare facilities with commercial uses.

These schemes deliver **economic vitality and social convenience**, making efficient use of valuable city-centre land. However, they also introduce a **unique set of fire safety challenges**, often exceeding the scope of prescriptive codes.

13.1.1 Why Mixed-Use Developments Are Fire Safety Challenges

- **Different Fire Risks**
    - Kitchens, restaurants, or retail units introduce **higher fire loads** than offices or residential flats.
    - Warehousing or delivery hubs can add large-scale fuel sources directly below or beside public occupancies.
- **Different Occupant Profiles**
    - Office workers and shoppers are generally mobile and can self-evacuate.
    - Residents, hotel guests, hospital patients, or schoolchildren may be **asleep, unfamiliar with the building, or dependent on staff**.
    - Fire strategies must reconcile these contrasting evacuation needs.

- **Shared Infrastructure**
    - Services such as plant rooms, car parks, atriums, and escape stairs are often shared across occupancies.
    - If not carefully compartmented, a fire in one zone may compromise escape or safety in another.
- **Management Complexity**
    - Mixed-use schemes typically involve multiple stakeholders (developers, landlords, managing agents, individual tenants).
    - Clear allocation of responsibilities under the **Regulatory Reform (Fire Safety) Order 2005** is essential to avoid gaps in fire safety management.

13.1.2 The Role of Fire Strategies

In these complex settings, the fire strategy must go beyond "single-use" building solutions. It must:

- Define **clear compartmentation** between occupancies.
- Address **shared vs dedicated escape routes**.
- Ensure **fire service facilities** are appropriately located and accessible to all uses.
- Coordinate **active systems** (sprinklers, alarms, smoke control) across different occupancies and ownership boundaries.
- Anticipate **management handover challenges** and provide a framework for ongoing compliance.

13.1.3 Real-World Examples

- A **railway station with integrated shopping mall** must coordinate evacuation between transient passengers and shoppers, while ensuring smoke control in concourse atriums.
- A **high-rise residential tower above a podium retail block** must ensure that residents' "stay put" policy is not compromised by fires in retail areas below.
- A **student accommodation complex with bars and restaurants** must balance late-night occupancy risks against sleeping accommodation above.

13.1.4 Scope of this Chapter

This chapter explores how fire strategies address mixed-use developments by examining:

1. **Regulatory Frameworks** – guidance from ADB, BS 9999, and BS 9991, and how they intersect.
2. **Compartmentation Principles** – ensuring fire separation between distinct occupancies.
3. **Means of Escape** – challenges of shared stairs, lobbies, and atriums.
4. **Fire Protection Systems** – coordinating alarms, sprinklers, and smoke control across occupancies.
5. **Management and Responsibility** – allocation of duties under fire safety law.
6. **Case Examples** – successful and failed strategies in real developments.

13.2 Key Risks in Mixed-Use Developments

Mixed-use developments combine distinct occupancy types in a single structure, creating **unique interactions between risks** that would not exist in standalone buildings. A fire strategy must therefore identify and address these key risks at the outset, since **failures in integration** have been a root cause of serious past incidents.

13.2.1 Compartmentation Challenges

- **Fire Separation Between Occupancies**
  - Residential accommodation must be completely compartmented from retail or leisure uses below.
  - Separation typically requires **60–120 minutes fire resistance** to prevent vertical spread.
  - Breaches through service risers, ducts, or poorly maintained fire-stopping can undermine protection.
- **Vertical Fire Spread**
  - Atriums, open stairs, or poorly designed service shafts can allow smoke and flame to bypass compartment floors.
  - The **Grenfell Tower tragedy** demonstrated how external cladding and façade systems can compromise intended separation.

- **Basement Risks**
  - Shared basements (e.g., parking or plant) can become the point of failure, with smoke migrating into residential or retail areas above unless strong separation and smoke control are provided.

13.2.2 Shared Spaces and Services

- **Car Parks**
  - Fires in underground or multi-storey car parks generate dense, toxic smoke.
  - If ventilation systems are shared with retail or residential areas, contamination is a major risk.
- **Atriums and Concourse Spaces**
  - Atriums offer visual connectivity but act as **chimneys for smoke spread**.
  - Fire curtains, smoke reservoirs, and engineered extract are essential, particularly when atriums link different uses.
- **Plant Rooms and Risers**
  - Centralised plant serving multiple occupancies introduces dependencies.
  - Service risers must be fire-stopped at every floor to prevent concealed vertical fire spread.
- **Lifts**
  - Shared lifts between retail and residential uses can spread smoke unless protected by **lobbies and pressurisation systems**.
  - Evacuation lifts must be clearly designated, with controls preventing unauthorised use during fire events.

13.2.3 Evacuation Conflicts

- **Residential "Stay Put" vs Public Evacuation**
  - In residential towers, a **stay-put policy** is often adopted, relying on robust compartmentation.
  - Retail, leisure, or office uses demand **simultaneous or phased evacuation**.

    - A fire strategy must ensure that **simultaneous alarms in public areas do not trigger panic in residential zones**, unless residents are directly at risk.
- **Phased vs Zoned Evacuation**
    - Offices above retail may require phased evacuation, while retail concourses require immediate clearance.
    - Evacuation models must test for congestion at shared stair cores and exits.
- **Mixed Populations**
    - Residents may be asleep or unfamiliar with exits.
    - Shoppers or commuters may be transient and less aware of building layouts.
    - Strategies must account for **different behavioural responses** and ensure voice alarms provide tailored messages.

13.2.4 Management Complexity

- **Multiple Stakeholders**
    - A single development may involve a **landlord, managing agent, commercial tenants, residential management companies, and retail operators.**
    - Without clear responsibilities, critical tasks such as alarm testing, smoke control maintenance, or evacuation drills may fall between the cracks.
- **Regulatory Compliance**
    - Under the **Regulatory Reform (Fire Safety) Order 2005**, a "Responsible Person" must be clearly identified for each area.
    - A single coordinating fire strategy is needed to integrate the responsibilities of all parties.
- **Fire Safety Culture**
    - Staff training, signage, and resident engagement vary widely between occupancies.
    - A consistent **management framework** is essential so that, during an emergency, occupants receive a coherent response rather than conflicting instructions.

13.2.5 Lessons from Past Incidents

- Fires in mixed-use complexes have highlighted how **poor separation and uncoordinated evacuation policies** can lead to tragedy:
  - In some shopping centres with residential towers above, smoke spread from retail fires has forced full residential evacuation, even when "stay put" was intended.
  - International case studies show mall–hotel complexes where alarms were not integrated, leading to delayed evacuation in one occupancy while another evacuated.

Summary

The **key risks in mixed-use developments** are not only technical but also organisational:

- Compartmentation and vertical fire spread must be rigorously controlled.
- Shared spaces and services demand robust engineering and separation.
- Conflicting evacuation strategies must be reconciled through careful planning.
- Clear management coordination is critical to ensuring the strategy works in practice.

By explicitly identifying and addressing these risks, fire strategies for mixed-use developments can provide a **defensible, integrated, and resilient approach** to life safety and property protection.

13.3 Regulatory Framework

Mixed-use developments sit at the **intersection of multiple regulatory regimes**. Because they combine distinct occupancies, no single standard is sufficient — instead, a coordinated application of **Approved Document B, British Standards, and performance-based engineering (BS 7974)** is usually required. Early consultation with regulators and stakeholders is also critical to achieving approvals and delivering a defensible fire strategy.

13.3.1 Approved Document B (ADB)

- **Compartmentation Requirements**
  - ADB requires that occupancies with different fire risks (e.g., residential above retail) are **compartmented by fire-resisting construction**.

    - Separation between uses is typically **60–120 minutes fire resistance**, depending on height, risk, and occupancy.
- **Limitations for Mixed-Use**
    - While ADB provides clear prescriptive rules for individual occupancies, it does not fully address the **interactions between them**.
    - For example, ADB gives rules for atriums, but not for atriums linking residential, retail, and transport uses in the same building.
    - A fire strategy must therefore adapt ADB principles and supplement them with engineering analysis.

13.3.2 BS 9999 – Risk-Based Design

- **Flexibility in Application**
    - BS 9999 provides a **risk-based approach**, adjusting design allowances (e.g., travel distance, exit capacity) according to management levels, fire protection systems, and occupancy risks.
    - This flexibility is particularly valuable in mixed-use schemes where **prescriptive ADB distances may be exceeded** (e.g., in large malls or atrium-linked concourses).
- **Management Emphasis**
    - BS 9999 places strong emphasis on **management coordination**, which is often the weakest link in mixed-use buildings.
    - For example, it recognises that effective evacuation in a shopping mall with residential use above requires **robust communication protocols and consistent training** across all occupancies.

13.3.3 BS 9991 – Residential Guidance

- BS 9991 is tailored to **blocks of flats, maisonettes, and residential towers**.
- In mixed-use developments, this standard is applied to the **residential elements** of the scheme, covering:
    - Stay-put strategies.

    - Fire-resisting separation between flats and communal areas.
    - Smoke ventilation for corridors and lobbies.
    - Sprinkler requirements in high-rise or complex residential layouts.
- The challenge for fire strategies is to ensure that these **residential provisions remain uncompromised** by adjacent occupancies such as retail or leisure spaces.

13.3.4 BS 7974 – Performance-Based Fire Engineering

- BS 7974 provides the framework for **fire engineering analysis**, which is almost always necessary in mixed-use projects.
- Common applications include:
    - **Atrium Smoke Control**: where a retail atrium passes through or adjoins residential floors.
    - **Evacuation Modelling**: where phased evacuation in office towers must be reconciled with simultaneous evacuation in adjacent retail zones.
    - **Structural Fire Engineering**: ensuring that a fire in one occupancy does not compromise the stability of another.
- This standard provides the **evidence base** to justify deviations from ADB, particularly in highly integrated or innovative designs.

13.3.5 Consultation and Approvals

- **Planning Authorities**
    - Local authorities may impose **planning conditions** specific to mixed-use projects, such as requirements for fire hydrants, access roads, or façade fire performance.
    - Early fire strategy input can prevent costly redesign at planning approval stage.
- **Building Control Bodies**
    - Whether public or Approved Inspectors, Building Control will scrutinise the **compartmentation strategy** and the interface between uses.

    - They often require **clear documentary evidence** of compliance where prescriptive guidance is exceeded.
- **Fire and Rescue Services**
    - Local fire services are heavily involved in mixed-use schemes, particularly regarding:
        - Firefighting shafts and access.
        - Smoke control systems in atriums and basements.
        - Pre-incident information for operational crews.
    - Engagement during design ensures the fire service can operate effectively if an incident occurs.

Summary

The regulatory framework for mixed-use developments demands a **layered approach**:

- **ADB** provides baseline prescriptive requirements.
- **BS 9999 and BS 9991** adapt these to commercial and residential contexts.
- **BS 7974 fire engineering analysis** fills the gaps where prescriptive rules fall short.
- **Consultation with planning authorities, Building Control, and Fire and Rescue Services** ensures that the fire strategy is not only compliant but also operationally viable.

In practice, few mixed-use projects achieve approval without some form of **hybrid strategy**, blending prescriptive compliance with performance-based justification.

13.4 Compartmentation Between Occupancies

Compartmentation is the **cornerstone of fire safety in mixed-use developments**, ensuring that fire, smoke, and heat remain contained within the area of origin long enough to protect occupants in adjoining occupancies. Without robust separation, the fire risks of one use — such as a retail unit, restaurant, or cinema — can compromise the safety of others, such as residents in flats above or office workers in adjacent wings.

13.4.1 Retail Below Residential

- The most common mixed-use arrangement is **residential apartments or hotels above ground-floor retail or leisure uses.**
- Building Regulations and best practice (ADB, BS 9991) typically require:
    - **Minimum 120 minutes fire resistance** for separating floor construction between the retail and residential uses.
    - This includes both **structural stability** and **fire-stopping of all penetrations** (ducts, services, lift shafts).
- Where atriums connect retail podiums with upper residential areas, **fire-rated glazing, smoke curtains, and mechanical smoke control** may be required to preserve separation.

13.4.2 High-Risk Occupancies (Kitchens, Bars, Cinemas)

- **Commercial Kitchens**
    - Present high fire loads due to oils, fats, and cooking appliances.
    - Must be separated from other occupancies by **120-minute fire-resisting walls and floors**.
    - Extract ducts require **fire-rated enclosures and dampers** to prevent fire spread.
- **Bars, Nightclubs, Cinemas, and Leisure Uses**
    - Generate dense occupancies with late-night or high-risk conditions.
    - Require **robust separation from residential and office uses** with minimum 60–120 minutes fire resistance.
    - Soundproofing measures often double as **fire-separating construction**, but must be tested and certified accordingly.
- **Plant Rooms and Back-of-House Areas**
    - Electrical rooms, generator rooms, and delivery bays are treated as **special fire risk compartments**.
    - Strict fire-stopping and smoke separation measures are required.

13.4.3 Service Risers and Shafts

- **Vertical risers** for mechanical, electrical, and plumbing services are one of the most common pathways for uncontrolled fire spread.
- Best practice requires:

- Fire-stopping at **every floor level** to maintain compartmentation.
- Use of **intumescent wraps, collars, and tested fire-stopping products** around pipes and cables.
- **Separate risers** for residential vs retail systems wherever possible, to prevent cross-contamination.

- **Lift Shafts**
    - Must be enclosed in **fire-resisting construction**, with protected lobbies where serving multiple occupancies.
    - **Lift pressurisation** may be required in tall buildings to prevent smoke ingress.

13.4.4 Façade and External Wall Design

- Fire can spread externally between occupancies, bypassing internal compartmentation.
- The **Grenfell Tower fire (2017)** demonstrated how combustible cladding systems can allow vertical fire spread across multiple floors and occupancies.
- For mixed-use developments:
    - **Non-combustible external wall systems (Euroclass A1 or A2-s1,d0)** are required for residential, healthcare, and institutional uses above 18 m.
    - **Cavity barriers** must be installed at compartment floors, around windows, and at edges of façade systems.
    - Special attention is needed where retail podium façades interface with residential towers above.

13.4.5 Integration with Design and Management

- Compartmentation is only effective if **maintained throughout the building's life**.
- Common failures include:
    - Contractors breaching walls/floors for new services without reinstating fire-stopping.
    - Poor maintenance of cavity barriers or cladding systems.

   - Refurbishments that compromise compartmentation between uses.
- Fire strategies must therefore include a **management plan for inspecting, certifying, and maintaining compartmentation integrity**.

13.4.6 Case Example (Illustrative)

A **mixed-use tower** combines a retail podium, five floors of offices, and 25 storeys of residential apartments above.

- The separating floor between the podium and offices is **120-minute reinforced concrete with all penetrations sealed**.
- Between offices and residential floors, a further **120-minute compartment floor** is provided, with separate HVAC and electrical risers for each occupancy.
- The façade uses **non-combustible rainscreen cladding with cavity barriers** at every floor level.
- During design review, evacuation modelling confirmed that a fire in the retail podium would not compromise the residential stay-put strategy.

Summary

Effective compartmentation between occupancies is essential in mixed-use buildings. Fire strategies must:

- Provide **robust fire separation** between retail, leisure, office, and residential uses.
- Ensure **service risers, shafts, and plant rooms** are fully fire-stopped.
- Address **façade risks and external fire spread**.
- Include provisions for **long-term inspection and maintenance**.

By doing so, compartmentation ensures that fires remain contained within their area of origin, allowing evacuation strategies and fire service intervention to succeed without exposing other occupancies to undue risk.

13.5 Means of Escape

In mixed-use developments, the **provision of safe and effective means of escape** is one of the most challenging aspects of the fire strategy. Each occupancy type may have **different evacuation needs, behaviours, and strategies**, meaning that escape routes must be carefully designed to prevent overlap, congestion, or conflict.

13.5.1 Independent Escape Routes

- **Residential vs Public Uses**
  - Residential stair cores should be **completely independent** of retail, office, or leisure stairs wherever possible.
  - This ensures that residents, particularly at night, are not forced to share escape routes with unfamiliar, high-density crowds from public occupancies below.
- **Security and Access Control**
  - Independent escape stairs also provide greater security by preventing unauthorised access from public areas into residential or office floors.
- **Exceptions and Engineering Justification**
  - Where full independence cannot be achieved, a **shared stair** may be permitted if:
    - It is enclosed in **120-minute fire-resisting construction**.
    - It is served by **protected lobbies** with smoke control.
    - The evacuation models confirm no **capacity or tenability conflicts** between different user groups.

13.5.2 Exit Discharge

- **Avoiding Shared Assembly Bottlenecks**
  - Escape stairs from different occupancies must not discharge into the **same small external courtyard or dead-end space**.
  - This prevents overcrowding and ensures that one fire cannot compromise all occupants at the point of final exit.
- **Design Best Practice**
  - Residential exits should discharge to **separate, secure assembly areas** away from retail loading bays or public gathering points.
  - Large complexes (e.g., shopping malls with residential towers) should provide **multiple, geographically separated discharge points**.

13.5.3 Evacuation Strategy Conflicts

- **Residential vs Public Evacuation**
    - Public occupancies (retail, leisure, offices) typically rely on **simultaneous or phased evacuation**.
    - Residential occupancies often adopt a **stay-put policy**, with evacuation only if directly affected.
    - Fire strategies must clearly define how these two approaches interact:
        - **Separate alarm systems** to prevent unnecessary evacuation of residents during a small retail fire.
        - **Clear protocols** for overriding "stay put" if compartmentation is compromised.
- **Office Phased Evacuation vs Retail Simultaneous Evacuation**
    - In mixed-use towers, phased office evacuation may conflict with retail areas requiring immediate clearance.
    - Performance-based modelling (RSET vs ASET) is often required to ensure both strategies can operate without conflict.

13.5.4 Wayfinding and Signage

- **Clarity of Routes**
    - Escape signage must clearly distinguish between **residential-only routes** and **public escape routes**.
    - Confusion during evacuation can lead residents to follow public routes, or vice versa, creating congestion or exposure to risk.
- **Accessible Design**
    - Signage should be supported by **emergency lighting, tactile indicators, and multilingual messaging** in large, diverse complexes such as airports or malls.
    - For residential occupants, wayfinding must reflect the potential for evacuation at night and in low-visibility conditions.

13.5.5 Case Example (Illustrative)

A **mixed-use high-rise** combines a shopping centre podium, ten floors of offices, and apartments above.

- Residential cores are provided with **dedicated staircases**, completely independent of retail and office escape stairs.
- Office floors use **phased evacuation**, while retail spaces rely on **simultaneous evacuation supported by voice alarms.**
- Exit discharge points are located on different façades to prevent crowding in a single assembly area.
- Signage is colour-coded: **green pictograms for public exits** and **blue-coded signage for residential escape routes**, ensuring clarity for all users.

Summary

Effective means of escape in mixed-use developments requires **careful separation and coordination.** Fire strategies must ensure that:

- Escape routes are **independent** wherever possible.
- Exit discharge points are **separated** to avoid bottlenecks.
- **Evacuation strategies do not conflict**, particularly between stay-put and simultaneous evacuation policies.
- Wayfinding is **clear, accessible, and differentiated** between occupancies.

By addressing these issues, fire strategies provide a defensible and practical framework for ensuring all occupants — regardless of occupancy type — can reach safety without conflict or delay.

13.6 Fire Protection Systems

In mixed-use developments, the design and coordination of **fire protection systems** is critical to ensuring that safety is maintained across all occupancies without creating conflicts between them. Detection, alarm, suppression, and smoke control must be integrated, yet also allow for **independent operation** where appropriate. Poorly coordinated systems can lead to unnecessary evacuations, system failures, or even escalation of risk.

13.6.1 Detection and Alarm Systems

- **Coverage Across All Occupancies**
    - All areas, including residential, retail, leisure, and plant spaces, must be fitted with suitable **automatic detection and manual call points.**

    - System design should align with **BS 5839-1 (non-domestic premises)** and **BS 5839-6 (domestic dwellings)**.
- **Independent Operation**
    - A fire alarm activation in one occupancy (e.g., a restaurant kitchen) should not automatically trigger full evacuation of unrelated areas (e.g., residential apartments).
    - Zoning and **cause-and-effect programming** are essential to prevent unnecessary disruption.
- **False Alarm Management**
    - Mixed-use buildings are particularly vulnerable to false alarms due to varied activities (cooking, maintenance, retail fit-outs).
    - Alarm verification systems, phased evacuation, and **differentiated voice messaging** help reduce inappropriate evacuation.

13.6.2 Suppression Systems

- **Sprinkler Requirements**
    - Sprinkler protection is increasingly **mandatory in high-rise residential developments and large retail spaces**.
    - Both areas may be served by a **shared water supply**, but must be independently zoned to avoid cross-activation.
- **Zoning and Control**
    - Retail, office, and residential zones should each have **dedicated sprinkler control valves** with clear indication in the fire control room.
    - System design must allow the fire service to **isolate affected zones without disabling protection elsewhere**.
- **Specialist Systems**
    - Commercial kitchens may require **wet chemical suppression**.
    - Data centres or plant areas may use **gaseous systems (e.g., Novec 1230, Inergen)** to avoid water damage.

13.6.3 Smoke Control Systems

- **Shared Atriums and Concourse Spaces**
  - Atriums linking retail, transport, and residential uses present major smoke challenges.
  - **Performance-based smoke control design** (using CFD modelling under BS 7974) is essential to demonstrate tenability for occupants until evacuation is complete.
- **System Types**
  - **Mechanical extraction**: Fans remove smoke from retail or concourse spaces while maintaining a tenable smoke layer.
  - **Pressurisation**: Keeps escape stairs and firefighting shafts clear of smoke.
  - **Natural venting**: May supplement mechanical systems in atriums or large open volumes.
- **Coordination Between Uses**
  - Smoke control systems must be carefully zoned so that smoke from one occupancy does not compromise another.
  - For example, a shopping centre fire should not draw smoke into a residential lobby via a shared shaft.

13.6.4 Integration and Interfaces

- **Cause-and-Effect Programming**
  - Fire alarm activation may trigger:
    - Release of magnetic door holders.
    - Activation of smoke vents or mechanical extract.
    - Grounding of lifts.
    - Operation of suppression systems.
  - These sequences must be designed and tested to ensure **interdependence does not cause failures**.
- **Fire Control Room Panels**
  - A single centralised fire control room is usually required, housing:
    - Fire alarm repeat panels.

    - Sprinkler and suppression controls.
    - Smoke control interfaces.
    - Communication systems for the fire service.
  - Controls must be **clearly labelled by occupancy** to avoid confusion during emergencies.

13.6.5 Case Example (Illustrative)

A **mixed-use development** includes a shopping mall, cinema, offices, and a residential tower:

- A fire in the cinema triggers sprinklers locally and smoke extract in the mall atrium.
- The residential tower remains on "stay put," with no alarm activation in apartments unless smoke enters the residential lobby.
- Voice alarm messages direct shoppers and cinema-goers to multiple exits while preventing unnecessary disruption to office and residential areas.
- The fire control room provides a single point of command for the fire service, with zone maps for each occupancy.

Summary

Fire protection systems in mixed-use developments must balance **integration and independence**:

- **Detection and alarms** should allow zoned evacuation while avoiding false-alarm disruption across occupancies.
- **Suppression systems** must be comprehensive, zoned, and tailored to specific risks.
- **Smoke control** must be performance-based and tested through CFD modelling to maintain tenability.
- **Interfaces and fire control rooms** must ensure that the fire service can operate systems safely and effectively.

A robust fire strategy ensures that these systems work **together, not against each other**, safeguarding occupants across different occupancies with diverse needs.

13.7 Shared Services and Infrastructure

In mixed-use developments, it is common for **building services** — such as plant rooms, ventilation ducts, lifts, and power systems — to be shared between multiple occupancies. While this approach brings economic and operational efficiency, it also creates **significant fire safety risks**. Unless carefully designed and managed, shared infrastructure can become a pathway for fire and smoke spread, or a single point of failure that compromises safety across the whole complex.

13.7.1 Plant Rooms and Mechanical Systems

- **Separation of Plant**
    - Plant rooms serving different occupancies (e.g., residential boilers vs retail HVAC) should, where possible, be **separated into independent compartments**, each with robust fire resistance (typically 120 minutes).
    - If sharing is unavoidable, fire strategies must ensure **compartmentation, fire detection, and suppression** within the plant space.
- **Fuel Storage and Generators**
    - Shared fuel sources (e.g., diesel tanks for standby generators) increase the risk profile.
    - Such storage should be **isolated, ventilated, and fire-resisting**, with automatic suppression if volumes are significant.

13.7.2 Ducts, Shafts, and Ventilation Systems

- **Fire Dampers and Smoke Control**
    - Where ducts cross compartment lines or serve multiple occupancies, **automatic fire dampers** (to BS EN 15650) must close on smoke/heat detection.
    - Dedicated **smoke control ducts** should not be shared between occupancies unless proven safe through CFD modelling.
- **Risks of Shared HVAC**
    - Shared ventilation can spread smoke between occupancies if not fitted with fire and smoke dampers at the interface.
    - Fire strategies should seek **dedicated HVAC systems** for residential and public occupancies.

13.7.3 Lifts

- **Occupancy Segregation**
  - Lifts should be **clearly designated** for residential, office, retail, or hotel use, with separate lobbies where possible.
  - Shared lifts create both security and fire safety concerns — for example, shoppers or office workers entering residential lobbies during evacuation.
- **Firefighting and Evacuation Lifts**
  - Mixed-use towers must provide **firefighting lifts** accessible to the fire service, designed to BS EN 81-72.
  - Where evacuation lifts are provided (e.g., for mobility-impaired residents), they must be clearly distinguished from firefighting lifts to avoid confusion in emergencies.

13.7.4 Emergency Power Provision

- **Combined Power Demands**
  - Fire strategies must ensure that **emergency power systems** (generators, battery backups) are sized for the **combined critical demands** of all occupancies.
  - This includes fire detection, alarm, suppression pumps, smoke control fans, emergency lighting, firefighting lifts, and evacuation systems.
- **Redundancy**
  - Dual supply systems or **backup generators** may be required for resilience, especially in healthcare, transport hubs, or tall residential towers.
  - Transfer switches and load shedding must be designed to **prioritise life safety systems** in the event of partial failure.

13.7.5 Case Example (Illustrative)

In a **mixed-use podium and tower development**:

- Retail units, offices, and apartments shared a central chiller plant in the basement.

- The fire strategy required **2-hour fire separation walls** around the plant room, with automatic suppression and smoke detection.
- Ventilation ducts serving retail and residential occupancies were separated, with fire dampers at all interfaces.
- Lifts were divided into **public (retail/office) and private (residential)** groups, each with their own protected lobbies.
- Emergency power was provided by **two diesel generators**, one dedicated to life safety systems and the other to essential business continuity loads.

This arrangement ensured that no single service failure could compromise all occupancies simultaneously.

Summary

Shared services and infrastructure are an inevitable feature of many mixed-use developments, but they represent **high-risk interfaces**. A robust fire strategy must:

- Provide **fire separation** between plants serving different occupancies.
- Include **automatic dampers** to prevent smoke transfer through ducts.
- Clearly **separate lifts** for residential vs commercial/public use.
- Ensure **emergency power** is sized and prioritised for all critical life safety demands.

By treating shared services as **critical nodes of risk**, the strategy ensures that the integrity of one occupancy is not compromised by failures or hazards originating in another.

13.8 Fire Safety Management

One of the most **complex challenges in mixed-use developments** is ensuring that fire safety management is coherent and coordinated across multiple occupancies. Even if the technical design of the building is robust, **fragmented management arrangements** can undermine fire safety provisions, leading to gaps in responsibility, inconsistent training, or conflicting evacuation procedures.

The **Regulatory Reform (Fire Safety) Order 2005 (RRO)** requires that every non-domestic premises has a **Responsible Person**, but in mixed-use schemes this often means that **multiple Responsible Persons** operate within the same building. Without central coordination, fire safety management can quickly become inconsistent, leaving occupants at risk.

13.8.1 The Responsible Person (RRO 2005)

- Each commercial tenant (e.g., a shop, restaurant, or office) has its own **Responsible Person**, typically the employer or leaseholder.
- The residential element of a building (e.g., flats or apartments) is normally covered by the **freeholder, managing agent, or Residents' Management Company (RMC)**.
- Each Responsible Person is legally required to carry out a **Fire Risk Assessment (FRA)** for their part of the building.
- The challenge arises when these FRAs are **not aligned**, leading to duplication, omissions, or conflicting recommendations.

13.8.2 Central Coordination

- A centralised **Fire Safety Manager** (appointed by the landlord, managing agent, or development consortium) is often necessary to coordinate responsibilities across all occupancies.
- In very large developments, a **Joint Fire Safety Committee** may be formed, with representatives from each occupancy type meeting regularly to ensure consistency.
- This central role should cover:
    - Maintenance of **shared fire systems** (alarms, sprinklers, smoke control).
    - Oversight of **fire strategy updates** following refurbishment or tenant changes.
    - Liaison with **Building Control and Fire and Rescue Services**.

13.8.3 Drills and Training

- **Coordinated Fire Drills**
    - All occupancies should participate in drills, with scenarios designed to test the interactions between different evacuation strategies (e.g., "stay put" in residential vs simultaneous evacuation in retail).
    - Failure to test these scenarios can result in confusion and conflict during a real incident.

- **Staff Training**
  - Security staff, receptionists, and building maintenance teams play a critical role in directing occupants during an evacuation.
  - In healthcare, leisure, or hospitality occupancies, **PEEPs (Personal Emergency Evacuation Plans)** must be rehearsed with staff to ensure vulnerable occupants are assisted safely.
- **Resident Engagement**

In residential elements, where formal drills may not be practical, residents should be provided with **clear written guidance** and periodic awareness sessions.

13.8.4 Tenant Fire Safety Guides

- Each tenant or occupier should receive a **Fire Safety Guide**, tailored to their part of the building, but consistent with the overall fire strategy.
- Guides should cover:
  - Evacuation routes and procedures.
  - Use of shared facilities (e.g., escape stairs, refuges).
  - Responsibilities for maintaining in-unit fire precautions (alarms, extinguishers).
  - Rules around high-risk activities (hot works, storage of flammable materials).
- In residential occupancies, guides should explain:
  - The building's evacuation policy ("stay put" or simultaneous).
  - What to do if smoke is detected in corridors or lobbies.
  - How to report concerns about fire doors, compartmentation, or alarms.

13.8.5 Common Failures in Management

- Lack of clarity about **who maintains shared systems** such as sprinklers or smoke extract.
- Tenants carrying out **fit-out works** that compromise compartmentation without informing the fire safety manager.
- **Inconsistent training** across occupancies leading to conflicting actions during evacuation.

- Fire strategies becoming **outdated** as tenants change, with no single party updating the master document.

13.8.6 Case Example (Illustrative)

In a **mixed-use podium and tower** with retail, offices, and residential units:

- Each shop carried out its own FRA, but none addressed the **shared atrium smoke control system**.
- Office management assumed that retail staff would guide the public in an evacuation, while retail assumed the reverse.
- A central fire safety manager was later appointed, creating a **unified evacuation plan**, updating the fire strategy, and ensuring that **joint training and drills** were carried out.

Summary

Fire safety management in mixed-use developments requires **coordination across multiple Responsible Persons**. A robust strategy must:

- Define clear responsibilities under the RRO.
- Appoint a central fire safety manager or joint committee.
- Conduct coordinated drills and provide consistent staff training.
- Issue tenant fire safety guides and engage residential occupants.

Without strong management, even the best-designed fire strategy can fail in practice. With it, complex mixed-use schemes can deliver safety, resilience, and confidence for all occupants.

13.9 Case Example (Illustrative)

A **city centre mixed-use tower** provides a useful example of how a coordinated fire strategy addresses the unique challenges of combining different occupancies. The development consists of:

- A **three-storey retail podium**, including shops, restaurants, and leisure units.
- **Eight floors of open-plan offices** above the podium.
- A **25-storey residential tower** rising above the offices.

Compartmentation and Structure

- The retail podium is separated from the office and residential elements by **120-minute fire-resisting floors and walls**, ensuring that a fire in the high-risk retail environment cannot compromise the tower above.
- Offices and residential occupancies are further separated by an additional **120-minute compartment floor**, with **independent service risers** to prevent cross-contamination.
- The external façade is designed to **Euroclass A2-s1,d0** with cavity barriers at each floor plate, reducing the risk of vertical external fire spread.

Means of Escape

- **Independent Escape Cores**
  - The podium, offices, and residential tower each have **dedicated stair cores**, preventing overlap of public and private evacuation routes.
  - Final exits discharge onto different street frontages, ensuring that large crowds from the mall do not mix with residents evacuating from the tower.
- **Evacuation Strategies**
  - **Retail:** Simultaneous evacuation supported by voice alarms.
  - **Offices:** Phased evacuation, beginning with the fire floor and adjacent levels.
  - **Residential:** Stay-put strategy, with evacuation only if directly affected.
  - These strategies were modelled using evacuation simulation to confirm **no conflict or congestion** at stair or exit levels.

Fire Protection Systems

- **Sprinklers** are installed throughout all occupancies:
  - **Retail:** Life safety and property protection sprinklers, designed to BS EN 12845.
  - **Offices:** Fully sprinklered open-plan layouts.
  - **Residential:** Sprinklers to BS 9251, independent from commercial systems.

- Each occupancy has **separate sprinkler control valves**, monitored in the central fire control room.
- **Detection and Alarm Systems**
    - Retail and offices share a BS 5839-1 compliant addressable fire alarm system.
    - Residential areas have a separate BS 5839-6 system, preventing false alarms in the commercial areas from unnecessarily evacuating residents.
- **Smoke Control**
    - The retail atrium is served by **mechanical smoke extract fans and smoke reservoirs**, supported by CFD modelling.
    - Office lobbies and stairwells are protected by **pressurisation systems**.
    - Residential corridors have natural smoke shafts sized to ADB/ BS 9991 guidance.

Firefighting Facilities

- **Firefighting Shafts** are provided to serve both the office and residential towers, with **firefighting lifts to BS EN 81-72**.
- **Dry risers** serve the offices, while **wet risers** supply the residential tower above 60 m.
- A **dedicated fire control room** is located at ground level adjacent to the main fire service access point, housing:
    - Alarm repeat panels for all occupancies.
    - Sprinkler valve status displays.
    - Smoke control and ventilation controls.
    - Pre-incident plans and building layouts for the fire service.

Management and Coordination

- A **central fire safety manager** oversees the whole development, ensuring that:
    - Tenant fit-outs in the retail podium do not compromise compartmentation.
    - Coordinated fire drills are carried out for offices and retail staff.

    - Residents are provided with clear written guidance on the stay-put policy.
- A **joint fire safety committee** meets quarterly with representatives from retail, office tenants, and the residential management company.

Performance Outcomes

- Evacuation modelling confirmed that:
    - Shoppers could clear the retail podium within **8 minutes**.
    - Office phased evacuation avoided congestion, maintaining stair descent times below **12 minutes**.
    - Residential stay-put policy remained valid with **120-minute compartmentation**.
- Smoke modelling demonstrated that a **tenable layer height of at least 2.5 m** was maintained in atrium evacuation routes for more than 30 minutes.

Summary of Lessons

This case illustrates the **integration of prescriptive guidance and performance-based design**:

- Prescriptive compartmentation ensured strong separation of occupancies.
- Performance-based smoke and evacuation modelling provided confidence that strategies could operate side by side.
- Independent escape routes and fire protection system zoning eliminated conflicts between occupancies.
- Central management ensured that the strategy remained a **living document**, not just a design-stage exercise.

13.10 International Perspectives

Mixed-use developments are increasingly common worldwide, particularly in dense urban areas where **land values demand vertical integration** of retail, office, leisure, and residential functions. While the principles of **compartmentation, evacuation, and coordinated fire systems** are universal, regulatory approaches vary significantly across jurisdictions. Examining these differences provides insight into alternative solutions and highlights best practice that can inform UK projects.

13.10.1 United States (NFPA 101 & International Building Code)

- **NFPA 101 (Life Safety Code)** and the **International Building Code (IBC)** are widely used across the US and in many international jurisdictions.
- **Occupancy Separation**
    - Strong emphasis is placed on separating distinct occupancies with fire-resisting construction.
    - However, NFPA codes allow for **shared facilities and systems** if designers can demonstrate equivalent safety using performance-based analysis.
- **Mixed-Use Flexibility**
    - The IBC classifies occupancies separately (e.g., Assembly, Residential, Business), requiring designers to apply the most restrictive provisions unless robust separation is provided.
    - **Atrium smoke control** is a common feature in large US mixed-use malls and transport hubs, typically supported by CFD modelling and large-scale testing.
- **Management Focus**
    - US codes emphasise **fire drills, staff training, and emergency plans**, particularly in malls, stadiums, and tall towers.

13.10.2 Asia (Singapore, Hong Kong, and Others)

- **Refuge Floors**
    - Both Singapore and Hong Kong commonly require **dedicated refuge floors** in tall mixed-use towers.
    - These are hardened, fire-resisting floors located at intervals (typically every 20–25 storeys) where occupants can gather safely before further evacuation or rescue.
- **Strict Compartmentation**
    - Local codes require **robust fire separation** between retail podiums and residential or office towers above, often exceeding European standards.

- **Evacuation Lifts**
  - In Singapore, lifts specifically designed for **evacuation of mobility-impaired occupants** are mandated in certain mixed-use high-rise buildings.
- **Dense Urban Contexts**
  - Due to extreme density, Asian cities emphasise **rapid fire service response** and heavily regulate **access, water supplies, and firefighting lobbies** in mixed-use projects.

13.10.3 Middle East (UAE, Qatar, Saudi Arabia)

- **International Code Adoption**
  - Many Middle Eastern jurisdictions adopt **IBC and NFPA provisions** almost wholesale, with local amendments.
- **Strict Zoning of Occupancies**
  - Fire strategies typically require occupancies to be **zoned independently**, with dedicated stairs, lifts, and fire systems for each use.
  - Retail podiums, hotels, and residential towers are treated as **separate fire compartments**, linked by atriums or shared basements only with rigorous smoke control measures.

Cultural and Climatic Factors

The use of large atriums, open-plan lobbies, and extensive façade glazing is common in the region, driving reliance on **performance-based smoke control**.

Hot climates necessitate careful design of HVAC/smoke systems to avoid conflicts between comfort cooling and life safety.

13.10.4 Europe (Comparative Note)

- Many EU states adopt **Eurocodes and EN standards**, with harmonisation around façade performance, smoke control (EN 12101), and fire-resisting classifications (EN 13501).
- Refuge floors are **less common than in Asia**, but compartmentation between occupancies is enforced through EN-based structural and material standards.
- A strong emphasis is placed on **integration with sustainability goals**, e.g., balancing fire safety with energy-efficient façades and ventilation systems.

Summary

International practice shows both **common principles and regional variations**:

- **USA**: flexible approach, permitting shared systems with performance-based justification.
- **Asia**: prescriptive requirement for **refuge floors** and strict compartmentation in tall towers.
- **Middle East**: adoption of IBC/NFPA with **strict zoning** of occupancies, tailored to climatic challenges.
- **Europe**: harmonisation around EN standards with a growing emphasis on **sustainability integration**.

For UK practitioners, awareness of these global approaches is valuable, especially when working on international projects or benchmarking UK strategies against best practice worldwide.

13.11 Summary

Mixed-use developments present some of the **most complex challenges in fire safety engineering**, due to the coexistence of occupancies with very different risk profiles, occupant behaviours, and regulatory requirements. Unlike single-use buildings, fire strategies in these environments must be designed to **balance and integrate multiple systems, policies, and responsibilities**, while ensuring that the actions taken in one part of the development do not compromise safety in another.

The core principles that underpin safe mixed-use schemes include:

- **Compartmentation Between Occupancies:** Robust fire separation between residential, commercial, office, and leisure elements is fundamental. Compartment floors, fire-stopped risers, and non-combustible façades are essential to prevent fire and smoke spread across occupancies.
- **Coordinated Evacuation Strategies:** Residential "stay put" approaches must not conflict with simultaneous evacuation of public areas, or phased evacuation in office towers. Fire strategies must model and test these approaches to demonstrate that **all occupants can reach safety without congestion or confusion**.

- **Independent or Zoned Fire Protection Systems:** Detection, alarm, suppression, and smoke control systems must be carefully zoned so that a fire in one occupancy does not unnecessarily trigger alarms or suppression in others. Central fire control rooms must provide a clear interface for the fire service, with systems clearly labelled by occupancy.
- **Shared Services and Infrastructure:** Where plant, ducts, lifts, or power systems are shared, they must be compartmented, fire-stopped, and designed with redundancy. Emergency power provision must cover the combined demands of all occupancies.
- **Management and Responsibilities:** Perhaps the most critical aspect of mixed-use fire safety is **coordination across multiple Responsible Persons** under the Regulatory Reform (Fire Safety) Order 2005. A central fire safety manager or joint committee should oversee drills, system maintenance, and tenant engagement to ensure the fire strategy remains a **living, enforceable document**.

Final Reflection

When properly designed and managed, mixed-use developments can be both **safe and flexible**, supporting vibrant city centres and integrated communities. However, success depends on:

- Early consultation with regulators and Fire and Rescue Services.
- Clear technical documentation of compartmentation, evacuation, and system interfaces.
- Ongoing management coordination that ensures the strategy remains effective throughout the building's life.

The next chapter will turn to **Underground and Sub-Grade Structures**, where the challenges of restricted access, complex ventilation, and reliance on engineered solutions introduce an entirely different set of risks for fire strategies.

Chapter 14

Underground and Sub-Grade Construction

14.1 Introduction

Underground and sub-grade structures represent some of the most complex environments for fire safety engineering. These include:

- **Basements** – used for storage, plant, or commercial spaces.
- **Underground car parks** – often extending across multiple basement levels.
- **Tunnels** – for road, rail, and pedestrian transport.
- **Transport hubs** – integrated stations and concourses located below ground.
- **Subterranean leisure and commercial areas** – gyms, cinemas, retail spaces, or data centres built below grade.

Unlike above-ground buildings, where natural escape routes and natural ventilation are available, underground spaces operate with **inherent disadvantages** that demand engineered solutions. Fires in these environments are among the most dangerous because of their rapid impact on **tenability, visibility, and egress.**

14.1.1 Key Challenges

1. **Absence of Natural Ventilation**
    - Smoke accumulates quickly in confined spaces.
    - Reliance on **mechanical smoke extraction systems** is critical, and failure of these systems can be catastrophic.
2. **Limited Escape Routes**
    - Escape is typically vertical (via stairs or lifts), often involving long travel distances.

- Congestion at stairheads can occur rapidly, especially in high-occupancy uses such as stations or car parks.

3. **Difficult Firefighter Access**
    - Firefighters must descend into the hazard, often facing heat build-up, poor visibility, and unknown fire development.
    - Long hose runs and lack of ventilation add to operational risk.
4. **High Fire Loads in Confined Volumes**
    - Car parks contain large quantities of plastics and fuel, producing toxic smoke.
    - Underground retail or leisure spaces may involve dense occupant loads and combustible finishes.
5. **Smoke and Toxicity**
    - In underground fires, **toxic gases (CO, HCN)** often present a greater life safety hazard than flames themselves.
    - Rapid smoke logging can compromise escape in minutes if not controlled.

14.1.2 Fire Strategy Objectives

A fire strategy for underground and sub-grade structures must address four interdependent priorities:

- **Safe Evacuation** – ensuring occupants can reach the surface or a protected refuge before conditions become untenable.
- **Effective Smoke Control** – mechanical and natural ventilation systems designed to maintain visibility and breathable air along escape routes.
- **Firefighter Intervention** – dedicated firefighting shafts, access routes, and rising mains to enable safe and rapid response.
- **System Resilience** – redundancy in power supplies, fans, and controls to reduce the risk of system failure during an incident.

14.1.3 Case Lessons

- The **King's Cross Underground fire (1987)** highlighted how smoke spread, poor communication, and inadequate evacuation planning can lead to catastrophic outcomes.

- Fires in **underground car parks** (e.g., Stavanger, Norway, 2019) have demonstrated the intensity and toxic smoke production associated with burning vehicles in confined spaces.
- Tunnel fires (Mont Blanc, 1999; Tauern, 1999) emphasised the difficulty of long-distance evacuation in smoke-filled underground structures.

14.1.4 Scope of This Chapter

This chapter explores the unique requirements of fire strategies for underground and sub-grade structures, including:

1. **Regulatory and Standards Framework** – ADB, BS 9999, NFPA 502, and international codes for tunnels and transport hubs.
2. **Compartmentation and Fire Resistance** – structural considerations below ground.
3. **Smoke Control Systems** – mechanical, natural, and hybrid solutions.
4. **Means of Escape** – vertical evacuation, refuges, and phased evacuation in transport hubs.
5. **Firefighting Access and Facilities** – shafts, rising mains, hydrants, and control rooms.
6. **Case Studies** – lessons from underground fires and engineered solutions.

14.2 Key Risks in Underground Structures

Underground and sub-grade buildings present a unique fire risk profile due to their **enclosed nature, restricted escape options, and dependence on engineered systems**. Unlike above-ground structures, where natural ventilation and multiple exit options are possible, underground environments impose inherent constraints that must be addressed through robust fire strategies.

14.2.1 Restricted Egress

- **Limited Escape Routes**
    - Basements and car parks typically have a small number of vertical escape routes, usually via stairs or evacuation lifts.
    - Travel distances are often extended, with occupants required to navigate **smoke-logged spaces before reaching a stair core**.

- **Bottleneck Effects**
    - Concentrating large occupant flows into a limited number of staircases can result in congestion and extended evacuation times.
    - This risk is exacerbated in transport hubs or leisure venues with **high transient populations**.
- **Vertical Evacuation Challenges**
    - Climbing stairs to reach the surface is physically demanding, particularly for vulnerable persons or those in high-rise sub-grade environments.
    - Fatigue and mobility impairments can lead to occupants becoming trapped partway through evacuation.

14.2.2 Smoke Movement

- **Buoyancy and Confinement**
    - Heat and buoyancy cause smoke to rise rapidly and fill upper portions of basements, shafts, or tunnels.
    - With limited natural leakage paths, smoke layers **descend quickly**, compromising tenability in minutes.
- **Longitudinal Smoke Spread**
    - In tunnels or concourses, smoke may travel horizontally for hundreds of metres, creating **low-visibility conditions far from the fire origin**.
- **Toxicity and Reduced Visibility**
    - Combustion products such as carbon monoxide (CO) and hydrogen cyanide (HCN) accumulate in higher concentrations underground.
    - Visibility can drop below 10 m within 2–3 minutes, impeding both self-evacuation and fire service intervention.

14.2.3 Limited Ventilation

- **Dependence on Mechanical Systems**
    - Unlike above-ground structures, underground spaces lack natural air paths. Effective smoke management relies almost

entirely on **mechanical extraction, pressurisation, or longitudinal jet fan systems.**

- System failure due to fire damage or power outage can result in catastrophic conditions.

- **Complex Control Requirements**
    - Ventilation systems often need **dynamic control**: for example, reversing jet fans in tunnels or switching smoke extract between zones in a shopping concourse.
    - Integration with fire alarm systems is essential, but complexity increases the risk of **operational errors or delays**.

14.2.4 Fire Service Access

- **Delayed Intervention**
    - Firefighters must descend into the hazard, often requiring extended hose runs through **smoke-filled, unfamiliar environments**.
    - Vertical descent increases time to reach the fire and exposes crews to higher levels of fatigue.
- **Operational Hazards**
    - Limited visibility, risk of backdraft in poorly ventilated compartments, and unfamiliar layouts increase operational risk.
    - Fire service reliance on **firefighting shafts, rising mains, and fixed installations** is far greater than in above-ground structures.

14.2.5 High Fire Loads

- **Vehicle and Fuel Fires**
    - Car parks and service basements often contain cars, fuel, and lubricants, leading to **high heat release rates** (>5–10 MW per vehicle).
    - Electric vehicle (EV) battery fires present emerging risks, with longer burning durations and toxic smoke production.
- **Storage Hazards**
    - Retail basements and logistics hubs may store **plastics, textiles, or flammable liquids**, creating extreme fire loads and rapid fire growth.

    - Concentrated storage can overwhelm suppression systems if not carefully designed.

- **Plant and Services**
    - Electrical switchrooms, boilers, and generators in basements introduce additional hazards, particularly if fuelled by diesel.

14.2.6 Psychological Effects

- **Stress and Panic**
    - Enclosed underground spaces intensify the psychological impact of fire alarms and smoke, increasing the risk of panic.
    - Occupants unfamiliar with the layout may become **disoriented**, leading them back towards the entry route rather than the nearest exit.
- **Claustrophobic Perception**
    - Even trained staff may hesitate to enter or move through smoke-filled tunnels or basements due to claustrophobic fear.
- **Communication Barriers**
    - Poor acoustics and limited radio coverage underground can reduce the effectiveness of **voice alarm systems** and fire service communication, worsening behavioural challenges during evacuation.

Summary

Underground fires are uniquely hazardous because of:

- **Restricted escape routes** requiring vertical evacuation.
- **Rapid smoke movement** filling confined volumes.
- **Dependence on engineered ventilation systems**.
- **Delayed fire service access**.
- **High fire loads** from vehicles, storage, and plant.
- **Psychological impacts** on occupants.

These risks underscore why underground and sub-grade structures demand **specialist fire strategies**, drawing on advanced smoke control design, resilient systems, and careful evacuation modelling.

14.3 Regulatory Framework

The design of fire strategies for underground and sub-grade structures is shaped by a combination of **UK regulations, British Standards, and international codes**. Because of the unique risks posed by below-ground environments — particularly in relation to smoke control, evacuation, and fire service access — performance-based methods are frequently required to complement prescriptive requirements.

14.3.1 Approved Document B (ADB)

- **Baseline Guidance**
    - Approved Document B (ADB) to the Building Regulations sets out the **minimum statutory requirements** for basements and underground car parks.
    - Prescriptive provisions include:
        - Escape route limits for basements (e.g., single-stair basements limited in size and depth).
        - Travel distance limits for basement compartments.
        - Minimum requirements for compartment walls, floors, and separation between basements and upper storeys.
        - Smoke ventilation provisions for underground car parks (natural or mechanical).
- **Limitations**
    - ADB guidance is often insufficient for large, deep, or complex sub-grade structures such as transport hubs or multi-level basements.
    - In such cases, reliance on **BS 9999 or performance-based fire engineering (BS 7974)** is almost inevitable.

14.3.2 BS 9999 – Code of Practice for Fire Safety in the Design, Management and Use of Buildings

- **Risk-Based Approach**
    - BS 9999 provides **flexibility beyond ADB**, particularly for basement escape and mechanical smoke ventilation.

    - It introduces risk-based factors that allow **extended travel distances** if management and suppression measures are in place.
- **Basement Guidance**
    - Explicit design advice on:
        - Number and arrangement of escape stairs.
        - Smoke ventilation systems, including requirements for both natural and mechanical extract.
        - Enhanced fire resistance ratings for basement enclosures.
- **Relevance**
    - BS 9999 is particularly useful in **large car parks, retail basements, and leisure facilities**, where prescriptive ADB provisions are too restrictive.

14.3.3 BS 7346-7 – Smoke Control in Car Parks

- **Specialist Guidance**
    - Provides a dedicated **code of practice for smoke and heat control systems in enclosed car parks**.
    - Covers both **natural ventilation systems** (using openings and grilles) and **mechanical ventilation systems** (fans and ducting).
    - Sets performance criteria such as:
        - Minimum free area for natural ventilation (2.5% of floor area, evenly distributed).
        - Extraction rates for mechanical systems (10 air changes per hour for smoke clearance, 6 ACH for pollution control).
        - Requirements for jet fan systems and CFD validation.
- **Integration**
    - Often applied alongside **BS EN 12101** standards for smoke and heat control systems.
    - Increasingly relevant given the rise of **EV car fires**, which require prolonged smoke management and robust ventilation design.

14.3.4 BS 7974 – Fire Safety Engineering Principles

- **Performance-Based Solutions**
  - BS 7974 and its supporting documents (PD 7974 series) are frequently essential for **underground projects** that exceed prescriptive limits.
  - Key applications include:
- Smoke movement modelling (using CFD tools such as FDS).
- Evacuation modelling for long travel distances.
- Structural fire engineering for underground walls, slabs, and linings.
- Reliability and resilience analysis of mechanical ventilation systems.
- **Why It Matters**
  - For deep basements, transit stations, or tunnels, prescriptive design alone cannot guarantee safety — BS 7974 provides the methodology for **quantitative risk analysis and system optimisation**.

14.3.5 International Standards

- **NFPA 502 (Road Tunnels, Bridges, and Other Limited Access Highways)**
  - Widely applied in tunnel projects worldwide.
  - Provides requirements for **longitudinal and transverse smoke ventilation**, fire service access, hydrants, and fire resistance of tunnel linings.
- **NFPA 130 (Fixed Guideway Transit and Passenger Rail Systems)**
  - Specific to underground and elevated rail stations.
  - Requires performance-based design to demonstrate safe evacuation under credible fire scenarios.
  - Strong focus on **smoke management in concourses and platforms**, and on integration with evacuation modelling.
- **Why UK Projects Reference Them**
  - While not statutory in the UK, NFPA codes are often **referenced in large infrastructure projects** such as metro systems, road tunnels, and airport terminals.

  - They provide a **benchmark for international best practice**, especially in projects with multinational design teams or where insurers require NFPA alignment.

Summary

The regulatory landscape for underground structures is **multi-layered**:

- **ADB** provides the statutory baseline.
- **BS 9999** offers a risk-based approach, especially useful for basements and car parks.
- **BS 7346-7** gives specialist guidance for car park smoke control.
- **BS 7974** provides performance-based tools essential for large or complex designs.
- **NFPA 502 and NFPA 130** supply international benchmarks for tunnels and transit stations.

A successful fire strategy will typically **combine prescriptive measures for basic compliance** with **performance-based analysis** to demonstrate safety in scenarios that prescriptive codes cannot adequately cover.

14.4 Means of Escape

The design of means of escape in underground and sub-grade structures is fundamentally different from above-ground buildings. In the absence of natural egress paths and the reliance on vertical ascent to reach open air, escape strategies must be robust, redundant, and tailored to the unique conditions of underground environments. Fire strategies for such buildings must demonstrate that **all occupants can reach a place of safety before conditions become untenable**, even under challenging fire scenarios.

14.4.1 Independent Exits

- **Minimum Provisions**
  - For multi-level basements, each level must have independent means of escape unless a fire-protected stair serves as a common escape route.
- **Practical Challenges**
  - Space constraints often make it difficult to provide multiple stairs. Where only one stair is feasible, fire engineering solutions

(e.g., smoke-control systems, sprinklers, and evacuation lifts) may be required to demonstrate equivalence.

- **Design Implication**
  - Independent escape stairs should discharge **directly to open air** at ground level without passing through upper floors, preventing occupants from re-entering potentially hazardous spaces.

14.4.2 Travel Distances

- **Prescriptive Limits**
  - For basements, ADB sets stricter limits than for above-ground floors due to increased risk:
    - **Dead ends:** maximum of 9 m.
    - **Two-way travel:** 18 m in assembly/entertainment use; up to 45 m in offices, subject to occupancy and suppression systems.
- **Performance-Based Extensions**
  - BS 9999 allows extended travel distances if risk factors (occupancy, management, fire load) are favourable.
  - Performance-based designs using **evacuation modelling** and **smoke control validation (CFD)** can justify longer travel distances, common in large underground car parks or retail concourses.
- **Design Risk**
  - Long escape paths without effective smoke control can result in untenable conditions within minutes, making **integration with smoke extraction** essential.

14.4.3 Protected Stairs

- **Enclosed, Fire-Resisting Staircases**
  - Escape stairs in basements must be enclosed in **fire-resisting construction (typically 60–120 minutes)** with FD30S/FD60S self-closing fire doors.
  - Stairs should be located remotely from likely fire hazards, ideally at opposite ends of the basement footprint.

- **Direct Discharge**
  - Stairs should discharge directly to the exterior or to a protected final exit passage, not into ground-floor lobbies or retail areas.
  - In transport hubs, stairs often link to **pressurised escape corridors** that guide occupants to surface exits.
- **Pressurisation Systems**
  - Pressurised stairs are often necessary in multi-storey basements or underground stations to keep them **smoke-free and tenable** during evacuation.

14.4.4 Refuges and Areas of Safety

- **Refuges in Complex Structures**
  - In very large underground environments such as metro stations or tunnels, it may not be feasible for all occupants to evacuate directly to the surface.
  - **Refuge areas** (fire-resisting spaces designed to hold occupants safely) provide interim protection until further evacuation or rescue is possible.
- **Key Requirements**
  - Minimum **30–60 minutes fire and smoke protection**.
  - Two-way communication systems linked to the fire control room.
  - Capacity sized for anticipated mobility-impaired or delayed evacuees.
- **Examples**
  - In long rail tunnels, **cross-passages** between parallel tunnels act as refuge areas.
  - In underground hospitals or healthcare facilities, **progressive horizontal evacuation** into protected compartments is standard.

14.4.5 Wayfinding and Occupant Support

- **Illuminated Exit Signage**
    - Clear, consistent signage is vital in underground structures where natural light and external references are absent.
    - Signs should be visible even in smoke-logged conditions, often incorporating **internally illuminated or photoluminescent signs**.
- **Floor-Level Lighting**
    - Smoke tends to accumulate at higher levels, so **low-mounted or floor-level lighting strips** provide guidance in reduced visibility conditions.
    - Particularly critical in transport hubs or car parks where disorientation is likely.
- **Audible Guidance**
    - Voice alarm systems (BS 5839-8) are often preferable to simple sounders, providing **clear spoken instructions** to direct occupants to safe exits.
    - Audible beacons or directional sound devices may be used in tunnels or long corridors.

Summary

Means of escape in underground and sub-grade structures must:

- Provide at least **two independent stairs** for larger basements and public areas.
- Limit **travel distances**, with performance-based design required where prescriptive limits are exceeded.
- Ensure **protected, fire-resisting stair enclosures** discharge directly to open air.
- Incorporate **refuge areas and communication systems** in long or complex underground routes.
- Use enhanced **wayfinding systems** including illuminated signage, floor-level lighting, and voice alarms.

These provisions collectively address the inherent challenges of underground evacuation, where time, visibility, and safe pathways are often in shortest supply.

14.5 Smoke Control

In underground and sub-grade environments, **smoke rather than flame** is the primary life safety hazard. Confined volumes, limited escape routes, and lack of natural ventilation mean that visibility and tenability can deteriorate in **minutes or even seconds**. An effective smoke control strategy is therefore a cornerstone of sub-grade fire design.

The objectives of smoke control in underground structures are to:

1. **Maintain tenable conditions** along escape routes for the duration of evacuation.
2. **Provide clear zones** for firefighter access and intervention.
3. **Delay or prevent flashover and backdraft,** which are more likely in confined, ventilation-limited fires.
4. **Support suppression systems** by improving oxygen balance and reducing smoke recirculation.

14.5.1 Mechanical Extract Systems

- **High-Capacity Fans**
    - Mechanical smoke extract is the default solution for basements and car parks, with fans sized to achieve rapid removal of hot gases and smoke.
    - Design extraction rates vary, but **10–12 air changes per hour (ACH)** for smoke removal is typical, with higher rates in high-risk areas.
- **Zoned Systems**
    - Large underground structures are divided into smoke control zones, each with dedicated fans and ductwork.
    - This prevents smoke being drawn across escape routes and allows targeted extraction near the fire.
- **Redundancy and Resilience**
    - Many designs adopt **n+1 arrangements**, where at least one additional fan is installed beyond the minimum requirement, ensuring performance if a unit fails.
    - Backup power supplies (generators or UPS) are essential, as loss of power would compromise life safety.

14.5.2 Natural Ventilation Shafts

- **Vertical Shafts**
    - In shallower basements, natural smoke ventilation may be feasible through vertical shafts, light wells, or open grilles.
    - Shafts must be carefully **sized to provide at least 2.5% of floor area** (as per BS 7346-7 for car parks) and evenly distributed.
- **Challenges**
    - Performance is heavily influenced by wind and stack effects, which may **reverse or diminish extraction** in adverse conditions.
    - Natural venting alone is rarely sufficient for large or deep sub-grade spaces, but it can supplement mechanical systems.

14.5.3 Pressurisation of Escape Routes

- **Stair and Lobby Protection**
    - Escape stairs, firefighting shafts, and lobbies in basements are often **pressurised with clean air** to prevent smoke ingress.
    - This ensures that occupants and firefighters can use stairs safely even if smoke fills the basement compartment.
- **System Design**
    - Airflow rates must be sufficient to counter smoke leakage but balanced to avoid **excessive door-opening forces (>100 N)**, which can hinder evacuation.
    - Pressure relief systems may be required to prevent over-pressurisation.
- **Reliability**
    - Fans, ducting, and controls must be robust, fire-resistant, and supplied with backup power. Regular testing is critical to ensure system readiness.

14.5.4 Long Tunnels and Transit Systems

- **Longitudinal Ventilation**
    - Road tunnels and metro systems often use **jet fans mounted on ceilings** to move smoke in one direction (usually downstream of evacuation routes).

    - This keeps upstream routes clear for evacuation and firefighting.
- **Transverse Ventilation**
    - In some transit systems, smoke is extracted through continuous ducts running alongside or above the tunnel, allowing more precise control.
- **Cross-Passages and Refuges**
    - To support evacuation, **cross-passages** to parallel safe areas are typically provided every 250–500 m in tunnels, acting as refuges and secondary escape routes.
- **CFD Modelling and Testing**
    - The complexity of tunnel smoke behaviour requires detailed modelling (e.g., FDS,ANSYS Fluent) and often **full-scale testing** before system approval.

14.5.5 System Integration

- Smoke control systems must be fully integrated with **detection and alarm systems**:
    - Automatic activation triggered by smoke detectors or sprinkler flow switches.
    - Manual override available to firefighters via the fire control room.
    - Sequenced operation (e.g., shutting down comfort ventilation to avoid spreading smoke).
- Clear **interfaces between zones** are critical — poor integration can cause smoke to be drawn into escape routes or unaffected areas.

14.5.6 Case Lessons

- **King's Cross Fire (1987)**: inadequate smoke ventilation in the Underground station allowed hot gases to accumulate, contributing to a catastrophic flashover.
- **Mont Blanc Tunnel Fire (1999)**: smoke travelled kilometres along the tunnel due to poor ventilation control, leading to 39 fatalities.
- **Stavanger Car Park Fire (2019)**: highlighted challenges of modern EV fires, which produce **intense, prolonged smoke** requiring powerful mechanical extraction.

Summary

Smoke control in underground structures requires:

- **High-capacity mechanical extract systems** with redundancy.
- **Supplementary natural ventilation shafts** where feasible.
- **Pressurised escape routes** to protect stairs and lobbies.
- **Longitudinal or transverse ventilation** in tunnels, supported by cross-passages.
- **Integration with detection, suppression, and fire service controls** for resilience.

Without effective smoke management, underground evacuation and firefighting become nearly impossible.

14.6 Suppression Systems

Active fire suppression is particularly critical in underground and sub-grade structures because of the **restricted access for firefighting crews** and the high risk that smoke will rapidly compromise evacuation. Unlike above-ground fires, where the fire service can often intervene early, sub-grade fires may burn unchecked for longer periods unless suppression is provided. A fire strategy must therefore set out the type, extent, and resilience of suppression systems, tailored to the building's use, fire load, and layout.

14.6.1 Sprinkler Systems

- **Car Parks and Basements**
  - Sprinklers are strongly recommended — and often required by insurers or local authorities — in underground car parks, multi-level basements, and underground retail areas.
  - In many jurisdictions, sprinklers are **mandatory in underground car parks over 20,000 m²** or in basements below certain depths.
  - They provide rapid control of fires, reduce smoke production, and limit the chance of fire spreading beyond the compartment of origin.
- **Design Standards**
  - **BS EN 12845** for commercial sprinkler systems.

  - **BS 9251** for residential or mixed-use underground spaces.
  - Underground designs often use **zoned control valves**, allowing isolation of affected areas while maintaining protection elsewhere.
- **Performance Benefits**
  - A single operating head can suppress a car fire, dramatically reducing the heat release rate.
  - By cooling smoke gases, sprinklers also **support smoke control systems** by slowing smoke descent.

14.6.2 Water Mist Systems

- **Principle**
  - Water mist uses high-pressure nozzles to generate fine droplets that **cool flames, displace oxygen, and absorb radiant heat**.
- **Applications Underground**
  - Suitable in spaces where conventional sprinklers may cause water damage, such as **data centres, control rooms, museums, or underground archives**.
  - Effective in **low-ceiling environments** where traditional sprinkler coverage may be less efficient.
- **Advantages**
  - Reduced water demand compared to sprinklers — useful where underground water storage is limited.
  - Less risk of water damage to sensitive electrical or cultural assets.
- **Limitations**
  - Effectiveness depends on correct nozzle placement and high-pressure pump reliability.
  - More complex to maintain than conventional sprinkler systems.

14.6.3 Deluge Systems

- **Principle**
  - Unlike sprinklers (which activate locally), **deluge systems discharge water across a wide area simultaneously** when triggered.

- **Applications in Sub-Grade Structures**
  - Used in **high-hazard plant areas**, such as transformer vaults, turbine halls, or fuel storage basements.
  - Applied in **road tunnels** to suppress vehicle fires, where rapid growth and multiple ignition points make sprinklers less effective.
- **Considerations**
  - Deluge systems require large water storage capacity and robust drainage to handle discharge volumes.
  - Their indiscriminate discharge may cause water damage but is justified in life-critical or high-hazard settings.

14.6.4 Integration with Other Systems

- Suppression systems should be **interlinked with detection and smoke control**:
  - Sprinkler or water mist activation may trigger **smoke extract fans** or **alarm escalation**.
  - Deluge activation should automatically alert the fire control room and isolate electrical equipment where necessary.
- Redundancy in **power supplies, pumps, and water storage** is particularly important below ground, where fire service replenishment is slower and less reliable.

14.6.5 Emerging Issues

- **Electric Vehicle (EV) Fires in Underground Car Parks**
  - Lithium-ion battery fires can produce **higher heat release rates** and prolonged burning, with risk of re-ignition.
  - Research is ongoing into whether **sprinklers or water mist** can fully control such events. Some projects are exploring **foam-water hybrid systems** or enhanced detection/suppression for EV charging areas.
- **Heritage or Commercial Basements**
  - In sensitive underground spaces (e.g., archives, art storage), water mist or **clean agent systems** may be preferred over sprinklers to prevent collateral damage.

Summary

Suppression systems are vital in underground structures because:

- **Sprinklers** provide reliable, proven protection in car parks, basements, and retail/storage uses.
- **Water mist** offers reduced water demand and minimal water damage, suitable for sensitive or low-ceiling environments.
- **Deluge systems** are reserved for **high-hazard areas** or long tunnels, where rapid and wide-area application is necessary.

In all cases, suppression must be supported by **redundant water and power supplies**, integrated with smoke control and alarm systems, and tailored to the specific fire load of the environment.

14.7 Fire Service Access

Underground fires are among the most hazardous operational environments for firefighters. Unlike above-ground incidents, crews must **descend into the hazard**, often through smoke-filled, unfamiliar routes with limited ventilation. This increases physical strain, extends response times, and makes crews heavily dependent on fixed firefighting facilities.

A fire strategy for underground and sub-grade structures must therefore prioritise **direct access routes, robust firefighting shafts, reliable lifts, and water supplies**. These provisions not only improve firefighter safety but also maximise the chances of successful rescue and fire suppression.

14.7.1 Direct Access Routes

- **Dedicated Stairs and Lobbies**
    - Firefighters require **protected, fire-resisting stairs and lobbies** that lead directly to underground levels.
    - Lobbies should provide **smoke separation** via FD60S doors and, where appropriate, be **pressurised** to prevent smoke ingress.
- **Layout and Wayfinding**
    - Access routes must be **clearly signed, illuminated, and wide enough** to accommodate hose deployment and breathing apparatus entry.

- Complex basements should include **firefighter orientation plans** posted at entrances to reduce disorientation.

14.7.2 Firefighting Shafts

- **Extension to Basements**
  - In multi-level basements, firefighting shafts should extend **continuously from ground level to the deepest basement**.
  - Shafts typically comprise:
    - A firefighting stair.
    - A firefighting lift.
    - A protected lobby, sized for equipment deployment.
- **Design Requirements**
  - Enclosed in **120-minute fire-resisting construction**.
  - Doors to basements and lobbies should be **FD60S minimum** with smoke seals.
  - Pressurisation or mechanical smoke ventilation should protect shafts from smoke contamination.

14.7.3 Firefighting Lifts

- **Depth Thresholds**
  - Firefighting lifts are required where occupied levels extend more than **10 m below fire service access level**.
  - They allow rapid deployment of crews and equipment and provide essential evacuation capability for mobility-impaired occupants.
- **Design Standards**
  - Must comply with **BS EN 81-72**, which governs load capacity (minimum 630 kg), fire protection, and firefighter control systems.
  - Lifts should only be accessible via **firefighting lobbies**, preventing smoke contamination of shafts.
- **Resilience**
  - Powered by **backup generators or dual supplies** to maintain operation during fire conditions.

- Emergency intercoms and override controls for exclusive firefighter use are mandatory.

14.7.4 Vehicle Access for Fire Appliances

- **Large Basements and Tunnels**
  - In very large or deep underground environments, fire appliance access may need to extend **directly into basement levels or tunnel portals**.
  - Examples include **road tunnels**, where fire appliances drive into the tunnel to attack fires.
- **Turning Circles and Staging Areas**
  - Where vehicle access is provided, **dedicated staging and turning areas** must be designed to accommodate the size and weight of fire appliances.

14.7.5 Water Supplies

- **Hydrants**
  - Underground hydrants should be located within **90 m of entrances**, as per ADB.
  - Where external hydrants are insufficient, **internal hydrants or landing valves** must be provided.
- **Rising Mains**
  - **Dry Risers**: required for basement levels where fire service access is less than 18 m vertically below access level.
  - **Wet Risers**: required for greater depths, supplied by pumps and tanks, ensuring immediate availability of water.
- **Dedicated Storage Tanks**
  - In deep or extensive underground structures, **dedicated water storage tanks** may be necessary to ensure prolonged firefighting capacity.

14.7.6 Case Lessons

- **King's Cross Underground Fire (1987)** highlighted the dangers of inadequate fire service access and poor smoke management in underground transport hubs.

- **Mont Blanc Tunnel Fire (1999)** demonstrated the operational difficulties of long hose runs and limited ventilation.
- **Underground Car Park Fires (e.g., Stavanger, 2019)** underline the importance of risers, hydrants, and clear firefighter access routes to contain EV-related fires.

Summary

Effective fire service access in underground structures requires:

- **Direct, protected access routes** with fire-resisting stairs and lobbies.
- **Firefighting shafts and lifts** extended to the deepest levels.
- **Vehicle access** for very large basements or tunnels.
- **Robust water supplies**, including hydrants, rising mains, and tanks.

These facilities are essential not only for **firefighter safety** but also for ensuring underground fires — which escalate rapidly and are difficult to reach — can be effectively controlled.

14.8 Evacuation Strategies

Evacuation in underground and sub-grade structures presents unique challenges compared to above-ground buildings. Restricted egress, longer vertical travel distances, and rapid smoke spread mean that evacuation strategies must be carefully matched to the **building type, occupancy profile, and engineered smoke control systems**. Unlike above-ground offices or residential blocks, **delay or partial evacuation underground is rarely viable without robust supporting systems**.

14.8.1 Simultaneous Evacuation

- **Definition**
    - All occupants are instructed to leave immediately once the alarm is triggered.
- **Applications**
    - Common in **basements, underground car parks, and smaller sub-grade retail units**.
    - Favoured where smoke spread can compromise escape routes rapidly and where there are limited compartments to hold occupants safely.

- **Advantages**
  - Simple, easy to communicate, and requires less reliance on complex alarm sequencing.
  - Ensures rapid clearance of hazardous spaces.
- **Limitations**
  - May create congestion at stair cores if large numbers attempt to escape simultaneously.
  - Not always practical in very large transport hubs with thousands of people present.

14.8.2 Zoned Evacuation

- **Definition**
  - Occupants are evacuated in stages, with alarms triggered sequentially in different areas or zones.
- **Applications**
  - Common in **large underground transport hubs, shopping concourses, and exhibition halls**, where simultaneous evacuation may overwhelm exits or cause panic.
  - Relies heavily on **voice alarm/public address systems (BS 5839-8)** to guide occupants and provide reassurance.
- **Design Considerations**
  - Smoke control systems must reliably contain smoke within the affected zone for the duration of the staged evacuation.
  - Requires accurate zoning of alarms, detectors, and suppression systems.
- **Risks**
  - Miscommunication or poor system performance can delay evacuation and compromise life safety.

14.8.3 Defend-in-Place

- **Definition**
  - Occupants remain in a designated fire-resisting compartment until fire service rescue or until the incident is controlled.

- **Applications Underground**
    - Rare, but may be justified in **highly protected service areas, control rooms, or safe havens** with robust fire resistance and independent ventilation.
    - Sometimes used in **underground hospitals or healthcare units**, where progressive evacuation of vulnerable occupants is prioritised.
- **Requirements**
    - Compartmentation providing **60–120 minutes fire resistance**.
    - Independent pressurisation or air supply.
    - Communication links to fire control rooms for reassurance and instructions.

14.8.4 Progressive Evacuation

- **Definition**
    - Occupants move in stages from one protected area to another, gradually reaching ultimate safety.
- **Applications**
    - Widely used in **long road or rail tunnels**, where direct evacuation to the surface may take too long.
    - Cross-passages between parallel tunnels or refuge areas every 250–500 m act as intermediate safe zones.
- **Design Factors**
    - Refuge areas must provide **adequate capacity, ventilation, and communication systems**.
    - Lighting, signage, and floor-level wayfinding are critical in smoky, disorienting environments.
    - Coordination with smoke control systems is essential to keep refuge paths clear.

14.8.5 Human Behaviour Considerations

- In underground fires, **psychological stress is amplified** by darkness, smoke, and enclosure.
- Occupants may be reluctant to move into smoke-logged routes, even if they remain tenable.

  - Research shows that clear, authoritative **voice instructions** significantly improve compliance in underground evacuations.

14.8.6 Case Lessons

- **King's Cross Underground (1987):** demonstrated the need for **rapid simultaneous evacuation** when smoke can quickly flood confined spaces.
- **Mont Blanc Tunnel (1999):** highlighted the role of **progressive evacuation via refuge points**, as long tunnel fires cannot be cleared quickly.
- **Channel Tunnel (1996 & 2008 fires):** showed the value of **cross-passages and compartmented rail carriages**, allowing staged evacuation into safe tunnels.

Summary

Evacuation strategies underground must be:

- **Simple and rapid** (simultaneous) for smaller basements and car parks.
- **Zoned** in large hubs where crowd management is essential.
- **Progressive** in long tunnels, supported by refuge areas.
- **Defend-in-place** only in exceptional cases with robust fire-resisting compartments.

The chosen approach must be underpinned by **detailed evacuation modelling, smoke control validation, and clear human behaviour assumptions**.

14.9 Human Behaviour in Underground Fires

Human behaviour in underground and sub-grade environments differs significantly from above-ground buildings. While panic is often overstated in fire safety literature, underground fires present **unique psychological and physiological stresses** that can accelerate disorientation, delay evacuation, and reduce compliance with instructions. Fire strategies must therefore integrate **realistic assumptions about behaviour** and provide systems and staff procedures that actively mitigate these risks.

14.9.1 Psychological Stress and Perception of Risk

- **Feelings of Entrapment**
    - Underground occupants often perceive themselves as trapped due to limited visibility of exits and the need to travel vertically to safety.
    - The lack of natural light and external reference points increases anxiety, even in well-designed structures.
- **Panic Potential**
    - True "panic" is rare, but underground fires create conditions where **loss of orientation and perceived helplessness** can trigger disorderly behaviour.
    - Research shows people are more likely to **follow crowds** in such conditions rather than seek the nearest exit, increasing congestion at familiar entry routes.

14.9.2 Impact of Smoke on Behaviour

- **Disorientation**
    - Smoke in underground environments accumulates quickly and often uniformly due to low ceiling heights or enclosed spaces.
    - This reduces occupants' ability to **navigate signage, locate exits, or follow familiar routes.**
- **Physiological Effects**
    - Carbon monoxide and irritant gases impair decision-making and slow physical movement.
    - Studies have shown that even low levels of visibility reduction lead to **slower walking speeds and increased reliance on tactile cues** (walls, handrails).

14.9.3 Importance of Voice Alarms and Communication

- **Voice Alarm Systems**
    - Research consistently shows that **clear, authoritative voice messages** improve compliance compared to simple bells or sirens.

- Instructions such as *"Use the nearest exit"* or *"Do not return to the platform"* help reduce hesitation and prevent backtracking.

- **Public Confidence**
  - In large underground stations, passengers often delay evacuation because they **wait for confirmation** that the alarm is real. Voice announcements from trusted staff or automated systems reduce this delay.

14.9.4 Role of Staff and Training

- **Staff-Led Evacuation**
  - Underground fires highlight the critical role of **trained staff** in guiding occupants, particularly in stations, malls, or leisure complexes.
  - Staff presence reassures the public, prevents disorder, and directs flow toward safe exits.
- **Scenario-Based Drills**
  - Regular exercises (including joint drills with Fire and Rescue Services) ensure staff can respond quickly, even when communications systems fail.
  - In transport systems, staff are trained to **direct passengers away from smoke and into refuges or cross-passages**, where applicable.

14.9.5 Human Behaviour Modelling

- **Limitations of Traditional Models**
  - Evacuation software often assumes ideal occupant behaviour (immediate response, use of nearest exit).
  - In underground fires, these assumptions are unrealistic and may **underestimate Required Safe Egress Time (RSET)**.
- **Advanced Modelling Approaches**
  - Incorporating **delayed reaction times, wayfinding errors, and group dynamics** provides more accurate predictions.
  - Scenarios should model **partial compliance** (e.g., some occupants attempting to re-enter to retrieve belongings).

14.9.6 Case Lessons

- **King's Cross Underground Fire (1987):** Many passengers initially ignored alarms and continued normal activities until smoke became visible, highlighting the importance of **early authoritative communication.**
- **Daegu Subway Fire, South Korea (2003):** Panic-like behaviour occurred when smoke overwhelmed passengers in enclosed platforms; the absence of clear instructions contributed to high fatalities (192 deaths).
- **Channel Tunnel Fires (1996, 2008):** Demonstrated the importance of **staff guidance and refuge areas**; passengers followed trained personnel into cross-passages with high compliance.

Summary

Human behaviour in underground fires is influenced by:

- **Heightened stress and feelings of entrapment.**
- **Rapid disorientation caused by smoke.**
- **Improved compliance when voice alarms and staff leadership are present.**
- **The need for realistic modelling of delays, confusion, and group behaviour.**

For fire strategies, this means evacuation provisions must not only be technically robust but also supported by **clear communication systems, trained staff, and behaviourally informed design assumptions.**

14.10 Case Examples

Case studies illustrate how fire strategies are applied in underground environments and highlight the importance of integrating **suppression, smoke control, means of escape, and human behaviour considerations.**

Case 1: Underground Car Park

A **three-level basement car park** beneath a mixed-use development was identified as a high-risk environment due to the presence of vehicles with combustible fuels, plastics, and increasingly, electric vehicles with lithium-ion batteries.

- **Smoke Control**
  - The car park employs **mechanical smoke extract fans** rated at **20 air changes per hour (ACH)**.
  - **Jet fans** supplement the system, pushing smoke towards extract points and preventing stratification in open areas.
  - Computational Fluid Dynamics (CFD) modelling confirmed that a smoke layer height of **2.5 m** could be maintained for at least **15 minutes** following ignition.
- **Suppression**
  - A full **sprinkler system** is installed to BS EN 12845, with design density increased for vehicle fire risks.
  - Sprinklers are zoned by level, each controlled via valves accessible from the fire service entrance lobby.
- **Firefighting Provisions**
  - **Dry risers** are installed in each stair core, with landing valves at every level to allow rapid hose deployment.
  - **Pressurised stair cores** ensure smoke-free conditions for both firefighters and evacuating occupants.
- **Evacuation Strategy**
  - Occupants (mostly drivers and maintenance staff) follow a **simultaneous evacuation policy**.
  - Escape stairs discharge directly to open air, with **photoluminescent signage and floor-level lighting** guiding evacuees.

Lessons Learned:

- Jet fan systems are highly effective in dispersing smoke in large open basements but must be carefully modelled to avoid smoke recirculation.
- EV fires present prolonged heat release challenges, requiring **enhanced sprinkler density** and additional cooling provisions.

Case 2: Metro Station

A large **underground transit station** comprises two concourses, three platforms, and interconnecting corridors, serving tens of thousands of passengers daily.

- **Smoke Control**
  - **High-capacity fans** provide extract from platforms and concourses, designed for redundancy (n+1 arrangement).
  - Fans are powered by **dual electrical supplies with generator backup**, ensuring resilience during grid failure.
  - CFD analysis confirmed that **tenable conditions (visibility >10 m, temperature <60°C at 2 m height)** could be maintained for at least 20 minutes in worst-case scenarios.
- **Evacuation Strategy**
  - A **zoned evacuation** approach is adopted. In the event of fire, only the affected platform and adjacent concourse are evacuated immediately, while unaffected areas remain open to avoid overcrowding.
  - Evacuation is supported by:
    - **Illuminated directional signage** throughout concourses.
    - **Floor-level path lighting** to counteract reduced visibility.
    - **Voice alarm systems** with pre-recorded and live instructions.
  - **Cross-passages** link to an adjacent parallel tunnel every 250 m, providing alternative escape routes and refuges.
- **Firefighting Access**
  - **Firefighting shafts** with pressurised stairs and firefighting lifts serve each platform.
  - A dedicated **fire control room** houses alarm panels, smoke control interfaces, and CCTV for incident management.

Lessons Learned:

- Zoned evacuation can prevent unnecessary disruption in high-volume stations but relies on **effective smoke control and reliable voice alarms** to maintain occupant confidence.
- Cross-passages between tunnels provide crucial resilience in large transit systems, enabling **progressive evacuation** when full surface evacuation is impractical.

Summary of Case Insights

- Underground car parks require **integrated suppression and smoke control**, with careful consideration of modern vehicle fire risks.
- Metro stations demand **redundant smoke ventilation, zoned evacuation, and strong communication systems** to manage very high occupant densities.
- Both cases illustrate that underground fire strategies must combine **engineering systems, robust evacuation design, and human factors** to achieve resilience.

14.11 Summary

Underground and sub-grade structures remain among the **most complex and unforgiving environments** for fire safety engineering. Unlike above-ground buildings, these settings offer occupants and firefighters **few natural escape routes, limited access to fresh air, and extended vertical travel distances**, all of which magnify the risks during a fire event. A well-developed fire strategy must therefore treat underground safety as a **specialist discipline**, where the failure of one system can quickly escalate into catastrophic consequences.

The main principles are clear:

- **Restricted Means of Escape**
    - Basements, tunnels, and sub-grade malls often rely on a small number of protected stair cores or cross-passages. Fire strategies must ensure escape routes are **protected, clearly marked, and supported by illuminated signage and wayfinding systems**.
- **Smoke Control as a Cornerstone**
    - In confined underground spaces, smoke poses the greatest hazard. **Engineered smoke control systems** — including high-capacity mechanical extract, pressurised lobbies and stairs, and in tunnels, longitudinal jet fan ventilation — are fundamental. Their performance must be validated through **CFD modelling and resilience testing**, ensuring they remain operational under worst-case scenarios.
- **Active Suppression for Rapid Intervention**
    - Because fire and rescue services face **delays and risks when entering underground spaces**, active suppression

is indispensable. Sprinklers, water mist systems, or deluge arrangements help **limit fire growth, buy evacuation time, and reduce smoke production**. For high-hazard areas, specialist systems such as ESFR sprinklers or foam may be necessary.

- **Firefighting Access and Facilities**
  - Effective design requires **robust firefighting shafts, protected lifts, reliable rising mains, and on-site water storage**. Where access for vehicles is provided (e.g., in tunnels), roadways and staging areas must be sized and protected to support fire appliance operations.
- **Human Behaviour Under Stress**
  - Underground environments increase psychological strain: feelings of entrapment, disorientation from smoke, and reliance on crowds or staff instructions. **Clear voice alarms, trained staff, and behaviour-informed evacuation planning** are essential to avoid confusion or delay.
- **Integrated Design Approach**
  - Success relies on **integration of passive measures, active systems, and management procedures**. Fire strategies must demonstrate how smoke control, suppression, evacuation, and fire service facilities complement one another, rather than working in isolation.

Forward Look

While underground environments demand solutions to overcome confinement and lack of natural ventilation, the next chapter turns to the **opposite challenge**: **Atriums and large open spaces**, where the risk lies in **vast interconnected volumes** that can act as chimneys for smoke movement. The fire strategy focus shifts from confinement to **managing smoke in large, complex spaces while maintaining tenable escape routes**.

Chapter 15

Atriums and Large Open Spaces

15.1 Introduction

Atriums and other large-volume spaces such as **airport terminals, sports arenas, convention centres, and shopping malls** have become defining features of contemporary architecture. Their purpose goes beyond functionality: they create a sense of openness, maximise natural light, and foster connectivity between multiple building levels and occupancies. From a design perspective, they are often seen as the "heart" of a development, forming both a social and visual focal point.

However, from a **fire safety perspective**, these spaces represent some of the **most challenging environments** to design for. Their open, interconnected nature undermines the fundamental principle of compartmentation, which is central to controlling fire and smoke spread in most other buildings. Instead, atriums and large open spaces can:

- **Act as chimneys**, drawing smoke and hot gases upwards through buoyancy-driven forces, threatening upper levels before lower levels are cleared.
- **Create long travel distances** across expansive floor plates, often exceeding prescriptive code allowances.
- **Combine multiple occupancies** (e.g., retail, leisure, transport), complicating evacuation planning and system coordination.
- **Introduce unusual architectural features** — such as glazed façades, roof lights, and suspension systems — which may impact fire resistance and smoke control design.

A fire strategy for atriums must therefore carefully balance **aesthetic ambition with robust safety measures**. It must integrate:

- **Smoke management systems** capable of maintaining tenable conditions across multiple levels.

- **Evacuation strategies** tailored to large crowds and diverse user groups, supported by voice alarms and clear wayfinding.
- **Structural fire protection** to ensure stability in large-span roof and façade systems.
- **Firefighting facilities** that allow safe intervention in tall or complex volumes.

In short, while atriums deliver architectural grandeur and functional flexibility, they also **magnify fire risk if not supported by carefully engineered strategies**. This chapter explores the principles, regulations, and design solutions specific to atriums and large open spaces, building on the foundation of compartmentation, means of escape, and fire service access developed in previous chapters.

15.2 Key Risks in Atriums and Large Open Spaces

Atriums and expansive public spaces present distinctive fire safety challenges that differ fundamentally from compartmented buildings. Their very qualities — openness, visibility across levels, and large volumes — can significantly magnify fire hazards if not addressed through engineered strategies. The key risks are outlined below.

15.2.1 Chimney Effect

Atriums, by their very geometry, can act as powerful vertical shafts during a fire, producing a chimney effect that greatly influences smoke movement and evacuation safety. This phenomenon arises from the buoyancy of heated gases, which rise rapidly and entrain cooler air from below, generating strong upward flows that can be difficult to predict and control.

Mechanism of the Chimney Effect

When a fire develops at a lower level of an atrium, the combustion process generates hot gases significantly lighter than the surrounding air.

These gases rise through the open void, pulling cooler air in from lower levels to replace the rising volume, thereby reinforcing the upward airflow.

The larger the vertical height of the atrium, the greater the potential for accelerated smoke movement and stratification at high levels.

Consequences in Tall Atriums

In atriums exceeding 20–30 metres, smoke can rapidly bypass intermediate balconies and accumulate in the upper reservoir before extraction systems have time to respond.

Occupants on upper levels may therefore be exposed to smoke spread from a fire several storeys below, even if the fire floor itself is relatively remote.

This vertical movement often challenges prescriptive approaches, which assume compartmentation and slower smoke spread.

Impact on Smoke Control Systems

- Poorly balanced or undersized extract systems may be overwhelmed by the volume and velocity of smoke generated.
- Natural buoyancy can stall mechanical extraction if insufficient make-up air is provided, leading to recirculation or spill of smoke into escape routes.
- In some cases, smoke curtains or reservoirs may be overtopped, compromising tenability across multiple storeys simultaneously.

Horizontal Spread Risks

- Once the rising plume interacts with structural elements — such as walkways, bridges, or balcony edges — smoke can spill horizontally into occupied circulation spaces.
- This can cut off escape routes, reduce visibility, and generate untenable conditions at exits far removed from the fire origin.
- Without adequate smoke reservoirs and barriers, the plume may flow unpredictably, complicating evacuation modelling and fire service intervention.

Design Implications

- Smoke management strategies must anticipate the chimney effect as a primary design driver in atrium fire safety.
- CFD (computational fluid dynamics) modelling is often required to predict smoke movement under different fire scenarios and to inform extract vent sizing, make-up air distribution, and the placement of smoke curtains.

- The design objective is to maintain a tenable layer of clean air for occupants during evacuation and to preserve conditions that enable firefighter access.

15.2.2 Exposure Across Multiple Storeys

One of the defining hazards of atriums is their ability to expose multiple storeys simultaneously to fire effects. Unlike single-storey compartments, where fire and smoke are largely contained, an atrium provides a direct pathway for vertical fire communication. This creates risks not only for the floor of origin but also for occupants located several levels above.

Vertical Fire and Smoke Spread

- A relatively small fire at ground level can result in untenable conditions at higher levels within minutes, due to the open vertical void.
- The rising smoke plume entrains air as it ascends, expanding in volume and spreading laterally when it meets intermediate floors, balconies, or walkways.
- Heat and smoke accumulation at higher levels can quickly compromise escape routes and threaten occupants remote from the fire source.

Compromise of Evacuation and Stay-Put Policies

- In compartmented high-rise buildings, a "stay-put" strategy may be feasible, as fire spread is limited by fire-resisting barriers.
- In atriums, however, the openness negates this approach: occupants several storeys above the fire may be exposed despite physical distance from the seat of fire.
- Evacuation strategies therefore require reconsideration, often involving simultaneous or phased evacuation rather than reliance on local containment.

Structural and Material Pathways for Spread

- Curtain walling, glazed façades, escalator voids, and open balustrades can act as conduits for flame and smoke spread.
- Unprotected glazing can fail rapidly under thermal stress, allowing fire to leap vertically between levels.
- Balconies or mezzanine edges, if unprotected by smoke curtains or fire-resisting construction, may enable smoke spill into circulation spaces.

Radiant Heat Effects

- In addition to smoke, radiant heat from a plume can directly affect occupants and combustible surfaces at higher levels.
- This exposure may trigger secondary ignitions or compromise structural elements not in direct contact with the fire floor.
- Radiant heat transfer is particularly critical where combustible decorative finishes, suspended features, or shopfronts are present within the atrium.

Design Implications

- Fire engineering solutions must be developed to prevent uncontrolled multi-storey exposure.
- Common measures include smoke reservoirs at each level, fire curtains to prevent spill, fire-resisting glazing, and suppression systems to limit fire growth.
- CFD modelling is essential to predict smoke spill patterns and to design extraction systems that can maintain tenable conditions across all levels during the evacuation phase.

15.2.3 Extended Travel Distances

Large open-plan layouts, such as those found in atriums, retail centres, airports, and arenas, frequently exceed the maximum travel distances permitted under prescriptive guidance. Approved Document B (ADB) sets limits for escape distances — for example, **18 metres in dead-end conditions** and **45 metres where two directions of travel are available in offices**. These benchmarks are rooted in conventional compartmented layouts and do not readily accommodate the scale and openness of atrium environments.

Challenges of Extended Travel

- Occupants may need to cross extensive concourses, open retail floors, or event spaces before reaching a protected stair or final exit.
- The lack of compartment boundaries means that smoke and heat can spread across large areas, potentially cutting off escape routes before occupants have travelled the required distance.
- Delays in recognising alarms in large public spaces — particularly where background noise or distractions are present — can further lengthen effective travel time.

Impact on Occupant Safety

- Longer travel distances increase exposure to smoke, reduced visibility, and toxic gases, all of which impair evacuation efficiency.
- Vulnerable occupants, such as children, the elderly, or persons with reduced mobility, may be disproportionately affected by extended egress requirements.
- In mass gatherings, crowd density can significantly slow movement speeds, amplifying the risk when travel distances are already above prescriptive limits.

Role of Performance-Based Design

- To justify longer distances, performance-based design is commonly employed, supported by **BS 9999** (risk-based approach) and **BS 7974** (application of fire engineering principles).
- These frameworks allow for extended egress routes, provided that compensatory measures — such as smoke control, early detection, suppression, and clear wayfinding — are incorporated.
- Computational evacuation modelling is often used to demonstrate that safe egress can be achieved within available safe egress time (ASET), even when distances exceed prescriptive thresholds.

Engineering Solutions

- **Smoke control systems**: Mechanical extraction, natural ventilation, or smoke curtains to maintain tenable escape conditions.
- **Alarm and communication systems**: Voice alarm systems, public address, and visual signals to ensure occupants receive and act upon evacuation cues quickly.
- **Refuge areas and progressive horizontal evacuation**: Providing intermediate safety points for vulnerable occupants in very large buildings.
- **Wayfinding strategies**: Clear signage, lighting, and intuitive layout design to reduce pre-movement time and confusion.

Design Implications

- Extended travel distances can only be justified where a holistic strategy demonstrates that occupants are protected by a combination of engineered measures.

- This requires robust coordination between architectural layout, fire engineering input, and management procedures, all validated by detailed modelling and risk assessment.

15.2.4 Occupant Familiarity and Behaviour

Atriums and large public spaces such as airports, shopping centres, transport hubs, and sports arenas are characterised by high volumes of transient occupants. Unlike office or residential buildings, where users may be familiar with escape routes, these environments are often populated by visitors who may be unfamiliar with the layout and emergency procedures. This introduces a distinct layer of risk during fire emergencies.

Transient and Unfamiliar Occupants

- Many users of atrium spaces are first-time or infrequent visitors with little awareness of the location of exits, refuge areas, or assembly points.
- In an emergency, occupants often attempt to retrace their entry route rather than using alternative, and often closer, exits. This behaviour can lead to congestion and extended evacuation times.
- Tourists, shoppers, or event-goers may also be distracted or encumbered by luggage, shopping bags, or personal belongings, further slowing their response.

Behavioural Dynamics in Emergencies

- Research into crowd behaviour shows that panic is rare; instead, hesitation, confusion, and reliance on group dynamics are more common.
- Occupants frequently wait for visual or social confirmation before responding to alarms, leading to **delays in pre-movement time**.
- People are more likely to follow others, even if those ahead are moving towards less optimal or more congested exits.
- Staff behaviour and instructions strongly influence crowd response. In environments such as airports and stadiums, trained staff providing authoritative guidance is critical to reduce uncertainty.

Congestion and Evacuation Flow

- In large atriums, wide-open spaces can paradoxically contribute to congestion, as crowds funnel towards the most obvious or familiar exits.

- Without effective directional signage and clearly visible escape routes, bottlenecks may form at entry doors, escalators, or main staircases.
- Where children, elderly people, or those with reduced mobility are present, evacuation flow can become uneven, further complicating crowd movement dynamics.

Communication and Wayfinding

- Clear, unambiguous communication is essential in unfamiliar environments. Voice alarm systems are more effective than bells or tones in prompting action, particularly when combined with multilingual or visual messaging.
- Dynamic signage and illuminated directional indicators can adapt to changing smoke conditions, guiding occupants away from compromised routes.
- Staff training is equally vital: security staff, retail employees, or stewards should be empowered to initiate and manage evacuation effectively.

Design Implications

- Fire safety strategies for atriums must account not only for the physical environment but also for **human factors and behaviour under stress**.
- Measures such as **enhanced voice alarm systems, intuitive wayfinding, and staff emergency response training** are as critical as smoke control or structural fire protection.
- Evacuation modelling should include behavioural variables — such as hesitation time, group movement, and exit choice — to provide realistic predictions of evacuation performance.

15.2.5 Evacuation Complexity

Large open spaces such as atriums, shopping centres, airports, and arenas can accommodate thousands of occupants simultaneously. While this scale supports commercial and social activity, it creates significant challenges in ensuring safe, efficient evacuation during a fire emergency. The interaction between vast volumes, varied occupant types, and complex layouts introduces evacuation risks that differ fundamentally from smaller, compartmented buildings.

Mass Crowd Dynamics

- In high-density environments, evacuation performance is dictated as much by crowd behaviour as by physical exit capacity.
- Bottlenecks commonly occur at **stair heads, concourses, escalators, and doorways**, where wide open areas converge into narrower circulation routes.
- Once congestion begins, movement slows exponentially, increasing the exposure time of occupants to smoke and heat.
- The risk is amplified in unfamiliar settings where visitors may hesitate, stop to orient themselves, or follow others rather than dispersing efficiently.

Complex Occupant Profiles

- Populations in atriums and stadia are highly diverse, including children, elderly people, and individuals with reduced mobility.
- In venues such as airports or retail centres, many occupants may be carrying luggage, pushchairs, or shopping trolleys, which slow evacuation flow.
- In arenas and stadia, alcohol consumption and heightened emotions can also affect crowd behaviour, reducing compliance with instructions.

Evacuation Strategies

Effective strategies for large spaces must go beyond simple provision of exits and incorporate **engineered management measures**, such as:

- **Voice alarm and public address systems**: Clear, multi-lingual, and authoritative instructions have been shown to reduce hesitation and improve compliance compared with simple bells or tones.
- **Phased or zoned evacuation**: Releasing occupants in controlled sequences reduces the risk of overloading circulation routes, particularly staircases. This is especially important in high-rise atriums or stadia with tiered seating.
- **Multiple, well-distributed exits**: Ensures occupants can disperse evenly rather than funnelling towards familiar entry routes. Exits must be clearly signed, illuminated, and accessible under low-visibility conditions.

- **Disabled access and inclusive design**: Evacuation strategies must incorporate refuge areas, evacuation lifts, or pitch-level egress routes to ensure safe evacuation for mobility-impaired occupants.

External Crowd Management

- Evacuation planning does not end at the building threshold. Large arenas and stadia can discharge tens of thousands of people onto concourses, plazas, or surrounding streets.
- Without coordinated external crowd management, secondary hazards such as crushing, queuing at transport interchanges, or obstruction of emergency access routes may occur.
- Collaboration with local authorities, transport providers, and emergency services is essential to ensure that external environments can safely accommodate discharged crowds.

Design and Modelling Implications

- Evacuation modelling must simulate crowd dynamics, not just individual egress times, to identify potential pinch points under different scenarios.
- Management procedures, staff training, and rehearsal exercises are critical in ensuring theoretical strategies work in practice.
- Ultimately, evacuation complexity in large open spaces can only be mitigated through a combination of **robust physical design, engineered systems, and active management protocols**.

15.2.6 Case Insights

Historical fire incidents provide critical lessons for understanding the risks inherent in atriums and large open spaces. While each case occurred in different contexts, the themes of rapid fire growth, smoke movement, occupant unfamiliarity, and evacuation challenges consistently emerge.

Bradford City Stadium Fire (UK, 1985)

- A discarded cigarette ignited litter beneath wooden seating during a football match. Within minutes, flames engulfed the entire stand, fed by combustible seating and a timber roof.
- Over **50 people lost their lives**, many trapped by congested exit routes that became impassable as the fire spread rapidly overhead.
- *Lessons*:

    - Open seating areas can enable rapid horizontal flame spread, overwhelming occupants before they can evacuate.
    - Exit doors, gates, and stairways must be sufficient in number, distributed evenly, and free from obstruction.
    - Fire growth in large open environments demands early detection and suppression to prevent evacuation routes being cut off.

Düsseldorf Airport Atrium Fire (Germany, 1996)

- A welding operation ignited polystyrene insulation in the ceiling of the terminal's atrium. The resulting fire produced dense toxic smoke that rose rapidly and spread across multiple levels due to the chimney effect.
- More than **15 fatalities** occurred, largely from smoke inhalation, with many others injured.
- *Lessons*:
    - Atrium fires can propagate vertically with alarming speed, bypassing intermediate storeys.
    - Inadequate smoke control systems can lead to rapid contamination of evacuation routes.
    - The event underscored the importance of integrating smoke extraction, make-up air supply, and fire-resisting construction in atrium design.

The Station Nightclub Fire (Rhode Island, USA, 2003)

- Though not an atrium, this incident remains a defining case study in occupant behaviour during fire. Pyrotechnics ignited foam insulation, creating a flashover within minutes.
- **100 fatalities** occurred as patrons attempted to escape via the main entrance, despite multiple available exits.
- Lessons:
    - Occupant unfamiliarity with layouts leads to reliance on familiar entry routes, resulting in fatal congestion.
    - Clear signage, visible exits, and staff guidance are critical in unfamiliar venues.
    - Performance-based evacuation design must incorporate human behaviour, not just theoretical exit capacity.

Cross-Cutting Insights

- These incidents collectively highlight the **compounding risks** in large open environments:
    - **Rapid fire spread** (Bradford).
    - **Vertical smoke movement** (Düsseldorf).
    - **Behavioural and congestion risks** (The Station).
- Effective fire strategy for atriums must therefore balance **engineering solutions** (smoke control, suppression, structural protection) with **human-centred measures** (signage, voice alarms, crowd management).

Summary

Atriums and large open spaces present a fire safety profile that diverges sharply from traditional compartmented buildings. The defining risks include the **chimney effect of vertical smoke movement**, **exposure of multiple storeys to both smoke and radiant heat**, and the **extension of travel distances** well beyond prescriptive code allowances. These physical hazards are compounded by **human factors** such as occupant unfamiliarity, crowd dynamics, and behavioural delays, which can turn otherwise manageable scenarios into catastrophic events.

The complexity of these environments demands that fire strategies extend beyond prescriptive code compliance. Engineered solutions must include:

- **Advanced smoke control systems** capable of managing vertical and lateral smoke movement.
- **Evacuation modelling and crowd flow analysis** to anticipate bottlenecks and behavioural patterns.
- **Robust management protocols**, including staff training, communication systems, and external crowd control.

In short, atrium fire safety can only be assured through a **performance-based approach**, underpinned by BS 9999, BS 7974, and international equivalents, where prescriptive rules are insufficient. This sets the foundation for the next section on the **regulatory framework** governing the design and management of these complex spaces.

15.3 Regulatory Framework

The distinctive risks posed by atriums and large open spaces often extend beyond the scope of prescriptive fire safety codes. While guidance documents establish important baseline requirements, they are frequently inadequate for complex, high-volume environments. As a result, fire safety strategies for atriums typically rely on a performance-based approach, supported by advanced modelling and risk justification. The principal regulatory and standards framework in the UK and Europe is outlined below.

Approved Document B (ADB)

- ADB provides prescriptive requirements for means of escape, compartmentation, and fire resistance. However, its provisions are generally intended for conventional compartmented buildings.
- Atriums challenge these assumptions: vertical openness, extended travel distances, and multi-storey exposure cannot easily be addressed within ADB's fixed parameters.
- While ADB does recognise atriums and permits certain solutions (e.g., smoke reservoirs and ventilation), it often necessitates **alternative approaches supported by fire engineering justification**.
- Designers must therefore demonstrate equivalence with ADB's life safety objectives through performance-based strategies.

BS 9999: Code of Practice for Fire Safety in the Design, Management and Use of Buildings

- BS 9999 provides a **risk-based framework** that allows for greater flexibility than ADB by adjusting design solutions according to occupancy characteristics, fire growth rates, and management systems.
- The standard recognises that in large open spaces, prescriptive limits (e.g., travel distance) may not be appropriate, and permits extended provisions where supported by compensatory measures such as smoke control, suppression, and robust evacuation management.
- For atriums, BS 9999 serves as a valuable intermediate standard, bridging prescriptive guidance and full performance-based design.

BS 7974: Application of Fire Safety Engineering Principles

- BS 7974 underpins the performance-based design of complex buildings, providing methodologies for applying fire engineering principles in practice.
- It includes guidance on:

    - **Computational Fluid Dynamics (CFD) modelling** to predict smoke and heat movement within large volumes.
    - **Evacuation modelling** to simulate crowd behaviour, exit choice, and travel times.
    - **Structural fire engineering** to assess the impact of elevated temperatures on atrium structures, such as glazed roofs and steel framing.
- In atrium design, BS 7974 is the primary standard used to justify departures from prescriptive codes and to demonstrate that life safety objectives can be achieved.

BS EN 12101: Smoke and Heat Control Systems

- This European standard specifies requirements for smoke and heat exhaust systems, including natural and mechanical ventilation, smoke barriers, and powered fans.
- For atriums, BS EN 12101 provides the technical basis for designing **smoke extraction, smoke curtains, and make-up air provisions**, all of which are critical to managing the chimney effect and maintaining tenable escape routes.
- Compliance with BS EN 12101 ensures that systems are designed, installed, and tested in line with European best practice, providing confidence in their performance during fire events.

Summary of the Framework

- **ADB** sets the baseline but is limited for atrium conditions.
- **BS 9999** introduces flexibility through a risk-based approach.
- **BS 7974** provides the performance-based tools for modelling and justification.
- **BS EN 12101** supplies the technical standards for smoke and heat control systems.

Together, these documents form the regulatory foundation for atrium fire safety strategies, enabling designers to move beyond prescriptive rules and apply engineered solutions tailored to the specific risks of large open spaces.

15.4 Smoke Control in Atriums

The central challenge in atrium fire safety is the **management of smoke movement** to preserve a tenable, smoke-free layer for evacuation and firefighting. Given the chimney effect and the scale of these spaces, smoke can quickly rise, spread laterally, and compromise escape routes across multiple levels. Effective smoke control systems must therefore be engineered to ensure visibility, breathable air, and protected evacuation paths.

Mechanical Smoke Extraction

- Mechanical extraction systems are the most reliable means of maintaining a clear lower layer during an atrium fire.
- **High-capacity extract fans** are designed to remove smoke at a rate that prevents descent of the smoke layer below safe levels, typically maintaining a **minimum clear height of 2.5 m above floor level** in occupied areas.
- These systems offer consistent performance independent of external weather conditions, unlike natural ventilation.
- Design considerations include:
    - **Redundancy**: Multiple fans should be installed to provide resilience in case of equipment failure.
    - **Emergency power supply**: Fans must remain operational under fire conditions, requiring protected cabling and connection to backup generators.
    - **Integration with CFD modelling**: Extraction rates should be determined based on plume characteristics, fire size, and anticipated occupant load.

Natural Ventilation

- Natural venting relies on **Automatic Opening Vents (AOVs)** located in the roof or upper façade to exhaust hot gases.
- While energy-efficient, performance is heavily dependent on external factors such as wind speed, direction, and ambient temperature.
- In tall atriums, rising smoke plumes entrain cooler air as they pass open balconies, mezzanines, or intermediate floors.
- This entrainment dilutes and **cools the plume**, reducing its buoyancy. In some cases, this can cause the smoke column to **stall before reaching**

the roof-level AOVs, leading to lateral smoke spill across occupied levels.

- As a result, **natural ventilation alone is rarely sufficient**, and it is typically used in combination with mechanical extraction to ensure reliability under varied conditions.

Compartmentation via Fire Curtains

- Deployable **smoke curtains** provide a critical means of subdividing the atrium space into smoke reservoirs, directing smoke toward extraction points.
- Typically concealed within ceiling lines, curtains deploy automatically upon activation of the fire alarm system.
- Curtains can be used to:
    - Protect circulation routes such as escalators, bridges, and balconies.
    - Create **smoke reservoirs** sized in accordance with BS EN 12101 to control smoke spill and maintain a stable smoke layer.
    - Reduce the volume of space requiring mechanical extraction, optimising fan performance.
- Curtains must be carefully coordinated with architectural features to avoid obstructing circulation during normal use.

Make-Up Air

- For smoke extraction to function effectively, **make-up air** must be introduced to replace the volume of gases being exhausted.
- This prevents negative pressure that could stall extraction systems or draw smoke into occupied zones.
- Best practice is to introduce make-up air at **low level**, avoiding turbulence that disrupts the smoke layer.
- Controlled inlet design (e.g., louvres, grilles) ensures airflow is steady and does not compromise visibility or tenability.

Design Integration

- Smoke control in atriums rarely relies on a single system. Instead, effective solutions typically integrate:
    - **Mechanical extraction** for primary smoke removal.

- **Natural venting** as a supplementary measure or failsafe.
- **Smoke curtains** to direct and contain plumes.
- **Balanced make-up air supply** to stabilise the system.

- The combined system must be validated through **fire engineering analysis and CFD simulation**, ensuring performance under credible worst-case fire scenarios.

15.5 Means of Escape in Atriums

The design of means of escape in atriums is one of the most challenging aspects of fire safety engineering. The openness, extended travel distances, and large occupant loads mean that prescriptive requirements from Approved Document B (ADB) are frequently exceeded. As a result, atrium escape design typically relies on performance-based justification, supported by engineered systems such as smoke control, suppression, and evacuation modelling.

Travel Distances

- Atrium layouts often require occupants to cross large concourses, retail floors, or event spaces before reaching a protected stair or final exit.
- These distances frequently exceed the **ADB limits** of 18 m for dead-end conditions and 45 m where two directions of travel are available in office-type occupancies.
- Extended travel distances can be justified where compensatory measures are in place, including:
    - **Automatic sprinkler protection**, which controls fire growth and limits smoke production.
    - **Engineered smoke control systems**, maintaining a tenable layer above escape routes.
    - **Evacuation modelling** that demonstrates all occupants can reach safety within the available safe egress time (ASET).

Exit Distribution

- Multiple exits should be **strategically distributed around the atrium perimeter**, reducing the risk of congestion and ensuring occupants can disperse evenly.

- Clear sightlines to exits are critical in large open volumes where occupants may otherwise default to the familiar entry route.
- Exits must be designed to handle the maximum expected occupant load, with sufficient width and capacity to prevent bottlenecks.
- Refuge points or horizontal exits should be considered for mobility-impaired occupants in very large or complex layouts.

Evacuation Strategy

- In most atrium occupancies, a **simultaneous evacuation strategy** is required due to the openness of the space and the potential for rapid smoke spread.
- In tall atriums with upper-level balconies or galleries, a **phased evacuation** approach may be acceptable, provided smoke control measures maintain safe conditions on non-evacuating levels.
- **Voice alarm and public address systems** are essential, providing clear, calm, and preferably multi-lingual instructions. These reduce hesitation, direct occupants to alternative exits, and prevent crowd congestion at familiar routes.
- Staff training is vital to support evacuation, especially in airports, shopping centres, and stadia where large numbers of unfamiliar occupants rely on guidance.

Design Integration

- Means of escape in atriums must be considered holistically, integrating:
    - **Architectural layout** (shortest possible routes, distributed exits).
    - **Fire engineering systems** (sprinklers, smoke extraction, fire curtains).
    - **Management procedures** (staff roles, evacuation drills, external crowd management).
- Computational evacuation modelling, validated against realistic crowd behaviours, is a critical tool to demonstrate compliance and optimise the design.

15.6 Structural Fire Protection

Atriums frequently incorporate ambitious architectural features such as exposed steelwork, glazed façades, and long-span roof structures. While these enhance openness and visual appeal, they also present challenges for structural fire protection. The failure of critical elements can compromise smoke control, evacuation routes, or even lead to progressive collapse. Effective fire strategy must therefore ensure that structural elements achieve appropriate levels of fire resistance and remain stable under design fire scenarios.

Fire Resistance Ratings for Primary Structure

- Supporting elements such as columns, beams, and trusses should be designed to achieve the fire resistance period required by building regulations (commonly **60 to 120 minutes**, depending on building height, use, and evacuation profile).
- Where architectural expression leaves structural steel exposed, passive protection systems such as **intumescent coatings, fire-rated board encasement, or spray-applied fire-resistive materials** must be specified.
- Performance-based assessments may be used to demonstrate structural adequacy under specific design fires, particularly where atriums are partially separated from occupied spaces by fire curtains or smoke barriers.

Fire-Resistant Glazing and Façades

- Atriums often rely heavily on glazing for natural light and visual connectivity. However, standard toughened or laminated glass fails rapidly under thermal stress, leading to breakage, smoke spill, and potential vertical fire spread.
- Fire strategy must therefore require the use of **fire-resistant glazing** for critical locations, such as:
    - Shopfronts opening onto the atrium.
    - Balustrades and balcony edges.
    - Compartment boundaries where fire separation must be maintained.

- Fire-rated glazing assemblies should be tested and certified to relevant standards (e.g., **BS EN 1364, BS EN 1365**) to ensure integrity and insulation for the specified duration.

Roof Protection and Smoke Vent Support

- Roof structures in atriums frequently support **Automatic Opening Vents (AOVs)**, smoke exhaust fans, or mechanical plant.
- If these structural elements fail prematurely in a fire, the entire smoke management strategy may be compromised.
- Protection of roof trusses, purlins, and framing should therefore be prioritised to maintain the functionality of smoke control systems throughout the evacuation and firefighting period.
- For lightweight roofs incorporating tensile fabric or non-fire-resisting materials, performance-based assessment must confirm that roof failure will not create unacceptable hazards (e.g., falling debris or uncontrolled ventilation paths).

Integration with Fire Engineering Analysis

- Structural fire protection in atriums should not be considered in isolation but integrated with **smoke control modelling and evacuation analysis.**
- CFD and structural fire modelling can be used to predict heat exposure and demonstrate whether reduced fire resistance (e.g., 30 minutes instead of 60) is adequate where suppression systems significantly limit fire growth.
- This risk-based approach aligns with BS 7974 principles, allowing design flexibility while maintaining safety objectives.

15.7 Fire Detection and Suppression

Early detection and effective suppression are critical in atriums, where fire and smoke can spread rapidly across large volumes. Conventional systems may be less effective due to the height and openness of the space, requiring tailored solutions that combine advanced detection technology with suppression systems designed to control fire growth and maintain smoke control performance.

Fire Detection

- **Challenges of point detection**:
    - Traditional point smoke detectors can be unreliable in atriums because smoke tends to stratify at high levels, potentially never reaching the detector heads.
    - In tall atriums, buoyancy-driven smoke plumes may entrain cooler air, slowing their ascent or causing them to stall before reaching ceiling-mounted detectors. This delay can critically reduce available evacuation time.
- **Aspirating Smoke Detection (ASD)**:
    - ASD systems (e.g., VESDA) are widely used in atriums. They actively draw air samples through a network of pipes and continuously monitor for smoke particles.
    - These systems offer higher sensitivity and earlier warning than conventional detectors, enabling intervention before smoke conditions become untenable.
    - ASD can be concealed within architectural features, providing both functional and aesthetic benefits.
- **Beam Detectors**:
    - Optical beam detectors are another option, particularly suited for wide spans, concourses, or tall voids.
    - They project a beam across the atrium, detecting smoke when the light path is obscured.
    - Beam detectors are cost-effective for covering large areas, though they can be prone to false alarms from dust, reflections, or maintenance activities.

Fire Suppression

- **Sprinkler Systems**:
    - Automatic sprinklers are essential in most atrium fire strategies, significantly reducing heat release rates and smoke production.
    - By controlling fire size, sprinklers enhance the effectiveness of smoke control systems and extend the time available for evacuation.

  - In shopping centres, airports, and transport hubs, sprinklers are often a regulatory requirement and a key justification for extended travel distances.
- **Water Mist Systems**:
  - Water mist offers an alternative or supplementary approach, particularly effective in spaces with sensitive contents or where reduced water damage is beneficial.
  - Mist systems cool the flame, displace oxygen locally, and absorb radiant heat, reducing the potential for fire spread.
  - They are particularly valuable in **special risk areas**, such as electrical plant rooms opening onto atriums, or heritage atriums where water damage is a concern.
- **Deluge Systems**:
  - Deluge suppression may be specified for **high-hazard zones**, such as stages, performance areas, or exhibition stands within atriums.
  - These systems release water simultaneously across a large area, rapidly cooling and suppressing fire spread.
  - Due to their high water demand, deluge systems must be carefully integrated with building services and drainage provisions.

Integration with Smoke Control and Means of Escape

- Detection and suppression systems must be coordinated with **smoke extraction and evacuation strategies**.
- Early detection enables faster system activation, while suppression reduces fire size and smoke output, directly supporting tenable escape conditions.
- Fire engineering analysis should model the interaction of detection and suppression with smoke control, ensuring a holistic and reliable safety strategy.

15.8 Human Behaviour in Large Spaces

Human behaviour is a critical factor in fire safety design for atriums, malls, arenas, and other large public spaces. While engineering solutions address smoke, fire spread, and evacuation capacity, the success of these strategies ultimately depends on how people respond under stress. Behavioural studies have shown that real evacuation patterns often diverge from idealised assumptions, with significant implications for fire strategy.

Exit Choice and Wayfinding

- Occupants often default to the **entrance route they used to arrive**, even when alternative exits are closer or less congested.
- This tendency can lead to overloading of a few main exits, while others remain underused.
- Clear sightlines, illuminated signage, and intuitive layouts are essential to encourage the use of multiple exits.
- Staff and stewards play a critical role in redirecting occupants away from compromised or congested routes.

Role of Communication Systems

- Large digital displays, voice alarm systems, and public address announcements significantly improve compliance with evacuation instructions.
- Research shows that **specific, clear, and authoritative messages** ("Use the nearest exit to your left") are more effective than generic alarms or sirens.
- Multi-lingual and visual communication is especially important in airports, stadia, and international venues where language barriers may impede understanding.
- Integration of communication systems with **dynamic signage** allows evacuation messages to adapt in real time, guiding crowds away from smoke-affected areas.

Crowd Dynamics and Bottlenecks

- Panic is rare, but **crowding, hesitation, and congestion** are common behaviours in large open spaces.
- Bottlenecks often form at stair heads, escalators, and main entrances, where wide concourses converge into narrow escape routes.
- Crowd movement speeds decrease exponentially once density exceeds approximately **3–4 people per square metre**, dramatically slowing evacuation in high-occupancy venues.

- Behavioural modelling within evacuation simulations (e.g., using social force models) is essential to capture these effects and design effective strategies.

Implications for Fire Strategy

- Human factors must be integrated into fire engineering analysis, not treated as secondary to physical systems.
- Strategies to address behavioural risks include:
    - **Distributed exits** to reduce reliance on entry routes.
    - **Voice alarms and digital messaging** to provide unambiguous evacuation instructions.
    - **Staff training and drills**, ensuring personnel can intervene effectively to manage crowd flow.
    - **Evacuation modelling** that accounts for pre-movement times, group dynamics, and exit choice behaviour.
- Ultimately, a successful fire strategy in atriums and arenas combines **engineering controls** with **behavioural insights**, ensuring that systems are designed to accommodate how people actually respond in emergencies.

15.9 Case Examples

Real-world atrium fire strategies demonstrate how engineered solutions, supported by performance-based design, can address the unique challenges of large open spaces. The following case examples illustrate typical approaches.

Case 1: Airport Terminal Atrium

- **Geometry and risk profile**: A terminal atrium with a vertical height of 40 metres serves as a central circulation space, handling large numbers of transient passengers unfamiliar with the layout.
- **Smoke control**:
    - Mechanical smoke extract is provided by high-capacity fans rated at **60 m^3/s**, ensuring sufficient extraction to maintain a clear layer above the occupied zone.
    - Deployable **smoke curtains** form smoke reservoirs at ceiling level, directing smoke into extract points and preventing uncontrolled spill into circulation routes.

- **Means of escape**:
  - Travel distances of up to **60 m** are required due to the scale of the concourse.
  - These are justified by a combination of **automatic sprinklers** (to limit fire growth) and **evacuation modelling**, which demonstrates that occupants can reach safety within available safe egress time (ASET).
  - The design demonstrates that tenable conditions are maintained for at least **20 minutes**, providing sufficient margin for full evacuation.
- **Management systems**:
  - Voice alarm systems are integrated with multi-lingual announcements, essential in an international environment.
  - Staff training ensures effective passenger redirection, preventing congestion at the main entry routes.

Case 2: Shopping Mall Atrium

- **Geometry and risk profile**: A four-storey shopping mall incorporates a glazed central atrium, providing daylight and visual connectivity across multiple levels.
- **Smoke control**:
  - Natural roof vents provide buoyancy-driven smoke exhaust, sized in accordance with **BS EN 12101**.
  - Automatic **sprinkler protection** throughout the atrium and adjacent retail units limits smoke production, reducing reliance on mechanical extraction.
  - CFD analysis confirms that a tenable smoke layer can be maintained across the central atrium floor plate.
- **Means of escape**:
  - A **simultaneous evacuation** strategy is adopted for the atrium floor, ensuring all shoppers and visitors at ground level begin egress immediately.
  - A **phased evacuation** is implemented for upper-level retail units, supported by smoke curtains to prevent smoke spill into protected lobbies and stair enclosures.

- **Communication and wayfinding**:
    - Digital signage, dynamic directional indicators, and a voice alarm system guide occupants to the nearest exits.
    - This reduces reliance on entry routes and distributes occupants more evenly across multiple exit points.
- **Management systems**:
    - Retail staff are trained to act as evacuation marshals, assisting in the direction of occupants and prioritising vulnerable individuals.

15.10 Summary

Atriums and large open spaces are among the most visually impressive features of modern architecture, yet they present some of the most complex challenges in fire safety design. Their openness, height, and multi-storey connectivity mean that conventional prescriptive approaches are often inadequate. Instead, they demand **performance-based solutions** that integrate engineering, management, and human behavioural considerations.

Key principles for safe design include:

- **Robust smoke control systems**: Mechanical extraction, natural venting, or hybrid solutions — often supported by deployable smoke curtains and carefully managed make-up air — to maintain a tenable smoke-free layer for evacuation and firefighting.
- **Travel distance management**: Extended escape routes, frequently exceeding ADB limits, can only be justified through sprinklers, smoke modelling, and evacuation analysis that demonstrate occupants can reach safety within available safe egress time.
- **Early detection and effective suppression**: Aspirating smoke detection (ASD), beam detectors, and automatic sprinklers play a central role in controlling fire growth, reducing smoke production, and activating control systems in time to protect occupants.
- **Clear evacuation procedures**: Voice alarm systems, digital signage, and staff guidance are essential to overcome occupant unfamiliarity, reduce congestion at bottlenecks, and improve compliance with evacuation strategies.

- **Structural fire resilience**: Fire-resisting protection of exposed steel, glazing, and roof elements ensures that smoke control systems and primary escape routes remain operational throughout the evacuation period.

Atrium fire strategies are therefore not standalone measures but **integrated systems**, validated by computational modelling and underpinned by robust management procedures. Their success lies in balancing the architectural aspirations of openness and visibility with the uncompromising need for life safety.

The next chapter will address **Buildings with Sleeping Accommodation**, where the focus shifts from public congregation and crowd management to early detection, alarm audibility, and assisted evacuation — critical factors for protecting vulnerable occupants during fire events.

Chapter 16
Buildings with Sleeping Accommodation

16.1 Introduction

Sleeping accommodation presents distinctive challenges in fire safety that require different priorities from those in workplaces, shops, or public atriums. The fundamental issue is that occupants are often **asleep, vulnerable, or unfamiliar with their surroundings**, all of which increase the time required to detect and respond to fire. These factors make early warning, robust compartmentation, and effective evacuation procedures essential components of fire strategy.

Key risk factors include:

- **Delayed recognition of alarms**: Occupants who are asleep will not respond immediately to conventional alarm tones. Research shows that arousal thresholds vary widely, with some individuals — particularly children, the elderly, or those with hearing impairments — failing to respond at all without additional measures such as vibrating alarms or low-frequency sounders.
- **Vulnerability due to age, illness, or disability**: Residential care homes, hospitals, and sheltered housing accommodate occupants who may require assistance to escape. In such buildings, evacuation cannot be assumed to be independent and must be carefully managed through staff intervention, refuge spaces, or progressive horizontal evacuation.
- **Unfamiliarity with the building**: Guests in hotels, hostels, or boarding houses are unlikely to know the location of exits or escape routes. In a fire situation, they may hesitate or attempt to return to the entrance route, similar to behaviours seen in large public spaces.

Strategic priorities for sleeping accommodation:

- **Early detection**: Automatic detection throughout sleeping areas and escape routes is essential to provide the earliest possible warning.

- **Alarm audibility and communication**: Systems must ensure all occupants, including those with sensory impairments, can recognise and respond promptly to alarms.
- **Compartmentation**: Sleeping accommodation buildings must incorporate robust fire-resisting construction to contain fire and smoke, protecting escape routes and enabling phased evacuation where appropriate.
- **Assisted evacuation**: In care settings, hospitals, and high-dependency occupancies, fire strategy must address staff availability, evacuation aids, and refuge areas to safeguard residents who cannot evacuate unaided.

The unique life risk profile of sleeping accommodation requires fire engineers to place life safety above all other design objectives. Unlike commercial occupancies where phased evacuation or stay-put approaches may be acceptable, sleeping risk demands **immediate and reliable warning, protection of escape routes, and contingency for vulnerable occupants**.

16.2 Regulatory Framework

Sleeping accommodation is subject to some of the most stringent fire safety requirements due to the heightened life risk associated with sleeping occupants. A range of statutory and best-practice guidance documents apply, covering building design, detection and alarm provision, evacuation planning, and management procedures. The key elements of the regulatory framework are outlined below.

Approved Document B (ADB)

- ADB sets the prescriptive baseline for fire safety in England, with specific provisions for residential, hotel, and institutional occupancies.
- Requirements address compartmentation, means of escape, fire resistance, and alarm audibility.
- ADB recognises that sleeping occupancies require more robust levels of protection than workplaces or shops, mandating:
    - **Automatic detection** in circulation spaces.
    - **Protected staircases** to provide safe escape routes.
    - **Increased fire resistance** for separating elements to safeguard sleeping occupants.

- For complex or high-rise sleeping buildings, ADB often requires enhancement through fire engineering solutions or reference to supplementary standards.

BS 9991: Fire Safety in the Design, Management and Use of Residential Buildings

- BS 9991 is the principal UK standard providing a comprehensive framework for residential fire safety.
- It builds on ADB but introduces a **risk-based approach**, considering building height, layout, and occupancy profile.
- Key features include:
    - Extended travel distances where supported by sprinklers or alternative measures.
    - Guidance on single-stair residential buildings.
    - Specific provisions for evacuation lifts, refuges, and stay-put strategies in apartment blocks.
- For sleeping occupancies, BS 9991 is often the primary reference document for performance-based solutions.

BS 5839-1 and BS 5839-6: Fire Detection and Alarm Systems

- **BS 5839-1**: Applies to commercial sleeping occupancies such as hotels, hostels, student accommodation, and care homes. It sets out requirements for system design, audibility, and monitoring.
- **BS 5839-6**: Applies to domestic dwellings, including houses, flats, and HMOs. It defines system grades (e.g., A to F) and categories of detection coverage (LD1, LD2, LD3).
- These standards ensure alarms are both **audible and reliable**, accounting for the reduced responsiveness of sleeping occupants. Provisions include low-frequency sounders, visual alarms for the hearing impaired, and monitoring for staff response in managed settings.

Healthcare and Care Settings Guidance

- The **Care Quality Commission (CQC)** requires providers to demonstrate compliance with fire safety standards in hospitals, care homes, and supported living environments.

- **NHS Health Technical Memoranda (HTMs)** and **Health Building Notes (HBNs)** provide specific fire safety requirements for healthcare facilities, including compartmentation, progressive horizontal evacuation, and staffing levels to support assisted evacuation.
- In these occupancies, fire strategy is inseparable from operational procedures, with reliance on trained staff to manage vulnerable residents.

Local Authority Licensing and Housing Standards

- Local authorities impose additional controls on **HMOs (Houses in Multiple Occupation)**, hostels, and student housing.
- Licensing conditions typically require enhanced detection and alarm systems, emergency lighting, and stricter management arrangements to reflect the risks of transient, high-density, or vulnerable occupants.
- Fire safety orders may also impose specific requirements following risk assessment under the **Regulatory Reform (Fire Safety) Order 2005**.

Summary

ADB establishes the prescriptive foundation.

BS 9991 provides detailed guidance for residential fire safety.

BS 5839 Parts 1 and 6 govern detection and alarm system design.

CQC and NHS guidance regulate healthcare and care occupancies.

Local authority licensing strengthens standards for HMOs, hostels, and student housing.

Together, these documents form a **multi-layered regulatory framework** that reflects the high life-risk profile of sleeping accommodation, ensuring that design, detection, and management are all robustly addressed.

16.3 Key Fire Risks in Sleeping Accommodation

Sleeping accommodation presents a higher life risk than most other building types because occupants are less able to detect, respond to, and escape from fire unaided. Fires in these environments frequently escalate rapidly due to the combination of ignition sources, combustible contents, and delayed occupant response. The following factors are particularly critical.

Ignition Sources

- **Smoking materials** remain a leading cause of fatal fires in sleeping occupancies, particularly where residents smoke in bedrooms or conceal smoking in prohibited settings such as care homes or hostels.
- **Cooking appliances**, especially in hotels, hostels, and HMOs, introduce risks of unattended cooking, grease fires, and electrical faults.
- **Portable heaters and electrical appliances** (e.g., fan heaters, overloaded extension leads, or defective chargers) are common ignition sources in bedrooms, often positioned close to combustible bedding or curtains.

Combustible Furnishings and Bedding

- Bedrooms and dormitories typically contain high loads of **soft furnishings, bedding, and upholstered furniture**, all of which can contribute to rapid fire growth and toxic smoke production.
- Fires in these contents often develop faster than those in office environments, with **flashover possible within minutes** if suppression or early detection is absent.
- Synthetic fabrics, foams, and plastics further increase the production of toxic gases such as carbon monoxide and hydrogen cyanide, posing life-threatening hazards to sleeping occupants.

Delayed Occupant Response

- Occupants who are asleep may take significantly longer to respond to alarms than awake individuals.
- Intoxication (alcohol or drugs), illness, or medication can impair perception and mobility, delaying escape further.
- Research shows that children, older adults, and those with sensory impairments may not wake to conventional alarms, necessitating the use of **low-frequency sounders, visual alarms, or vibrating devices** in certain contexts.

High Occupant Density

- In **hostels, student housing, and HMOs**, sleeping accommodation is often provided at high density, with multiple occupants per room or shared facilities.

- This increases the risk of ignition sources being introduced and the likelihood of fire spreading rapidly between occupants.
- High density also creates challenges for evacuation, as staircases and corridors may quickly become congested.

Dependency on Staff in Managed Settings

- In **care homes, hospitals, and other institutional occupancies**, many residents may be unable to self-evacuate.
- The fire strategy must therefore assume dependency on staff for assisted evacuation, often requiring **progressive horizontal evacuation, refuge areas, and evacuation aids** such as ski sheets or evacuation chairs.
- Staffing levels, training, and night-time procedures are critical determinants of safety, as fire development times may outpace the ability of staff to move large numbers of residents without robust planning.

Summary

The principal fire risks in sleeping accommodation arise from a combination of **ignition sources**, **fuel load**, and **delayed or impaired occupant response**. These are exacerbated in high-density or managed occupancies where evacuation may be slow or staff-dependent. Fire strategies must therefore prioritise **early detection, effective alarm systems, suppression, and robust compartmentation** to mitigate these elevated risks.

16.4 Early Detection and Warning

In sleeping accommodation, where occupants are often unconscious, impaired, or otherwise slow to respond, **early detection and clear alarm warning are paramount**. Unlike workplaces or public assembly spaces, where individuals are alert and mobile, sleeping occupants require additional time and more robust warning systems to initiate safe evacuation.

Detection Systems

- **Comprehensive coverage** is essential, with smoke detectors provided in all **bedrooms, circulation corridors, stairways, and common areas**. Early detection within sleeping rooms ensures that fires are identified before they spread to escape routes.

- **Heat detection** should be used in kitchens, plant rooms, boiler houses, and similar areas where smoke detection would generate unwanted alarms.
- In **high-risk or high-dependency occupancies** such as care homes and hospitals, **Aspirating Smoke Detection (ASD)** systems may be employed. These continuously sample air for smoke particles, offering earlier warning than standard point detectors and enabling staff intervention before conditions become critical.
- All systems should be designed and installed in accordance with **BS 5839-1** (non-domestic buildings) or **BS 5839-6** (domestic premises such as flats and HMOs).

Alarm Systems

- Alarm audibility is critical. In bedrooms, sounders must achieve **at least 75 dB (A) at the bedhead**, ensuring that occupants are woken promptly, even with doors closed.
- In accessible bedrooms, **visual alarm devices (VADs)** such as flashing beacons must supplement audible alarms to ensure inclusion for occupants with hearing impairments.
- For large or complex occupancies (e.g., hotels, hostels, student residences), **voice alarm systems** provide clear, directive instructions that improve compliance and reduce hesitation compared to simple tone alarms. Messages should be concise, multi-lingual where appropriate, and tailored to the building's evacuation strategy.
- Where phased evacuation is adopted (e.g., in tall hotels), the alarm system must support zoned activation, ensuring that only the relevant areas are evacuated while others remain protected by compartmentation.

Staff Alerts

- In **care homes and hospitals**, general alarm sounding may be inappropriate, as many residents require staff-assisted evacuation.
- Instead, systems may trigger **staff pagers, electronic staff call systems, or silent alarms** that notify staff without unnecessarily alarming vulnerable occupants.
- This enables staff to implement **progressive horizontal evacuation (PHE)** or staged movement of residents in a calm and controlled manner.

- Such systems must be rigorously maintained and supported by **staff training and rehearsed procedures**, ensuring that staff respond immediately and consistently.

Integration with Management Procedures

- Early detection and alarm systems only succeed if coupled with effective management. Fire drills, staff training, and clear procedures are vital in ensuring that alarms lead to prompt and coordinated action.
- System design should also consider **night-time conditions**, where staffing levels are lower and reliance on automated systems becomes greater.

Summary

Early detection and effective alarm design form the **first line of defence** in sleeping accommodation. By combining comprehensive detector coverage, robust alarm audibility, inclusive warning for all occupants, and staff alert systems in care environments, fire strategies can significantly reduce the life risk posed by delayed occupant response.

16.5 Compartmentation and Fire Resistance

Robust compartmentation is the cornerstone of fire safety in sleeping accommodation. Because occupants are likely to be asleep, vulnerable, or dependent on staff, immediate evacuation is not always possible. Instead, the fire strategy must ensure that fire and smoke spread are contained long enough to allow safe escape or staff-assisted movement. Structural fire protection is therefore fundamental to life safety in these settings.

Bedroom Separation

- All bedrooms must be separated from corridors and escape routes by **fire-resisting doors**, typically **FD30S** (30 minutes fire resistance with smoke seals).
- These doors prevent early smoke leakage into corridors, protecting escape routes during the crucial initial stages of a fire.
- In higher-risk occupancies (such as care homes or high-rise hotels), **FD60S** doors may be specified to provide additional resilience.

Protected Corridors and Escape Routes

- Corridors serving bedrooms must be designed as **protected escape routes**, enclosed by fire-resisting construction.
- Ventilation ducts, service penetrations, and recessed storage must be avoided or fitted with fire-stopping and dampers to maintain integrity.
- Where corridors serve multiple sleeping rooms, they should connect directly to protected staircases, which form the primary escape routes from upper levels.

Compartment Floors and Walls

- Larger sleeping buildings, such as hotels, hostels, or hospitals, require subdivision into **fire compartments**.
- Compartment floors and walls should typically provide **60 minutes of fire resistance**, though higher ratings (90–120 minutes) may be necessary in high-rise or complex layouts.
- These barriers prevent fire spreading unchecked through multiple storeys or across large wings of a building, ensuring that evacuation can proceed safely from unaffected areas.

Sub-Compartmentation for Progressive Horizontal Evacuation

- In **care homes, hospitals, and other institutional occupancies**, residents may not be able to evacuate immediately.
- To address this, **sub-compartmentation** is introduced, dividing each floor into smaller fire-resisting zones.
- This enables **progressive horizontal evacuation (PHE)**, where residents are moved in stages from one sub-compartment to the next, remaining in relative safety until full evacuation is possible.
- Fire resistance of at least **30 minutes** is generally required for sub-compartment walls, with higher ratings where dependency levels are greatest.

Cavity Barriers and Concealed Spaces

- Modern sleeping accommodation buildings often include complex roof and wall constructions with concealed cavities.
- **Cavity barriers** are required to prevent unseen fire spread within roof voids, service risers, or cladding systems.

- This is particularly critical in the wake of high-profile cladding fires, where uncontrolled fire spread through concealed routes has caused catastrophic consequences.
- Regular inspection and maintenance of cavity barriers is essential to ensure ongoing effectiveness.

Integration with Evacuation Strategy

- Compartmentation and fire resistance are not isolated measures; they directly support evacuation strategies.
- By containing fire to the room or sub-compartment of origin, they buy crucial time for detection, alarm, staff response, and occupant movement.
- In residential care settings, compartmentation is the enabler of safe staff-assisted evacuation, often making the difference between survivable and fatal outcomes.

Summary

Compartmentation in sleeping accommodation provides **time and protection** where occupants cannot respond instantly to fire. From FD30S bedroom doors to 60-minute compartment floors and progressive horizontal evacuation in care facilities, structural fire resistance forms the backbone of a safe fire strategy, ensuring that escape routes remain viable until all residents are moved to safety.

16.6 Evacuation Strategies

The choice of evacuation strategy for sleeping accommodation depends on the **building type, occupancy profile, and level of dependency of residents**. Unlike workplaces or public buildings, where occupants are generally alert and mobile, sleeping occupancies often require tailored approaches that account for delayed response, vulnerability, and staff involvement.

Simultaneous Evacuation

- In small premises such as **guest houses, hostels, and small student halls**, a **simultaneous evacuation** strategy is normally adopted.
- On activation of the fire alarm, all occupants are expected to evacuate immediately and make their way to a place of safety.

- This approach is only appropriate where:
    - The number of occupants is relatively low.
    - Escape routes are short and direct.
    - Staff can supervise and assist evacuation quickly if required.

Phased Evacuation

- In larger sleeping buildings, such as **hotels or large student accommodation blocks**, **phased evacuation** may be used.
- Occupants on the **fire floor and immediately adjacent floors evacuate first**, while those on more remote levels await further instruction.
- This reduces congestion in staircases and ensures orderly movement, particularly in high-rise buildings with large populations.
- Phased evacuation requires:
    - **Robust compartmentation** to ensure the safety of occupants awaiting evacuation.
    - **Reliable communication systems**, typically voice alarms, to provide clear instructions and maintain occupant confidence.

Stay-Put Policy

- In some **purpose-built residential blocks**, a **stay-put strategy** is adopted, allowing occupants in unaffected flats to remain safely in place while only the fire-affected flat is evacuated.
- This strategy relies heavily on:
    - **High-quality compartmentation** between flats.
    - **Fire-resisting construction** to contain fire and smoke within the flat of origin.
- While effective in theory, public confidence in stay-put policies has been undermined by high-profile incidents (e.g., Grenfell Tower), making **robust management, resident communication, and reassurance** essential.

Progressive Horizontal Evacuation (PHE)

- In **care homes, hospitals, and healthcare facilities**, occupants often cannot self-evacuate and may depend entirely on staff assistance.

- **Progressive horizontal evacuation (PHE)** is the standard approach, moving residents in stages:
    - From the fire-affected compartment into an **adjacent sub-compartment** protected by fire-resisting construction.
    - If conditions deteriorate, residents are moved progressively further until they reach a final exit.
- PHE reduces reliance on vertical evacuation, which is impractical for residents in beds or wheelchairs.
- Sub-compartmentation, staff training, and evacuation aids are critical to its success.

Assisted Evacuation

- In care environments and some residential institutions, **staff-assisted evacuation** is a core requirement.
- Staff must be trained and regularly drilled in evacuation procedures, ensuring that they can safely move residents under time pressure.
- **Evacuation aids** such as chairs, sleds, or beds fitted with evacuation sheets enable staff to move immobile residents quickly and with minimal manpower.
- Adequate night staffing levels are particularly important, as fire incidents often occur when staff numbers are lowest.

Summary

Evacuation strategies in sleeping accommodation must be carefully matched to the building's risk profile: **simultaneous evacuation** for small premises, **phased evacuation** for large or high-rise buildings, **stay-put policies** where compartmentation is reliable, and **progressive horizontal evacuation** for healthcare and care settings. All approaches require **robust management, clear communication, and trained staff** to ensure that delayed or assisted occupants are moved to safety in time.

16.7 Refuge Areas and Special Provisions

In sleeping accommodation, particularly where vulnerable or mobility-impaired occupants are present, evacuation cannot always be immediate or independent. Fire strategies must therefore incorporate **refuge spaces and specialist provisions** to ensure all residents can reach safety, even where assisted or staged evacuation is required.

Refuge Spaces

- Refuges are **protected areas**, typically located within stair lobbies, protected corridors, or dedicated compartments, where occupants can safely await staff assistance or use of evacuation lifts.
- They must provide at least **30 minutes fire resistance** (though 60 minutes is often recommended in high-risk occupancies such as care homes or hospitals).
- Refuges should be large enough to accommodate a wheelchair user without obstructing circulation routes, typically requiring a **minimum clear space of 900 mm × 1400 mm**.
- In progressive horizontal evacuation strategies, refuges form part of the **sub-compartmented zones**, allowing phased movement of dependent occupants.

Two-Way Communication Systems

- Each refuge should be fitted with a **two-way intercom system**, linking directly to the fire control room, reception desk, or other staffed location.
- These systems enable occupants (or staff assisting them) to confirm their presence and receive instructions, providing reassurance during evacuation.
- Communication systems must be robust, **fire-protected, and backed up by emergency power supplies** to remain operational throughout an incident.
- Their design should comply with **BS 5839-9**, which governs emergency voice communication systems, including refuge intercoms.

Disabled Evacuation Lifts

- While conventional passenger lifts are not permitted for evacuation during fire, **dedicated evacuation lifts** may be provided where designed to the appropriate standards.
- These must comply with **BS 9999** and **BS EN 81-70**, ensuring:
  - Location within a **fire-resisting shaft**, protected for at least the same duration as the escape routes it serves.
  - **Independent or backup power supply**, ensuring continued operation in case of mains failure.

  - **Direct communication** between the lift car and the fire control point.
  - Signage and procedures that ensure the lift is used safely and only as part of an agreed evacuation plan.
- Evacuation lifts are especially valuable in hospitals, care homes, and high-rise hotels where occupants cannot be moved down staircases quickly or safely.

Integration with Evacuation Planning

- Refuge spaces, communication systems, and evacuation lifts must be embedded within a **holistic evacuation strategy**.
- Their effectiveness depends on **trained staff**, clear procedures, and regular drills, ensuring that vulnerable occupants can be moved in a calm and controlled manner.
- Fire strategies must also account for **night-time conditions**, when staff numbers may be reduced, by ensuring that refuges provide sufficient protection time for delayed evacuation.

Summary

Refuge areas and special provisions are essential safeguards in sleeping accommodation where immediate evacuation is not possible. **Protected refuges, reliable two-way communication, and purpose-built evacuation lifts** provide the infrastructure necessary for safe assisted evacuation, while ensuring compliance with modern standards of accessibility and inclusivity.

16.8 Active Fire Protection

Active fire protection systems are critical in sleeping accommodation, where delayed response and occupant vulnerability significantly increase life risk. Detection and alarm systems initiate the response, but suppression, emergency lighting, and other active measures are essential to contain fire, protect escape routes, and support evacuation under adverse conditions.

Sprinklers

- Sprinkler protection has proven to be one of the most effective life safety measures in sleeping occupancies, both for reducing fire growth and limiting smoke production.

- In **England**, sprinklers are **mandatory in all new residential buildings over 11 m in height** and in **all new care homes**, regardless of height.
- Residential systems are designed in accordance with **BS 9251**, while larger and more complex occupancies (e.g., high-rise hotels) may adopt **BS EN 12845** standards for commercial sprinkler systems.
- Benefits of sprinkler systems include:
    - Rapid fire control or suppression, often extinguishing fires before the fire service arrives.
    - Reduced heat release rate, thereby enhancing the performance of smoke control systems and protecting structural integrity.
    - Demonstrable justification for extended travel distances or reduced compartment sizes under performance-based design.

Suppression in High-Risk Areas

- Specific fire risks in sleeping accommodation require targeted suppression measures:
    - **Kitchens**: Deep-fat fryers and catering facilities should be protected by **wet chemical suppression systems**, capable of controlling oil-based fires that water cannot extinguish.
    - **Laundry facilities**: Tumble dryers, often implicated in residential fires, can be protected by sprinklers or dedicated suppression systems designed to mitigate ignition of lint or overheating equipment.
    - **Plant and service rooms**: Depending on the building, additional protection such as water mist or clean agent suppression may be justified, particularly where electrical risks are significant.

Emergency Lighting

- For sleeping occupants suddenly awoken at night, reliable emergency lighting is essential to avoid disorientation and ensure safe evacuation.
- **Corridors, staircases, and refuge areas** must be fitted with emergency lighting designed to **BS 5266-1**, providing:
    - Sufficient illuminance for safe movement under mains failure conditions.

   - Coverage of escape route signs, changes in direction, and final exits.
   - A minimum duration of **three hours**, sufficient for prolonged evacuation or delayed staff-assisted movement in care settings.
- In larger premises such as hotels or student halls, illuminated escape route signage must also be supported by dynamic or adaptive systems where smoke conditions could compromise visibility.

Integration with Overall Fire Strategy

- Active systems such as sprinklers, suppression, and emergency lighting are not standalone provisions; they directly support evacuation and structural fire protection.
- Together, they ensure that occupants are alerted promptly, escape routes remain tenable, and suppression limits fire growth to allow safe evacuation or relocation of vulnerable residents.
- Regular maintenance, testing, and integration with the building's fire safety management plan are essential to ensure reliability during an incident.

Summary

Active fire protection systems provide **frontline defence** in sleeping accommodation. **Sprinklers** deliver rapid suppression, **targeted suppression systems** protect high-risk areas such as kitchens and laundries, and **emergency lighting** ensures occupants can safely navigate escape routes in darkness. Collectively, these measures bridge the gap between detection and evacuation, underpinning a comprehensive fire safety strategy.

16.9 Fire Safety Management

The success of any fire strategy in sleeping accommodation depends not only on structural and engineered measures but also on **effective fire safety management**. Because occupants may be asleep, vulnerable, or unfamiliar with the building, the actions of staff and the reliability of management systems are critical to ensuring safe evacuation. Ongoing training, communication, and maintenance routines form the backbone of operational fire safety.

Staff Training

- Staff in hotels, care homes, hospitals, and student residences must receive **regular training** in fire safety procedures.
- Training should cover:
    - Operation of **detection and alarm systems**.
    - Familiarity with **evacuation equipment** such as evacuation chairs, sleds, or bed-mounted evacuation sheets.
    - Procedures for **assisted evacuation**, including the safe handling of residents with limited mobility, medical dependencies, or cognitive impairments.
    - Roles and responsibilities during evacuation, ensuring all staff understand their tasks in both day and night conditions.
- **Regular drills** should be carried out, reflecting realistic fire scenarios and including night-time staffing levels where appropriate.

Resident Communication

- Occupants unfamiliar with the building require clear instructions to act effectively in an emergency.
- **Fire action notices** should be prominently displayed in bedrooms, corridors, and communal areas, providing simple guidance on evacuation procedures.
- In student halls and long-term residential settings, **induction sessions** should explain alarm signals, evacuation routes, and assembly points at the start of each residency period.
- In care settings, communication must be tailored to residents' abilities, with emphasis on reassuring and supporting vulnerable individuals during both drills and actual incidents.

Routine Checks and Maintenance

- Daily inspection of **escape routes** is essential to ensure that corridors and stairways remain clear of obstructions such as furniture, equipment, or stored materials.
- Fire doors should be checked regularly to ensure they remain **closed, unblocked, and functional**, with self-closing devices in good working order.

- **Weekly alarm tests** must be conducted in accordance with **BS 5839**, ensuring sounders are audible in all areas and that staff understand their roles in responding to activations.
- **Monthly emergency lighting tests** should confirm that luminaires operate correctly under mains failure, as required by **BS 5266-1**.
- Records of all tests, inspections, and maintenance activities should be kept in a **fire safety logbook** for audit and compliance purposes.

Integration with Management Systems

- Fire safety management should form part of the building's **overall safety culture**, not just a compliance exercise.
- Responsible persons under the **Regulatory Reform (Fire Safety) Order 2005** must ensure that risk assessments are reviewed regularly, particularly following building alterations or changes in occupancy profile.
- In healthcare and care settings, fire safety procedures should be fully integrated into **staffing policies and care planning**, recognising the dependency of residents.

Summary

Fire safety management in sleeping accommodation relies on **trained staff, informed residents, and reliable maintenance routines**. Daily vigilance, regular drills, and clear communication complement the building's physical fire protection measures, ensuring that strategies function effectively when required. In sleeping occupancies, management is often the decisive factor in whether occupants are evacuated safely and without loss of life.

16.10 Case Examples

Real-world applications of fire safety principles in sleeping accommodation demonstrate how regulatory requirements, engineered systems, and management practices combine to protect occupants. The following examples illustrate contrasting strategies for hotels and care homes.

Case 1: Hotel

- **Building profile**: A 12-storey city-centre hotel accommodating several hundred guests, many of whom are transient and unfamiliar with the layout.
- **Evacuation strategy**: A **phased evacuation** approach is adopted. On activation of the alarm, the fire floor and the floors immediately above and below are evacuated first. Remaining floors await instruction, relying on robust compartmentation to maintain safety until evacuation is initiated.
- **Detection and alarm**: Each bedroom is fitted with a smoke detector, linked to a central addressable fire alarm panel. The system is supported by a **voice alarm system**, broadcasting clear instructions to guests in multiple languages. This reduces hesitation and prevents guests from defaulting to the entry route.
- **Compartmentation**:
    - **Compartment floors** provide **60 minutes fire resistance**, limiting vertical fire spread between levels.
    - **Stair cores** are enclosed in **120-minute fire-resisting construction**, ensuring protected escape routes for the duration of evacuation.
- **Staff provisions**: Staff are trained to manage evacuation, including assisting mobility-impaired guests. **Refuge areas** are provided in protected lobbies adjacent to stair cores, each equipped with two-way communication systems for coordination with the fire control point.
- **Outcome**: This layered strategy ensures that large numbers of guests can be evacuated in an orderly manner without overwhelming stair capacity, while maintaining life safety for those awaiting evacuation.

Case 2: Care Home

- **Building profile**: A single-storey care home housing elderly residents, many of whom require assisted evacuation.
- **Evacuation strategy**: The building is divided into four **30-minute sub-compartments**, enabling **progressive horizontal evacuation** (PHE). Residents can be moved in stages from the fire-affected compartment into an adjacent safe area without reliance on vertical movement.

- **Suppression and detection**:
    - **Sprinklers** are installed throughout in accordance with **BS 9251**, ensuring rapid fire suppression and limiting smoke production.
    - A comprehensive **smoke detection system** provides early warning, with detectors in bedrooms, corridors, and common areas.
- **Staff training and equipment**: Staff receive **quarterly fire safety training** covering evacuation procedures, alarm response, and use of evacuation aids such as sleds and sheets. Training is tailored to night-time staffing levels to reflect the most challenging conditions.
- **Outcome**: The combination of sub-compartmentation, suppression, and trained staff ensures that vulnerable residents can be relocated to safety in a controlled manner, with sufficient time margins even in the event of a rapidly developing fire.

Summary

These examples demonstrate the adaptability of fire strategies for different types of sleeping accommodation:

- **Hotels** prioritise phased evacuation, voice alarm systems, and robust compartmentation to manage large numbers of transient occupants.
- **Care homes** rely on progressive horizontal evacuation, suppression, and staff training to safeguard residents with limited mobility or dependency.

Both approaches illustrate how **engineering solutions and management systems** must be integrated to address the specific risks of sleeping environments.

16.11 Summary

Sleeping accommodation represents one of the **highest life-risk categories** in fire safety due to the delayed response of occupants, their vulnerability, and their unfamiliarity with building layouts. Fires in such settings have historically resulted in disproportionate loss of life, underscoring the need for a **holistic, performance-based fire strategy** that integrates engineering controls with management practices.

Key principles include:

- **Early detection and warning systems**: Comprehensive coverage with smoke detection in bedrooms, circulation routes, and common areas; heat detection in kitchens; and tailored alarm solutions such as low-frequency sounders, voice alarms, or staff alert systems in care environments.
- **Robust compartmentation**: Fire-resisting construction, from FD30S bedroom doors to 60-minute (or greater) compartment walls and floors, ensures fires are contained and escape routes remain viable. Sub-compartmentation is particularly vital in care settings to enable progressive horizontal evacuation.
- **Tailored evacuation strategies**: From simultaneous evacuation in small hostels, to phased evacuation in large hotels, stay-put approaches in residential blocks, and progressive horizontal evacuation in care homes and hospitals — strategies must reflect occupancy type and dependency levels.
- **Sprinklers and suppression systems**: Mandatory in many sleeping occupancies, these control fire growth, limit smoke production, and protect structural integrity, supporting extended travel distances and phased evacuation strategies.
- **Fire safety management and staff training**: The effectiveness of all engineered measures depends on trained staff, clear resident communication, routine checks, and regular drills — particularly in care homes and institutions where assisted evacuation is essential.

Together, these measures create a layered defence that ensures sleeping occupancies are afforded the **earliest possible warning, the longest possible time for escape, and the highest possible chance of survival**.

The next chapter will address **Assembly and Recreation Buildings**, where the challenges shift towards managing **large crowds, occupant unfamiliarity with exits, and the dynamics of crowd behaviour under stress**.

Chapter 17
Assembly and Recreation Buildings (Schools, Cinemas, Theatres, Arenas)

17.1 Introduction

Assembly and recreation buildings encompass a wide variety of occupancies where large numbers of people gather for education, entertainment, leisure, or worship. Examples include:

- **Educational premises** such as schools, colleges, and universities.
- **Entertainment venues** including cinemas, theatres, and concert halls.
- **Sports arenas and stadia**, accommodating tens of thousands of spectators.
- **Places of assembly** such as churches, mosques, community centres, and multi-purpose halls.

These buildings share a common feature: **large and diverse populations concentrated in relatively confined spaces**, often for prolonged periods. Unlike residential or workplace settings, the occupants may include children, elderly persons, visitors unfamiliar with the premises, or crowds in heightened emotional states due to events, performances, or sporting contests.

Fire safety challenges in assembly and recreation buildings include:

- **High occupant density**: Crowds can lead to rapid congestion at exits, staircases, and concourses, with evacuation times greatly influenced by crowd dynamics.
- **Occupant unfamiliarity**: Visitors may not know the location of exits, and research shows a tendency to return to familiar entry routes rather than dispersing to alternative exits.

- **Varied behavioural responses**: While panic is rare, hesitation, group behaviour, and crowd surges under pressure can create serious risks if not managed.
- **Complex evacuation routes**: Buildings such as stadiums and theatres often rely on tiered seating, concourses, and vomitories, requiring carefully coordinated evacuation strategies.
- **Special uses and functions**: Assembly spaces may host performances, religious services, or examinations, where occupants may initially resist evacuation until prompted by staff or alarms.

Strategic priorities for fire safety in these buildings are therefore:

- To provide **sufficient exit capacity** for the maximum anticipated occupancy, with well-distributed exits to avoid bottlenecks.
- To incorporate **voice alarm and communication systems** capable of directing large numbers of people with clarity and authority.
- To design **robust smoke control and compartmentation**, particularly in multi-level or enclosed venues.
- To ensure **staff training and crowd management procedures** are in place, recognising that staff play a decisive role in directing safe evacuation.

Assembly and recreation buildings thus require a **dual focus**: managing the **technical risks** of fire and smoke spread while also addressing the **behavioural dynamics** of large, unfamiliar crowds under stress. The following sections examine the key risks, regulatory framework, and engineered solutions that underpin safe design in these challenging occupancies.

17.2 Key Risks in Assembly and Recreation Buildings

Assembly and recreation buildings present a distinctive fire risk profile arising from large, diverse populations, complex layouts, and specialised functions. Unlike workplaces or residential settings, occupants are typically gathered in high numbers, may be unfamiliar with the building, and are influenced strongly by crowd behaviour. These factors can transform even a relatively small fire into a significant life safety emergency if not properly managed.

High Occupant Loads

- Theatres, sports arenas, and concert venues frequently accommodate **thousands of people simultaneously**.
- Evacuation must be planned for peak occupancy, not average use, with exit capacity and circulation routes sized accordingly.
- Inadequate provision can lead to dangerous congestion and extended evacuation times, particularly where events conclude and large numbers attempt to leave simultaneously.

Unfamiliarity with Layouts

- The majority of visitors are **first-time or occasional users**, meaning they are often unaware of exit locations and evacuation routes.
- Studies show that occupants frequently attempt to **exit via the entry point**, even when nearer or safer exits exist.
- Without effective wayfinding, signage, and staff guidance, unfamiliarity can lead to serious delays and imbalances in crowd flow.

Crowd Dynamics

- Large groups moving under pressure behave differently from individuals. **Bottlenecks** typically occur at doorways, stair heads, vomitories, or narrow concourses.
- Once density exceeds about **3–4 persons per square metre**, crowd movement slows dramatically, increasing exposure time to smoke and heat.
- Crowd surges or uneven flows can cause trip and fall hazards, particularly for children, elderly spectators, or those with mobility limitations.

Special Hazards

- Entertainment and assembly spaces often include **unique ignition and fuel risks**:
  - **Stage effects** such as pyrotechnics, lighting rigs, and sound systems.
  - **Electrical distribution** for performances and exhibitions.
  - **Commercial kitchens and concessions** in arenas or multi-purpose halls.

    - **Combustible decorations, drapes, or props**, which may not always meet fire-retardant standards.
- These can produce **rapid flame spread** or **dense smoke conditions**, overwhelming escape strategies if not adequately controlled.

Multiple Age Groups and Vulnerable Occupants

- Occupant profiles vary by building type:
    - **Schools and universities**: Children and young adults may require supervision and guidance.
    - **Sports arenas and places of worship**: Elderly or mobility-impaired spectators may be present in large numbers.
    - Fire strategies must therefore account for **slower movement speeds** and the potential need for **assisted evacuation** of vulnerable groups.

Potential for Panic and Disorder

- While true panic is rare, **miscommunication, delayed alarms, or lack of crowd control** can trigger anxiety, disorder, or stampedes.
- The risk is heightened in emotionally charged environments such as sporting events, concerts, or festivals.
- Clear voice communication, staff leadership, and visible exit routes are essential to maintaining order and ensuring safe, controlled evacuation.

Summary

The key risks in assembly and recreation buildings arise from a combination of **large occupant numbers, unfamiliarity with layouts, complex crowd dynamics, and special hazards** unique to performance and gathering spaces. Fire strategies must therefore combine **robust engineering measures** with **crowd management and communication systems** to ensure safe evacuation under pressure.

17.3 Regulatory Framework

Assembly and recreation buildings fall under a wide range of prescriptive and performance-based regulations, reflecting the high life-risk posed by large, often unfamiliar, occupant groups. Fire strategies for these occupancies must address both **statutory building requirements** and **sector-specific guidance**, including education, entertainment, and sports facilities.

Approved Document B (ADB)

- ADB sets out **baseline fire safety provisions** for assembly and recreation occupancies in England, including schools, theatres, cinemas, and places of worship.
- Requirements address:
    - **Compartmentation** to limit fire spread.
    - **Travel distances** and exit capacity relative to maximum occupancy.
    - **Fire resistance ratings** for structural elements.
    - **Detection and alarm systems**, with emphasis on audibility in large spaces.
- While ADB provides prescriptive benchmarks, its provisions may be restrictive in large or complex venues such as arenas and stadia, where performance-based justification is often required.

BS 9999: Code of Practice for Fire Safety in the Design, Management and Use of Buildings

- BS 9999 provides a **risk-based, performance-oriented framework** for assembly occupancies, offering flexibility in travel distances, exit provisions, and compartmentation.
- It allows designers to vary prescriptive requirements where compensatory measures such as sprinklers, smoke control, or robust management systems are in place.
- For large entertainment venues or stadia, BS 9999 is frequently used alongside ADB to justify engineered solutions that meet life safety objectives without compromising architectural intent.

Building Bulletins – Education Premises

- **BB 100 (Design for Fire Safety in Schools)** is the principal guidance for fire safety in educational buildings.
- It sets requirements for:
    - **Means of escape** for pupils and staff.
    - **Compartmentation and protected routes** to reflect pupil dependency.
    - **Special risks** such as science labs, design/technology workshops, and assembly halls.

- For universities and higher education buildings, fire safety generally follows ADB and BS 9999, though large lecture theatres may require additional risk-based justification.

Licensing Requirements for Entertainment Venues

- Theatres, cinemas, nightclubs, and concert halls are often subject to **local authority licensing**, which may impose specific fire safety conditions in addition to building regulations.
- Licensing authorities typically require:
    - Sufficient exit capacity and distribution.
    - Voice alarm systems.
    - Staff training and evacuation procedures.
    - Maintenance of fire safety systems as part of continued operation.
- Compliance is generally enforced through regular inspections and licence renewals.

Sports Grounds Safety Authority (SGSA)

- For sports stadia and arenas, fire safety sits within the broader remit of the **SGSA Green Guide** (*Guide to Safety at Sports Grounds*).
- The Green Guide provides detailed recommendations for:
    - **Safe capacities and exit design** based on crowd modelling.
    - **Fire precautions** for concourses, seating tiers, and ancillary areas.
    - **Emergency procedures**, including coordination with police and emergency services.
- SGSA guidance is widely used in the UK and internationally as a benchmark for stadium safety management.

Summary

- **ADB** establishes prescriptive requirements for assembly and recreation occupancies.
- **BS 9999** introduces flexibility through risk-based design.
- **BB 100** provides school-specific fire safety guidance.

- **Licensing conditions** regulate theatres, cinemas, and entertainment venues.
- **SGSA's Green Guide** governs fire and crowd safety in sports grounds.

Together, these frameworks provide a **multi-layered regulatory foundation**, balancing prescriptive requirements with performance-based solutions tailored to large and complex public venues.

17.4 Means of Escape

In assembly and recreation buildings, means of escape design is one of the most critical elements of the fire strategy. With occupant numbers often in the hundreds or thousands, small deficiencies in exit provision or distribution can have catastrophic consequences. Travel distances, exit widths, and distribution must therefore be carefully calculated, with consideration given not only to prescriptive code requirements but also to crowd behaviour and performance-based modelling.

Travel Distances

- Approved Document B (ADB) sets baseline travel distance limits for assembly and recreation occupancies:
    - **18 m** in dead-end conditions.
    - **45 m** where two directions of travel are available.
- In large buildings such as arenas, shopping centres, or theatres, these limits may be exceeded where justified by **performance-based design** under **BS 9999**.
- Risk factors such as sprinklers, smoke control, or reduced fire growth potential may allow extended travel distances, provided that evacuation modelling demonstrates occupants can reach safety within the available safe egress time (ASET).

Exit Widths and Numbers

- Exit capacity must be sufficient to accommodate the maximum anticipated occupant load.
- ADB and BS 9999 provide calculation methods based on occupant flow rates, typically:
    - **5 mm per person** for door exits serving more than 220 persons.

- Exits must be sized and numbered to prevent congestion, especially in **arenas, stadia, and theatres**, where crowds funnel through vomitories or tiered exits.
- In practice, wide exits are more efficient than multiple narrow doors, reducing the likelihood of blockages and improving crowd flow.

Exit Distribution

- Exits should be **distributed around the perimeter** of the assembly space, ensuring occupants can disperse evenly during evacuation.
- Exit distribution is critical in venues with high-density seating, such as stadiums or concert halls, where reliance on a single or limited number of exits can lead to severe crowding.
- Exits must be clearly visible and directly accessible, with **unobstructed sightlines** that encourage use of the nearest available route rather than reliance on familiar entry points.
- Secondary exits should be equally prominent to reduce dependence on main entrances.

Design Considerations for Large Venues

- **Arena vomitories and concourses** must be sized to prevent bottlenecks where large numbers of spectators leave simultaneously.
- **Tiered seating arrangements** require sufficient staircases and radial gangways, designed in accordance with SGSA's *Green Guide*.
- **Horizontal and vertical circulation routes** must be free of obstructions and supported by emergency lighting to maintain flow under power failure conditions.
- Evacuation modelling is increasingly used to test exit capacity, crowd dispersal, and the impact of localised hazards (e.g., blocked exits or smoke spread).

Summary

Means of escape in assembly and recreation buildings must address **large occupant numbers, unfamiliar users, and complex circulation patterns**. Travel distances must be justified through risk-based design, exit widths calculated against full occupancy, and exits distributed to encourage balanced crowd flows.

Ultimately, safe evacuation depends on both the **engineering of exit capacity** and the **behavioural design of exit visibility and distribution**.

17.5 Evacuation Strategies

Evacuation strategies for assembly and recreation buildings must account for **large occupant numbers, varied user profiles, and the dynamics of crowd behaviour**. The choice of strategy depends on building type, scale, and management arrangements, with success relying on both engineered systems and trained staff.

Simultaneous Evacuation

- The most common approach in **theatres, cinemas, and schools** is simultaneous evacuation, where all occupants leave the building together upon alarm activation.
- This strategy is suitable for buildings where:
    - Occupant numbers are relatively moderate.
    - Escape routes are short and well-distributed.
    - Compartmentation cannot guarantee extended protection for phased approaches.
- In schools, simultaneous evacuation is straightforward as pupils are familiar with drills and guided by teachers. In cinemas and theatres, patrons rely heavily on staff instructions and clear voice alarm systems to prompt swift and orderly evacuation.

Phased Evacuation

- In very large venues, such as **arenas, stadia, or high-rise educational buildings**, a **phased evacuation** strategy may be adopted.
- Occupants in the fire-affected zone evacuate first, followed by adjacent areas, while those in remote compartments await instruction.
- This reduces congestion in staircases, concourses, and vomitories, ensuring smoother crowd flow.
- Phased evacuation requires:
    - **Robust compartmentation** to protect occupants who remain in place temporarily.

  - **Reliable voice communication systems** capable of issuing clear, authoritative instructions to thousands of people simultaneously.
  - Close coordination with stewards, marshals, or teaching staff to manage crowd dispersal effectively.

Staff-Assisted Evacuation

- Staff play a vital role in ensuring evacuation strategies function as intended.
- In **schools**, teachers are directly responsible for guiding pupils to assembly points, checking roll calls, and ensuring vulnerable children receive additional support.
- In **arenas and stadia**, trained stewards and marshals are deployed throughout seating areas, concourses, and exits to direct spectators towards safe routes and prevent bottlenecks.
- Staff training is critical: without effective leadership, crowds may hesitate or surge towards familiar entry routes, undermining the strategy.

Integration with Communication Systems

- Both simultaneous and phased evacuation strategies depend on **effective communication**, typically through voice alarm/public address systems supported by clear signage.
- In large or acoustically complex spaces, digital displays and visual indicators may supplement voice systems, guiding crowds towards the nearest safe exits.
- Regular drills and rehearsals, particularly in schools and licensed venues, reinforce evacuation strategies and ensure staff competence.

Summary

Evacuation strategies in assembly and recreation buildings range from **simultaneous evacuation in smaller venues** to **phased evacuation in very large or complex occupancies**, with staff guidance forming the critical link between alarm activation and safe crowd movement. Successful strategies rely not only on exit capacity and compartmentation but also on **well-trained staff and reliable communication systems** to maintain order under pressure.

17.6 Voice Alarm and Communication Systems

In high-occupancy venues such as schools, theatres, sports arenas, and places of worship, **clear and authoritative communication** is critical to safe evacuation. Traditional alarm tones or sirens are often insufficient for large or unfamiliar crowds, where hesitation, confusion, or inappropriate actions may delay egress. Voice alarm and communication systems ensure that occupants receive specific instructions, reducing uncertainty and improving compliance.

Voice Alarm Systems

- Unlike simple sounders, **voice alarms deliver spoken instructions**, guiding occupants towards safe actions.
- Research has shown that people respond more quickly and reliably to **clear voice messages** than to bells or sirens, especially in crowded or acoustically complex environments.
- Voice alarm systems are typically integrated with the fire detection and alarm panel, automatically broadcasting messages on alarm activation.
- Standard messages may include:
    - "Attention please, leave the building by the nearest exit."
    - Zone-specific instructions for phased evacuation.
- In large or international venues, multi-lingual messages may be programmed to reflect the diversity of the occupant population.

Integration with Public Address (PA) Systems

- In theatres, cinemas, and sports arenas, voice alarm systems are often integrated with the existing **public address (PA) infrastructure**, allowing alarms to override background music, announcements, or entertainment audio.
- This ensures immediate clarity, preventing the alarm from being masked by performance sound systems or crowd noise.
- In some venues, a central control room allows operators to issue **live announcements**, tailoring messages to the developing situation and providing reassurance to occupants.

Multi-Language and Inclusive Communication

- International airports, sports arenas, and concert venues often attract occupants from multiple nationalities.

- In these settings, **multi-language voice messages** or **pictogram-based signage** may be required to ensure comprehension across diverse audiences.
- Supplementary visual systems, such as **dynamic digital displays or illuminated directional signage**, provide additional clarity in environments where noise, acoustics, or panic may impair the effectiveness of voice messages.
- For hearing-impaired occupants, integration with **visual alarm devices (VADs)** ensures that emergency messages remain inclusive.

System Reliability and Management

- Voice alarm systems must be designed to the standards of **BS 5839-8**, ensuring reliability, audibility, and clarity under fire conditions.
- Systems require:
  - **Backup power supplies**, maintaining functionality during mains failure.
  - **Zoned messaging capability**, allowing selective evacuation of specific areas while others remain in use or await instruction.
  - Regular **testing and staff training**, ensuring both technical reliability and human familiarity with system operation.
- Integration with crowd management plans ensures that staff instructions and voice messages reinforce one another, maintaining order and minimising hesitation.

Summary

Voice alarm and communication systems are indispensable in high-occupancy assembly and recreation buildings. By providing **clear, specific, and inclusive instructions**, they reduce confusion, guide crowds towards safe exits, and support both simultaneous and phased evacuation strategies. Properly integrated with PA systems, signage, and staff procedures, they are among the most effective tools for managing large-scale evacuations.

17.7 Fire Protection Systems

Assembly and recreation buildings demand carefully tailored fire protection systems, balancing the need for robust life safety with the operational and architectural characteristics of venues such as theatres, stadiums, schools, and places of worship. Systems must not only detect and suppress fire but also manage smoke in ways that maintain tenable conditions for evacuation and fire service intervention.

Detection and Alarm

- **Comprehensive coverage** is essential, with a combination of **smoke and heat detectors** installed throughout auditoria, foyers, circulation spaces, and back-of-house areas.
- In theatres and performance venues, conventional smoke detectors can be prone to false alarms from **stage effects** such as smoke machines, pyrotechnics, or theatrical lighting.
    - In such cases, **aspirating smoke detection (ASD)** is often specified, providing highly sensitive, reliable detection that distinguishes real smoke from special effects.
- Detection systems must be fully integrated with **voice alarms and public address systems** to provide clear instructions, particularly in crowded auditoria or multi-level concourses.

Suppression Systems

- **Sprinklers** are typically provided in high-risk areas such as:
    - **Stages**, where electrical equipment, scenery, and drapes present significant fuel loads.
    - **Projection rooms**, where equipment operates at high temperatures.
    - **Commercial kitchens and concessions**, where deep-fat fryers or cooking appliances introduce ignition risks.
- **Water mist or gaseous suppression** systems may be appropriate in **heritage theatres, museums, or archives**, where traditional sprinklers could damage sensitive materials.
- In sports arenas and exhibition centres, suppression systems may also be extended to large concourses, storage rooms, and plant areas to control ignition at multiple risk points.

Smoke Control

- **Smoke vents** in auditoria are critical to prevent smoke accumulation in large seating areas, protecting both escape routes and fire service access.

These may be natural roof vents or mechanically powered fans sized to BS EN 12101.

- In modern arenas or large foyers, which often resemble **atriums**, smoke movement can be complex. Computational Fluid Dynamics (**CFD analysis**) is frequently required to model plume dynamics, reservoir sizing, and make-up air provision, ensuring tenable escape conditions are maintained for the duration of evacuation.
- **Smoke curtains** may be used to direct plumes towards extract points, particularly where open concourses connect with seating bowls or multi-level lobbies.

Integration with Building Design

- Fire protection systems must be coordinated with the architectural and operational use of the building. For example:
    - Systems in theatres must operate without interfering with performances, yet remain reliable in emergencies.
    - In schools, systems must be robust against tampering or accidental activation while still providing comprehensive coverage.
    - In stadia, systems must integrate with large-scale PA, stewarding, and external crowd management plans.

Summary

Fire protection systems in assembly and recreation buildings must be **multi-layered and occupancy-specific**: reliable detection and alarm systems for early warning, targeted suppression in high-risk zones, and smoke control strategies tailored to complex geometries. By combining these measures, designers can ensure tenable conditions are preserved long enough for safe evacuation, even in the most challenging large-scale environments.

17.8 Human Behaviour and Crowd Safety

The safety of assembly and recreation buildings is shaped as much by **human behaviour** as by engineering design. With thousands of people often gathered in one place, crowd psychology, wayfinding, and flow dynamics can determine whether evacuation is safe and orderly or chaotic and hazardous. Fire strategies must therefore integrate behavioural insights alongside physical systems.

Flow and Congestion

- People rarely evacuate as isolated individuals; they tend to move in **groups**, whether as families, school classes, friendship circles, or sports teams. This can slow movement, as groups often wait for one another rather than dispersing efficiently.
- **Bottlenecks** commonly occur at stairheads, concourses, or exit gates, where wide open areas funnel into narrower escape routes. Once densities exceed about **3–4 persons per m^2**, crowd movement slows dramatically, with "stop-start" flows developing.
- In stadiums and arenas, **vomitories** are critical pinch points, requiring careful sizing to avoid dangerous crowding when large sections attempt to evacuate simultaneously.

Wayfinding

- Occupants often attempt to **exit by the same route they entered**, even when alternative exits are closer and safer.
- This behaviour can cause dangerous overloading of main entrances, while other exits remain underutilised.
- **Signage, lighting, and steward direction** are therefore essential to redistribute occupants evenly. Dynamic signage and digital displays, when integrated with alarm and smoke control systems, can redirect flows away from compromised or congested routes.
- In schools or theatres, where occupants are more familiar with procedures or led by staff, reliance on entry routes is reduced but not eliminated.

Panic vs. Orderly Evacuation

- Contrary to popular belief, **panic is rare**. Most people behave rationally, seeking information and following others. However, when exits are blocked or communication is poor, confusion can escalate into **disorder or crowd surges**, increasing the risk of crushing or trampling.
- Clear **voice alarm systems**, supported by visible staff and stewards, greatly reduce these risks by providing authoritative reassurance and direction.
- The presence of trained staff has been shown to anchor behaviour, turning uncertain crowds into orderly flows by reinforcing safe routes and preventing bottlenecks.

Design and Management Implications

- Fire strategies must assume that crowd behaviour will not align perfectly with engineered capacities. Real-world flows are slower, more uneven, and influenced by group dynamics.
- Evacuation modelling should incorporate behavioural variables, such as pre-movement times, exit choice, and group movement, rather than relying solely on theoretical flow rates.
- Staff training, communication, and drills are as important as physical systems, particularly in large venues with diverse and unfamiliar populations.

Summary

Human behaviour in assembly and recreation buildings is characterised by **group movement, reliance on familiar routes, and vulnerability to congestion at bottlenecks**. While panic is rare, poor communication can cause disorder. Effective fire safety therefore depends on **clear voice alarms, visible staff, intuitive wayfinding, and crowd-aware design**, ensuring that evacuation strategies reflect how people actually behave in emergencies.

17.9 Management and Staff Training

In assembly and recreation buildings, where large numbers of people may be present and many are unfamiliar with the premises, the **role of management and staff** is fundamental to fire safety. Even the most carefully engineered fire protection systems cannot ensure safe evacuation without trained personnel to lead, supervise, and reassure occupants. Licensing and regulatory authorities increasingly view staff competence as central to compliance.

Schools

- Fire drills must be conducted **regularly**, ensuring that pupils and staff are familiar with alarm signals, evacuation routes, and assembly points.
- Teachers are responsible for **leading classes to safety**, supervising movement along escape routes, and maintaining calm.
- **Roll calls** at assembly points are critical to confirm all pupils are accounted for and to identify any missing persons immediately.
- School fire strategies must also include **provisions for pupils with disabilities**, ensuring assisted evacuation procedures are rehearsed and reliable.

Theatres and Cinemas

- Ushers and front-of-house staff are often the first point of contact with audiences during an evacuation.
- Training must equip them to:
    - **Guide audiences quickly and calmly** to exits, using clear and authoritative instructions.
    - Direct patrons away from congested routes, reducing reliance on familiar entry doors.
    - Reassure occupants, minimising anxiety and preventing disorder.
- Staff must be familiar with the location of all exits, refuge points, and communication systems, and drills should reflect scenarios where visibility is reduced by smoke or lighting failures.

Sports Arenas and Stadia

- With crowds often numbering in the tens of thousands, effective evacuation relies on the deployment of **trained stewards**.
- Each steward is typically assigned to a **specific zone** — seating areas, concourses, or vomitories — ensuring full coverage of the venue.
- Stewards are supported by **CCTV monitoring and central control rooms**, which allow supervisors to monitor crowd flows, detect congestion, and coordinate live announcements.
- Regular rehearsals and event-day briefings reinforce procedures and maintain readiness. Integration with police and emergency services is also vital for large-scale events.

Licensing and Regulatory Requirements

- Local authorities and licensing bodies often require **demonstrations of evacuation procedures** as a condition of operating licences for theatres, cinemas, nightclubs, and concert halls.
- Documentation of staff training, drill records, and fire safety management plans may be subject to inspection during licensing renewals.
- This ensures that fire safety management is not only a design requirement but an **ongoing operational responsibility**.

Summary

Effective fire safety in assembly and recreation buildings depends on **well-trained staff supported by robust management systems**. Whether it is teachers supervising pupils, ushers guiding audiences, or stewards managing sports crowds, staff provide the human leadership that transforms alarm activations into safe, orderly evacuations. Regulatory oversight reinforces this, ensuring that fire strategies are actively maintained and rehearsed, not left as static design documents.

17.10 Case Examples

Real-world examples of fire strategy in assembly and recreation buildings demonstrate how regulatory provisions, engineering measures, and staff training come together to ensure life safety. The following case studies highlight contrasting approaches across a theatre, a sports stadium, and a school.

Case 1: Theatre Fire

- **Building profile**: A 1,200-seat theatre with a large stage, fly tower, and backstage facilities.
- **Fire protection systems**:
  - Comprehensive **smoke detection** installed throughout the auditorium, lobbies, and backstage areas, linked to a central alarm panel.
  - A **voice alarm system** provides clear, zone-specific evacuation messages to patrons.
  - Stage areas are fitted with **sprinklers and deluge curtains**, protecting against the rapid spread of fire from scenery, drapes, and lighting rigs.
- **Means of escape**:
  - Exit widths total **8 m**, distributed evenly around the building, avoiding reliance on the main entrance.
  - Exits lead directly to external assembly areas, with illuminated signage ensuring visibility in low-light conditions.
- **Management and staff**: Ushers are trained to guide patrons quickly and calmly to the nearest exit, reducing the tendency to return via entry doors.

- **Outcome**: This layered approach ensures that even a full-capacity audience can evacuate safely within the available safe egress time (ASET).

Case 2: Football Stadium

- **Building profile**: A modern stadium accommodating 40,000 spectators across multiple seating tiers.
- **Means of escape**:
  - Multiple **vomitory exits** serve each seating tier, designed in accordance with the **SGSA Green Guide** to provide sufficient exit capacity.
  - Wide concourses allow circulation without dangerous bottlenecks.
- **Fire protection systems**:
  - Covered concourses are fitted with **mechanical smoke control systems** to maintain tenability in case of fire.
  - Detection and suppression systems are provided in kitchens, plant rooms, and storage areas.
- **Evacuation strategy**:
  - **Phased evacuation** may be employed in certain scenarios, allowing controlled dispersal of spectators without overwhelming exits.
  - A central **control room with CCTV monitoring** oversees crowd movement and can issue targeted announcements via a **public address override system**.
- **Management and staff**: Trained stewards are assigned to zones, directing spectators towards exits and maintaining order. Coordination with police and emergency services supports external crowd management.
- **Outcome**: The combination of engineered exit capacity and trained staff ensures safe dispersal of tens of thousands of spectators under pressure.

Case 3: School Building

- **Building profile**: A three-storey secondary school with multiple classrooms, assembly halls, and laboratories.
- **Evacuation strategy**: A **simultaneous evacuation** approach is adopted. On alarm activation, each class exits under teacher supervision.

- **Means of escape**:
  - **Travel distances** are restricted to a maximum of **30 m in multi directions**, in compliance with ADB.
  - Multiple protected **stair cores** provide safe vertical escape routes, enclosed in 60-minute fire-resisting construction.
- **Fire protection systems**:
  - Smoke detection in classrooms, corridors, and common areas provides early warning.
  - Heat detection is installed in laboratories and kitchens to reduce false alarms.
- **Management and staff**: Teachers play a pivotal role, leading pupils to external assembly points and conducting roll calls to ensure accountability.
- **Outcome**: Frequent fire drills ensure pupils are familiar with evacuation procedures, supporting calm and efficient egress during an emergency.

Summary

- The **theatre** example highlights the integration of sprinklers, deluge systems, and voice alarms with trained staff.
- The **stadium** case demonstrates the scale of design and management required to move tens of thousands of spectators safely.
- The **school** case shows how familiar drills and staff-led evacuation can offset the risks associated with young occupants.

Together, these examples illustrate the **adaptability of fire strategies** to different forms of assembly and recreation, all built on the common foundation of **detection, suppression, means of escape, and trained staff leadership**.

17.11 Summary

Assembly and recreation buildings present some of the most complex challenges in fire safety. Their defining features—**large occupant numbers, high-density crowds, and varied levels of occupant familiarity**—demand strategies that go beyond prescriptive design and place equal emphasis on management, communication, and behavioural factors.

Key principles include:

- **Multiple, well-distributed exits**: Exit capacity must be sufficient for peak occupancy, avoiding bottlenecks at doors, concourses, or vomitories. Exits should be visible and located around the perimeter to encourage even dispersal.
- **Clear voice alarm and communication systems**: Voice messages are more effective than bells or sirens, reducing hesitation and guiding crowds towards safe routes. Integration with PA systems and digital signage is essential in large or noisy environments.
- **Early detection and targeted suppression**: Comprehensive smoke and heat detection, supplemented by aspirating systems in theatres or arenas, ensures timely alarm activation. Sprinklers, water mist, or gaseous suppression should protect high-risk areas such as stages, projection rooms, kitchens, and plant spaces.
- **Trained staff and stewards**: Human leadership is critical. Teachers in schools, ushers in theatres, and stewards in stadia provide the authority and guidance that transform alarms into safe, orderly evacuations.

The success of fire strategies in assembly and recreation buildings lies in the **integration of engineered systems with crowd-aware management**. By addressing both the physical and behavioural aspects of evacuation, designers and operators can ensure that even the largest gatherings of people can be moved to safety under pressure.

The next chapter will address **Healthcare and High Dependency Occupancies**, where evacuation presents an even greater challenge due to mobility limitations, medical dependencies, and the reliance on staff-assisted movement.

Chapter 18

Healthcare and High Dependency Occupancies

18.1 Introduction

Healthcare premises and high-dependency occupancies—including **hospitals, care homes, mental health units, hospices, and rehabilitation centres**—present some of the most complex challenges in fire safety design and management. Unlike most other building types, where evacuation can be assumed to be independent and relatively rapid, many occupants in these environments are **non-ambulant, highly dependent, or medically vulnerable**, requiring staff intervention for movement.

Unique challenges in healthcare and high-dependency occupancies include:

- **Limited mobility**: Many patients and residents are unable to self-evacuate due to age, illness, injury, or disability. Evacuation must often be carried out using evacuation chairs, stretchers, or bed-mounted evacuation aids.
- **Dependence on staff assistance**: The speed and effectiveness of evacuation are directly linked to staffing levels, training, and the availability of equipment. Night-time staffing ratios are particularly critical, as they may be significantly reduced compared to daytime.
- **Continuous medical support**: Some patients rely on life-sustaining equipment (e.g., ventilators, monitors, or infusion pumps) that cannot be easily disconnected during evacuation. Strategies must balance immediate life risk from fire with the clinical risks of moving patients.
- **Infeasibility of simultaneous evacuation**: Evacuating all occupants at once is rarely possible in hospitals or care facilities. Instead, **progressive evacuation** is adopted, moving occupants in stages from the fire-affected compartment into adjacent safe areas, supported by robust compartmentation.

Strategic priorities for healthcare fire safety include:

- **Progressive evacuation strategies**, particularly **progressive horizontal evacuation (PHE)**, where patients are moved horizontally into adjacent compartments before any vertical evacuation is attempted.
- **Robust compartmentation**, ensuring each ward, treatment area, or sub-compartment can contain fire and smoke long enough to permit safe staff-assisted movement.
- **Early detection and alarm systems**, including addressable fire alarm panels and staff alert functions, to ensure rapid response before conditions deteriorate.
- **Suppression systems**, such as sprinklers or water mist, to control fire growth and reduce smoke production, thereby extending available safe egress time (ASET).
- **Staff training and preparedness**, with regular drills, clear role allocations, and contingency planning for both day and night operations.

Healthcare and high-dependency occupancies therefore demand a **performance-based approach**, supported by prescriptive standards and rigorous management procedures. Here, the **human factor**—the role of trained staff—is as critical as any engineering solution.

18.2 Regulatory Framework

Healthcare and high-dependency occupancies are governed by a **multi-layered regulatory framework**, reflecting their high life-risk profile and the need for tailored fire safety solutions. Compliance requires not only adherence to baseline building regulations but also to healthcare-specific standards and operational guidance.

Approved Document B (ADB)

ADB sets the **baseline requirements** for fire safety in all building types, including hospitals, care homes, and mental health units.

It establishes minimum standards for compartmentation, travel distances, means of escape, fire resistance, and alarm audibility.

While ADB provides a starting point, healthcare occupancies often require **enhanced measures** beyond ADB due to occupant dependency and the impracticality of simultaneous evacuation.

Health Technical Memoranda (HTMs)

- The **Department of Health's Health Technical Memoranda (HTM 05 series)** provide detailed guidance specific to NHS and healthcare premises.
    - **HTM 05-02: Firecode – Guidance in Support of Functional Provisions (Fire Safety in the Design of Healthcare Premises)** sets design standards for compartmentation, evacuation strategy, detection, suppression, and fire-fighting access.
    - **HTM 05-03: Operational Provisions** covers management, maintenance, staff training, and fire drills, ensuring strategies are embedded in daily operations.
- HTMs form the **core reference** for hospital fire strategies and are widely recognised as best practice in both public and private healthcare facilities.

BS 9999: Fire Safety in the Design, Management and Use of Buildings

- BS 9999 provides a **risk-based approach** to fire safety design, offering flexibility compared to ADB.
- In healthcare premises, it can be used to justify alternative solutions (e.g., extended travel distances, reduced stair provision) where supported by sprinklers, robust compartmentation, or staff management systems.
- BS 9999 is often applied in private care homes or smaller healthcare facilities outside the direct scope of NHS HTMs.

BS 5839-1: Fire Detection and Alarm Systems

- This standard governs the design, installation, and maintenance of fire detection and alarm systems in non-domestic buildings, including hospitals and care homes.
- It requires:
    - **Comprehensive smoke and heat detection**, with zoning aligned to evacuation compartments.
    - **Staff alert systems**, enabling staff to be notified discreetly before a general alarm is sounded.
    - **Integration with voice alarm systems**, essential in larger facilities for phased or progressive evacuation.

BS 9251: Residential Sprinkler Systems

- Residential sprinklers, designed to **BS 9251**, are mandatory in new care homes in England and strongly recommended for retrofitting in existing facilities.
- Sprinklers provide rapid suppression, slowing fire growth and extending available safe egress time (ASET).
- In hospitals, sprinklers may be installed in high-risk areas (e.g., treatment rooms, plant spaces), supplementing compartmentation and PHE strategies.

Care Quality Commission (CQC)

- In England, the **CQC** regulates health and social care providers, ensuring that fire safety standards are embedded in daily operations.
- The CQC requires care homes and hospitals to demonstrate compliance with fire safety legislation, risk assessment under the **Regulatory Reform (Fire Safety) Order 2005**, and adherence to HTM/BS guidance.
- Inspections may include review of fire safety policies, staff training records, and maintenance logs for detection and suppression systems.

Summary

The regulatory framework for healthcare and high-dependency occupancies combines:

- **ADB** for baseline compliance.
- **HTM 05-02 and HTM 05-03** for detailed NHS standards.
- **BS 9999** for performance-based flexibility.
- **BS 5839-1 and BS 9251** for detection, alarm, and suppression systems.
- **CQC oversight** to ensure operational compliance.

Together, these frameworks establish a comprehensive approach that integrates **design, engineering, and management** to protect highly vulnerable occupants.

18.3 Key Fire Safety Objectives

The overarching goal of fire safety in healthcare and high-dependency occupancies is to protect **life, care continuity, and building resilience** in environments where immediate full evacuation is often impractical. Strategies must therefore account for patient vulnerability, staff dependency, and the need to maintain essential medical functions during and after a fire incident.

Protect Patients, Residents, and Staff

- The primary objective is the **preservation of life**, particularly for patients and residents who cannot self-evacuate.
- Fire safety systems must ensure that conditions remain tenable long enough for staff to relocate or evacuate occupants safely.
- Protection extends not only to patients and residents but also to staff, who must remain in the building to perform evacuation and life-sustaining duties.

Support Progressive Horizontal Evacuation (PHE)

- In hospitals and care homes, **progressive horizontal evacuation** is the default strategy.
- Instead of attempting immediate total evacuation, occupants are moved in stages to **adjacent sub-compartments** with fire-resisting construction.
- This staged movement allows staff to relocate patients safely without the risks or impracticalities of vertical evacuation, particularly for bed-bound or high-dependency individuals.

Maintain Continuity of Care During Fire Incidents

- Healthcare occupancies must continue to provide **essential medical care** even in the event of fire.
- Fire strategies should minimise disruption to critical functions such as intensive care, operating theatres, and high-dependency units.
- Systems must be designed so that life-support equipment can continue to operate, often requiring **protected power supplies, medical gas shut-off controls, and resilient infrastructure**.

Provide Safe, Smoke-Free Refuges

- Fire strategies must incorporate **refuge areas** within each compartment or sub-compartment, allowing patients and staff to remain protected until onward evacuation is possible.

- These areas must be kept **smoke-free** through compartmentation, fire doors, and—where appropriate—smoke control systems.
- Refuges provide the critical time margin needed for staff to prepare assisted evacuation in a calm and controlled manner.

Ensure Resilience Through Redundancy in Systems and Compartmentation

- Healthcare facilities must be resilient against single points of failure.
- **Redundancy** is required in critical systems such as detection, alarm, suppression, and emergency power to ensure continued functionality under fire conditions.
- **Compartmentation** provides structural redundancy, ensuring that even if one area becomes untenable, the building can continue to function safely in other zones.
- This resilience is particularly important for **large hospitals**, where total evacuation may take several hours and services must remain operational for unaffected patients.

Summary

Key fire safety objectives in healthcare and high-dependency occupancies go beyond life safety alone.

They encompass:

- **Protection of vulnerable patients and staff**.
- **Support for progressive horizontal evacuation** (PHE).
- **Continuity of medical care** during fire incidents.
- **Provision of safe refuges** for delayed or staged evacuation.
- **System and compartment resilience** through redundancy.

These objectives recognise the unique challenges of healthcare settings, where fire safety must balance immediate life protection with the ongoing delivery of critical medical services.

18.4 Compartmentation and Sub-Compartmentation

Compartmentation is the **backbone of fire safety** in healthcare and high-dependency occupancies. Since simultaneous evacuation of all occupants is rarely feasible, fire and smoke must be contained within the area of origin long enough to permit **progressive horizontal evacuation (PHE)**. This requires robust structural fire protection, carefully designed compartmentation, and reliable fire doors to maintain separation.

Compartment Floors and Walls

- Healthcare buildings are divided into **compartments**, separated by walls and floors designed to provide **60–120 minutes fire resistance** depending on building size, function, and risk profile.
- These compartments act as fire barriers, limiting the spread of heat, flames, and smoke between wards, treatment areas, and circulation spaces.
- Service penetrations, ductwork, and cable trays must be appropriately sealed with fire-stopping to maintain compartment integrity.

Sub-Compartmentation

- Within each compartment, **sub-compartments** are created to provide intermediate levels of protection.
- Sub-compartment walls are typically designed to provide **30 minutes fire resistance**, sufficient to enable staff to relocate patients horizontally into an adjacent safe zone.
- This staged approach allows controlled movement without requiring immediate vertical evacuation, which is often impractical for bed-bound or high-dependency patients.

Maximum Compartment Sizes

- To ensure evacuation remains manageable, **compartment sizes are strictly controlled**.
- HTM 05-02, for example, recommends maximum areas of **750 m²** in hospitals, though this may vary depending on layout and risk assessment.
- Limiting compartment size reduces the number of patients and staff affected by any single fire incident, ensuring staff resources can cope with progressive evacuation.

Fire Doors

- Compartment and sub-compartment boundaries rely on **fire doors** fitted with **self-closing devices, smoke seals, and vision panels**.

- Doors must be easy for staff to operate, allowing quick passage of beds, wheelchairs, and evacuation aids while maintaining their fire-resisting function when closed.
- Fire doors form a critical component of **progressive evacuation routes**, enabling separation between affected and safe areas.

Example in Practice

- A typical hospital ward may be divided into two or more **sub-compartments**, separated by 30-minute fire-resisting construction.
- In the event of fire, patients can be moved horizontally into the adjacent safe area without using stairs or leaving the floor.
- If conditions worsen, further horizontal or eventual vertical evacuation can be undertaken in a staged manner, maintaining patient safety throughout.

Summary

Compartmentation and sub-compartmentation provide the **structural framework** that enables progressive evacuation in healthcare occupancies. With 60–120 minute compartments, 30-minute sub-compartments, and well-maintained fire doors, staff gain the vital time margin needed to relocate patients safely while preventing smoke and fire spread. This layered protection underpins every other aspect of healthcare fire safety strategy.

18.5 Evacuation Strategies

Evacuation in healthcare and high-dependency occupancies is fundamentally different from most other building types. Because many patients cannot self-evacuate and staff resources are finite, evacuation must be carefully staged and supported by compartmentation. The choice of strategy depends on the building design, patient dependency levels, and the availability of suppression and smoke control systems.

Progressive Horizontal Evacuation (PHE)

- **PHE is the primary strategy** for hospitals, care homes, and similar facilities.
- Patients are moved from the fire-affected area into an **adjacent fire-resisting sub-compartment** on the same floor.

- Each move places occupants into a **safe, smoke-free refuge**, allowing staff to regroup and continue patient care while planning further relocation if necessary.
- If conditions deteriorate, evacuation may continue **progressively into further compartments**, and eventually, if unavoidable, towards a final exit.
- PHE is only effective if supported by **robust compartmentation** (typically 30 minutes for sub-compartments and 60–120 minutes for compartments), and if staff are adequately trained in evacuation procedures.

Vertical Evacuation

- Vertical evacuation is **used only as a last resort** in healthcare settings, due to the challenges of moving bedbound or medically dependent patients down staircases.
- It becomes necessary if fire or smoke spreads beyond the compartment or sub-compartment of origin and further horizontal relocation is not viable.
- Evacuation aids such as **ski sheets, evacuation chairs, or specially adapted beds** are essential to facilitate vertical movement.
- Staircases serving healthcare facilities must therefore be wide, unobstructed, and protected to at least 60 minutes fire resistance, recognising that vertical evacuation may be slow and resource-intensive.

Defend-in-Place

- In certain **highly protected areas**, such as intensive care units (ICUs), operating theatres, or high-dependency wards, it may be safer to keep patients in place until the fire is controlled by suppression systems or fire service intervention.
- Defend-in-place requires:
    - **Very high levels of fire resistance**, typically 120 minutes or more for surrounding construction.
    - **Automatic suppression systems**, such as sprinklers or water mist, to limit fire growth.
    - **Smoke control measures**, including pressurisation or high-specification air-handling systems to keep the area tenable.

- This strategy must be underpinned by **detailed risk assessment**, recognising that keeping patients in place demands absolute reliability of fire protection systems.

Integration with Staff Procedures

- All evacuation strategies depend on **staff training, drills, and resource planning**.
- Daytime and night-time procedures must be separately rehearsed, recognising reduced staffing levels at night.
- Staff must be proficient in the use of evacuation aids and in prioritising patients based on dependency and mobility.
- Coordination with fire and rescue services is critical, ensuring responders are familiar with compartment layouts and evacuation strategies.

Summary

Evacuation strategies in healthcare occupancies are built around **progressive horizontal evacuation (PHE)**, supported by compartmentation and staff training. **Vertical evacuation** is a last resort due to its complexity, while **defend-in-place** may be adopted in highly protected, critical care areas. All approaches rely on robust fire protection systems and well-trained staff to ensure the safety of vulnerable, non-ambulant occupants.

18.6 Fire Detection and Alarm Systems

Early detection and clear communication are critical in healthcare and high-dependency occupancies, where patients may be unable to self-evacuate and staff must take the lead in managing progressive evacuation. Fire detection and alarm systems must therefore provide rapid, reliable warning to staff while avoiding unnecessary disruption to patients.

Comprehensive Coverage

- Healthcare facilities require **extensive smoke and heat detection coverage**, including:
    - **Wards, bedrooms, and day rooms**, ensuring fires are detected at the earliest stage.
    - **Corridors, circulation spaces, and protected escape routes**, to maintain safe egress.

    - **Plant rooms, service risers, and kitchens**, where electrical or mechanical faults pose ignition risks.

- Detection systems must be sensitive enough to provide early warning while being designed to avoid false alarms from activities such as cooking or medical equipment use.

Zoned Systems

- Fire alarm systems in healthcare occupancies are divided into **zones aligned with compartments and sub-compartments**.
- This enables staff to quickly identify the **exact location of a fire** and determine the appropriate evacuation response.
- Zoned alarms also allow fire service crews to locate and tackle the fire more efficiently, reducing disruption to unaffected areas.

Two-Stage Alarms

- Many healthcare premises use a **two-stage alarm system**:
    - **Stage 1 – Staff Alert**: On initial detection, staff are alerted discreetly (via panels, pagers, or sounders in staff-only areas). This provides a short **investigation period** (commonly 2–3 minutes) to verify the cause of the alarm.
    - **Stage 2 – General Alarm**: If the alarm is confirmed, or if no action is taken within the investigation period, the system escalates to a full evacuation alarm in the affected zone.
- This approach minimises unnecessary disturbance to patients, reducing anxiety and avoiding the risks associated with unwarranted evacuations.

Staff Paging Systems

- Staff paging systems or **silent alarms** are particularly valuable in care homes and hospitals.
- On activation, messages are sent directly to staff pagers, handheld devices, or nurse stations, allowing trained personnel to respond quickly without alarming residents or patients unnecessarily.
- This supports a calm, controlled evacuation process and ensures that patients are only moved when necessary.

Integration with Voice Alarm and Communication Systems

- In larger hospitals and healthcare complexes, voice alarm systems may be integrated to provide **clear instructions to staff and visitors**.
- Messages can be **zoned**, allowing instructions to be tailored (e.g., general alarm in one compartment while adjacent areas remain in staff-alert mode).
- This ensures that evacuations remain proportionate, avoiding unnecessary disruption to unaffected wards or treatment areas.

Summary

Fire detection and alarm systems in healthcare occupancies must deliver **early, reliable warning** while supporting the **progressive evacuation strategy**. Comprehensive coverage, zoning by compartments, two-stage alarms, and staff paging systems all work together to ensure fires are detected promptly and managed without unnecessary disruption to patients.

18.7 Active Fire Protection

Active fire protection systems play a central role in safeguarding healthcare and high-dependency occupancies. Because many occupants cannot self-evacuate, suppression and emergency systems are essential for slowing fire growth, preserving tenable conditions, and buying staff the time needed to carry out progressive evacuation.

Sprinklers

- Sprinklers are **strongly recommended in all new care homes and healthcare facilities**, and are increasingly mandated by national guidance.
- Designed to **BS 9251 (residential)** or **BS EN 12845 (commercial/industrial)** depending on building type, sprinklers offer multiple benefits:
    - Automatic suppression or control of fire growth, often extinguishing fires before the fire service arrives.
    - Reduction in smoke and toxic gases, directly protecting vulnerable patients who cannot self-rescue.
    - Extension of available safe egress time (ASET), supporting progressive horizontal evacuation strategies.

- Retrofitting sprinklers into existing hospitals and care homes is becoming more common, especially following high-profile fire incidents in healthcare settings.

Water Mist Systems

- In **heritage hospitals, mental health units, or buildings with sensitive equipment**, **water mist systems** may be preferable to conventional sprinklers.
- They use less water, minimising collateral damage to historic interiors, medical records, or electronic equipment.
- Water mist also improves survivability by rapidly cooling the fire plume and reducing smoke particle density.

Portable Extinguishers

- Portable fire extinguishers provide first-aid firefighting capability, but their effectiveness depends on **staff training and confidence**.
- Staff should be trained to use extinguishers appropriately, balancing early intervention against the need to prioritise evacuation.
- Extinguishers must be **readily accessible but not obstruct escape routes**, ensuring they support rather than hinder safe movement.
- In healthcare settings, staff training typically focuses on small fire containment, allowing suppression in its incipient stage without endangering staff.

Emergency Lighting

- **Emergency lighting**, designed in accordance with **BS 5266-1**, is essential in healthcare environments, particularly in wards and long corridors.
- Key functions include:
  - Maintaining illumination during power failures to support evacuation.
  - Providing clear visibility for staff moving patients on stretchers, wheelchairs, or beds.
  - Highlighting exit signage, staircases, and critical equipment areas.

- A minimum duration of **three hours** is recommended, reflecting the possibility of prolonged evacuation in large hospital complexes or care facilities.

Integration with Fire Strategy

- Active fire protection measures must be coordinated with detection, alarm, and compartmentation systems.
- Suppression systems reduce fire growth, while emergency lighting ensures staff can evacuate patients safely under adverse conditions.
- Together, these measures create a **layered defence**, recognising that active systems in healthcare are not only life safety measures but also operational enablers for staff-assisted evacuation.

Summary

Active fire protection in healthcare occupancies combines **sprinklers, water mist systems, portable extinguishers, and emergency lighting** to slow fire development, protect vulnerable patients, and extend the time available for safe evacuation. These systems form a vital layer of protection in environments where delayed evacuation and staff dependency are unavoidable.

18.8 Staff Training and Management

In healthcare and high-dependency occupancies, **staff are the most critical element of fire safety**. Even with robust compartmentation, early detection, and suppression systems, the safe evacuation of patients and residents ultimately depends on the skill, preparedness, and availability of staff. Without well-trained personnel, even the best-designed fire strategy will fail.

Regular Drills

- Staff must participate in **regular fire drills**, reflecting both day and night conditions.
- Drills should practise **progressive horizontal evacuation (PHE)**, with staff rehearsing the movement of patients into adjacent sub-compartments and refuges.
- Night-time drills are especially important, as reduced staffing levels may require different strategies and role allocations.

- Records of drills should be logged and audited to ensure compliance with regulatory expectations (e.g., CQC inspections).

Evacuation Aids

- Staff must be trained in the correct use of **evacuation aids**, including:
  - **Ski sheets and evacuation sleds**, fitted beneath mattresses for rapid bed-bound patient relocation.
  - **Evacuation chairs**, enabling staff to move mobility-impaired patients down protected staircases if vertical evacuation becomes necessary.
  - **Bed evacuation systems**, allowing entire beds to be wheeled into sub-compartments or refuges where space permits.
- Regular hands-on training ensures that staff are confident and efficient under pressure, reducing delays during real emergencies.

Compartment Familiarity

- Staff must be thoroughly familiar with the **layout of compartments and sub-compartments**, including alternative refuge areas and escape routes.
- Training should emphasise the importance of **closing fire doors promptly** to maintain compartment integrity during evacuation.
- Maps, signage, and local induction for new staff members should reinforce awareness of safe zones and evacuation paths.

Communication Protocols

- Effective evacuation depends on **clear and coordinated communication** between staff.
- Hospitals and care homes should implement systems such as:
  - **Two-way radios or wireless headsets** for ward-to-control communication.
  - **Fire alarm panels** with zoned indicators for precise fire location.
  - **Central control rooms**, where larger facilities can coordinate staff response, manage voice alarm systems, and liaise with the fire service.

- Protocols must ensure that staff receive instructions quickly and avoid conflicting messages that could delay evacuation.

Management Responsibilities

- Senior staff and fire safety managers must ensure:
    - Adequate **night-time staffing ratios**, recognising that fires often occur at night when patient dependency is high.
    - Ongoing **training programmes** to maintain staff competence.
    - Regular review and updating of the **fire risk assessment**, particularly after building modifications or changes in patient dependency.

Fire safety must be embedded into the **operational culture** of healthcare settings, treated as a continuous process rather than a periodic compliance exercise.

Summary

Staff training and management form the **operational backbone of fire safety in healthcare occupancies**. Regular drills, competence in evacuation aids, familiarity with compartment layouts, and clear communication protocols ensure that staff can act decisively under pressure. Without trained staff and robust management, even the most sophisticated fire safety systems cannot guarantee the safe evacuation of vulnerable patients.

18.9 Special Considerations

Healthcare and high-dependency occupancies encompass a wide range of facilities, each presenting unique challenges that go beyond standard hospital or care home design. Fire strategies must therefore be **tailored to specific contexts**, balancing fire safety with patient safety, medical continuity, and security requirements.

Mental Health Units

- Mental health facilities require a careful balance between **fire safety** and **security considerations**.
- Doors and escape routes must be designed to allow safe evacuation while also preventing absconding or harm to patients.

- Locking arrangements must comply with **fire safety regulations**, ensuring staff can override locks quickly in an emergency.
- Materials must be robust, tamper-resistant, and designed to minimise ligature risks, without compromising compartmentation or means of escape.
- Staff training is particularly important, as patient cooperation may vary during an evacuation.

Operating Theatres

- Fires in operating theatres present exceptional challenges, as patients may be **anaesthetised and mid-surgery**, making immediate evacuation impossible.
- Fire strategies in these areas often rely on **defend-in-place measures**, including:
    - **High levels of compartment fire resistance** (often 120 minutes or more).
    - **Automatic suppression systems**, such as sprinklers or water mist.
    - **Smoke control and air-handling systems** designed to prevent smoke spread into adjacent areas.
- Where possible, patients may be transferred to an adjacent protected operating theatre or recovery room once life-critical procedures allow.
- Staff training must include protocols for continuing essential surgery under fire alarm conditions while preparing for safe transfer.

Care Homes

- Care homes present some of the **highest fire risks**, with large numbers of elderly or immobile residents concentrated in relatively small buildings.
- **Progressive horizontal evacuation (PHE)** is essential, with wards or bedrooms divided into sub-compartments that allow staff to relocate residents without immediate vertical movement.
- **Sprinklers** are strongly recommended—and now mandatory in new care homes in England—as they significantly reduce fire growth and smoke production, buying critical time for staff-assisted evacuation.
- Staff training, night staffing ratios, and the availability of evacuation aids are vital to ensure that all residents can be relocated safely.

Summary

Special considerations in healthcare fire safety require strategies that account for **security in mental health units, continuity of care in operating theatres, and the extreme vulnerability of residents in care homes**. These occupancies demand **enhanced compartmentation, suppression systems, and staff preparedness**, tailored to their unique risks and operational constraints.

18.10 Fire Service Access and Support

Fire and rescue service access is especially critical in healthcare and high-dependency occupancies, where **progressive evacuation** and **defend-in-place strategies** often depend on the ability of firefighters to intervene quickly. Because occupants may not be fully evacuated before the arrival of the fire service, building design must ensure safe, reliable access and clear operational support.

Firefighting Shafts and Lifts

- Large hospitals and multi-storey care facilities must provide **firefighting shafts** in accordance with **Approved Document B (ADB)** and **BS 9999**.
- Shafts typically include:
    - A **firefighting stair**, enclosed in fire-resisting construction (at least 60 minutes).
    - A **firefighting lift** capable of carrying crews and equipment to upper levels.
    - A **firefighting lobby** providing a safe bridgehead from which crews can deploy hoses and plan operations.
- Firefighting lifts must comply with **BS EN 81-72**, with protected power supplies, water ingress protection, and fire service override controls.

Compartmentation Plans for Fire Service Use

- Healthcare premises must provide the fire service with **up-to-date compartmentation and evacuation plans**.
- Plans should identify:
    - **Compartment and sub-compartment boundaries**.
    - **Refuge areas** and evacuation routes.
    - Locations of **fire doors, suppression systems, and smoke control zones**.

- These plans enable fire crews to prioritise intervention, assess the spread of fire, and coordinate with staff conducting progressive evacuation.

Fire Control Rooms

- In **large hospitals or healthcare campuses**, a **dedicated fire control room** may be required.
- Fire control rooms should include:
    - The **fire alarm panel** and repeater displays.
    - Controls for **smoke ventilation and suppression systems**.
    - Two-way communication systems for **refuges and evacuation lifts**.
    - Compartmentation plans and building schematics.
- Control rooms allow joint coordination between the fire service and hospital management, ensuring effective response to complex incidents.

Smoke Control in Escape Routes

- Corridors and stairwells form critical routes for both **evacuation and fire service access**.
- Smoke control systems must therefore be provided, which may include:
    - **Natural or mechanical smoke ventilation** in stairwells.
    - **Pressure differentials** (pressurisation systems) to prevent smoke entering protected routes.
    - **Automatic smoke curtains** to limit horizontal spread in long corridors.
- Maintaining smoke-free access routes ensures firefighters can reach the fire quickly while patients continue to be evacuated or defended in place.

Summary

Fire service access and support are essential to healthcare fire strategies, enabling intervention while vulnerable patients remain in the building. **Firefighting shafts and lifts, accurate compartmentation plans, fire control rooms, and smoke-free escape routes** give firefighters the tools and information they need to protect life in complex, high-dependency environments.

18.11 Case Examples

Practical examples from healthcare and high-dependency occupancies illustrate how compartmentation, evacuation strategies, suppression systems, and staff management combine to provide resilience and life safety in complex environments.

Case 1: NHS Hospital

- **Building profile**: A large acute hospital serving a major city, accommodating several hundred patients across multiple wards and treatment areas.
- **Compartmentation**:
    - The building is divided into **60-minute fire-resisting compartments**, each ward or treatment block forming a distinct compartment.
    - Within these, **30-minute sub-compartments** enable progressive horizontal evacuation (PHE), allowing patients to be moved one zone at a time.
- **Evacuation strategy**:
    - **PHE is the default approach**, with staff relocating patients horizontally into adjacent sub-compartments.
    - If fire spread continues, patients can be moved progressively into further compartments or eventually towards final exits.
- **Fire protection systems**:
    - A **voice alarm system** delivers clear, zoned evacuation instructions to staff and visitors.
    - A **staff paging system** provides discreet alerts to nursing staff, enabling investigation and action before a full general alarm.
- **Outcome**: This layered strategy ensures patients remain protected while staff coordinate evacuation in a controlled manner until the fire service arrives to contain the incident.

Case 2: Care Home

- **Building profile**: A 40-bed, single-storey care home for elderly residents, many of whom are mobility-impaired.

- **Compartmentation**:
    - The care home is divided into **wings**, each forming a **30-minute sub-compartment**.
    - This allows staff to relocate residents horizontally into adjacent wings without needing vertical evacuation.
- **Fire protection systems**:
    - The entire building is protected by a **sprinkler system** in accordance with **BS 9251**, ensuring rapid suppression of any fire outbreak.
    - Smoke detection is provided in all bedrooms, corridors, and communal areas, linked to both audible alarms and staff pagers.
- **Staff training**:
    - Staff receive **quarterly training** in the use of evacuation aids, including ski sheets and evacuation chairs.
    - Night-time staffing arrangements are tested in drills to ensure that residents can be moved effectively even with reduced personnel.
- **Outcome**: The combination of sprinklers, sub-compartmentation, and trained staff provides a robust safety net, ensuring that residents can be moved progressively to safety with minimal disruption.

Summary

These case studies demonstrate the importance of integrating **compartmentation, suppression, detection, and trained staff** in healthcare occupancies:

- In **hospitals**, large-scale compartmentation and voice/paging systems underpin PHE for hundreds of patients.
- In **care homes**, sprinklers, sub-compartmentation, and staff competence ensure safe relocation of highly vulnerable residents.

Both examples highlight how effective fire strategies in healthcare must be **multi-layered, resilient, and tailored to patient needs**, providing protection until fire service intervention is achieved.

18.12 Summary

Healthcare and high-dependency occupancies demand **specialist fire strategies** that prioritise the protection of life above all other considerations. Unlike most building types, where evacuation can be assumed to be rapid and independent, patients and residents in these facilities are often immobile, dependent on staff, or reliant on continuous medical support. This makes **progressive evacuation, robust compartmentation, and staff preparedness** central to fire safety.

Key features of healthcare fire strategies include:

- **Strong compartmentation**: Compartment and sub-compartment boundaries provide 60–120 minutes and 30 minutes fire resistance respectively, enabling **progressive horizontal evacuation (PHE)** without immediate reliance on stairs.
- **Early detection and staged alarms**: Zoned detection, two-stage alarms, and staff paging systems ensure that fires are identified quickly and acted upon before conditions escalate, while minimising unnecessary disturbance to patients.
- **Active fire protection**: Sprinklers, suppression systems, water mist, portable extinguishers, and emergency lighting provide critical layers of defence, controlling fire growth and maintaining tenable conditions for staff-assisted evacuation.
- **Staff training and management**: Trained staff are the decisive factor in safe evacuation. Regular drills, familiarity with evacuation aids, and clear communication protocols ensure that staff can respond effectively, even under night-time or reduced staffing conditions.
- **Fire service consultation**: Firefighting shafts, fire control rooms, smoke control systems, and detailed compartmentation plans support realistic and timely intervention by fire crews while patients remain in situ.

Together, these measures create a **multi-layered and resilient safety system**, ensuring that even the most vulnerable occupants can be protected during a fire emergency. Healthcare fire safety is therefore as much about **management and staff capability** as it is about engineering solutions.

The next chapter will turn to **Fire Safety Management**, exploring the **ongoing policies, responsibilities, and procedures** that ensure strategies remain effective long after design and construction are complete.

Chapter 19
Fire Safety Management

19.1 Introduction

Fire safety management forms the **operational backbone of any fire strategy**. While passive measures (such as compartmentation and fire-resisting construction) and active measures (such as detection, alarms, and suppression systems) provide the physical framework for protection, they are only effective if properly **managed, maintained, and integrated** into daily operations.

In practice, it is **people, policies, and procedures** that determine whether a fire strategy works in reality as well as in design. A building equipped with sophisticated fire protection can still present significant life risks if escape routes are obstructed, staff are untrained, or fire protection systems are neglected. Conversely, even relatively simple fire precautions can be effective when combined with **robust management and well-trained personnel**.

Key elements of fire safety management include:

- **Clear responsibilities and accountability**: Identifying the *Responsible Person* under the Regulatory Reform (Fire Safety) Order 2005, and defining roles for staff, managers, and contractors.
- **Planned maintenance and testing**: Ensuring fire alarm systems, emergency lighting, suppression systems, and fire doors are inspected and tested to the required standards.
- **Policies and procedures**: From evacuation drills and permit-to-work systems for hot works, to housekeeping arrangements that keep escape routes clear.
- **Training and awareness**: Equipping staff, residents, or building users with the knowledge and confidence to act appropriately in the event of fire.
- **Integration with the fire strategy**: Management arrangements must reflect and support the technical design, ensuring that evacuation

strategies, compartmentation, and fire service access provisions can be relied upon in practice.

This chapter explores the **principles of fire safety management**, the **legal and regulatory responsibilities of duty holders**, and the **practical arrangements** that keep strategies effective throughout the life of a building.

19.2 Legal Framework

Fire safety management in the UK is governed by a framework of legislation that places clear duties on building owners, employers, and managers. These laws establish not only the **requirement for fire precautions** but also the **responsibility for ongoing management and compliance**.

Regulatory Reform (Fire Safety) Order 2005 (RRO)

- The RRO is the principal piece of legislation governing fire safety in non-domestic premises in England and Wales.
- It places responsibility on the **"Responsible Person"** (typically the employer, building owner, or managing agent) to ensure that fire safety measures are in place, suitable, and maintained.
- Key duties include:
    - Completing a **suitable and sufficient Fire Risk Assessment (FRA)**, which must be kept up to date and reviewed regularly.
    - Implementing appropriate **fire precautions**, covering detection, alarm, means of escape, and firefighting equipment.
    - Providing **training and fire drills** for staff, ensuring all personnel know what to do in an emergency.
    - Ensuring **maintenance and testing** of fire safety systems, with records kept for inspection.
- The RRO adopts a **risk-based approach**, requiring that precautions are proportionate to the level of life risk in the building.

Health and Safety at Work etc. Act 1974

- This Act sets out the **general duty on employers** to protect the health, safety, and welfare of employees and others affected by their work activities.
- While not specific to fire, it underpins the RRO by reinforcing the requirement for employers to provide a safe working environment, including adequate fire precautions.

- Senior managers and directors may also bear personal liability under this Act if they neglect their duties.

Enforcement

- Compliance with the RRO is enforced by the **local Fire and Rescue Authority**.
- Inspectors have powers to:
  - Conduct inspections and audits of fire risk assessments and management systems.
  - Issue **alterations notices** (requiring changes in use or layout to be notified).
  - Serve **enforcement notices** where failures are identified, requiring improvements within a set timeframe.
  - Issue **prohibition notices** where premises present an immediate risk, preventing their use until deficiencies are corrected.
- Serious or repeated non-compliance can result in **prosecution**, with unlimited fines or custodial sentences for responsible persons who breach their duties.

Summary

The legal framework for fire safety management places responsibility squarely on building owners, employers, and managers. The **RRO 2005** requires a suitable Fire Risk Assessment, ongoing management, and maintenance of precautions, while **enforcement by fire and rescue services** under the RRO ensures compliance, with significant penalties for those who fail in their duties.

19.3 The Responsible Person

At the heart of the Regulatory Reform (Fire Safety) Order 2005 (RRO) is the requirement for every building to have a clearly identified **Responsible Person (RP)**, sometimes referred to as the duty holder. This role is fundamental to ensuring that fire safety is not only designed into a building but also **maintained and managed throughout its operational life**.

Core Duties of the Responsible Person

The RP carries the ultimate responsibility for implementing and overseeing fire safety measures. Their duties include:

- **Implementing the fire strategy**: Ensuring that the fire strategy developed during design is translated into day-to-day management procedures and is actively maintained.
- **Commissioning and reviewing Fire Risk Assessments (FRAs)**: Making sure that a **suitable and sufficient FRA** is carried out by a competent person, reviewed regularly, and updated whenever the building use, layout, or risk profile changes.
- **Ensuring staff training and competence**: Providing fire safety induction, ongoing training, and evacuation drills so that staff understand their roles and responsibilities in an emergency.
- **Maintaining fire safety systems and equipment**: Arranging for regular inspection, servicing, and testing of fire detection and alarm systems, extinguishers, emergency lighting, fire doors, sprinklers, and suppression systems.
- **Keeping records**: Maintaining clear and auditable records of FRAs, drills, staff training, equipment tests, and maintenance, which may be reviewed during inspections by the enforcing authority.

Multi-Occupied Premises

- In complex or shared premises, such as **shopping centres, office towers, healthcare campuses, or mixed-use developments**, multiple organisations may each act as Responsible Person for their part of the building.
- In such cases, fire safety management must be **coordinated** to ensure that responsibilities are not duplicated or overlooked.
- The RRO places a duty on all Responsible Persons to **cooperate and share information**, for example:
    - Agreeing a joint evacuation strategy.
    - Sharing details of fire safety systems, compartmentation, and maintenance responsibilities.
    - Coordinating fire drills and staff training across multiple occupiers.
- Landlords, managing agents, and tenants must work together to provide a seamless approach to fire safety that protects all occupants, regardless of which part of the building they occupy.

Accountability

- The Responsible Person role carries significant **legal accountability**. Failure to meet these duties can result in enforcement action, including fines or prosecution.
- In practice, the RP may delegate day-to-day tasks to competent staff or contractors, but **legal responsibility cannot be delegated**—the RP remains ultimately accountable.

Summary

The Responsible Person is the **cornerstone of fire safety management**, ensuring that strategies are implemented, risks are assessed, staff are trained, and systems are maintained. In single-occupancy premises this is straightforward, but in multi-occupied buildings, cooperation and coordination are essential to provide a unified and effective fire safety regime.

19.4 Fire Safety Policy

A written **fire safety policy** provides the foundation for effective fire safety management within an organisation. It sets out the organisation's overall **commitment, structure, and procedures** for maintaining a safe environment, ensuring that staff, contractors, and visitors understand both the importance of fire safety and their individual responsibilities.

Purpose of the Fire Safety Policy

- The policy acts as a **formal statement of intent**, demonstrating management's commitment to life safety, property protection, and compliance with legislation.
- It ensures that fire safety is embedded in the organisation's broader health and safety culture, rather than being treated as a standalone requirement.
- A clear policy also provides evidence of due diligence in the event of enforcement action or legal proceedings.

Key Components of a Fire Safety Policy

- **Commitment to fire safety**: A statement by senior management that fire safety is a priority, outlining how risks will be managed and resources allocated.

- **Defined roles and responsibilities**: Clear allocation of duties to the Responsible Person, fire wardens, facilities managers, and contractors, ensuring accountability.
- **Maintenance and testing**: An outline of the arrangements for routine inspection and maintenance of detection systems, extinguishers, fire doors, sprinklers, and emergency lighting, with reference to statutory testing intervals.
- **Training and drills**: Details of staff induction, refresher training, and the frequency of evacuation drills (typically at least annually, with additional drills for high-risk premises such as schools, hospitals, and care homes).
- **Evacuation arrangements**: A summary of the building's evacuation strategy (e.g., simultaneous, phased, progressive horizontal), including procedures for mobility-impaired persons and visitors.
- **Contractor management**: Requirements for contractors to comply with fire safety procedures, including hot works permits, housekeeping standards, and emergency arrangements.

Communication and Accessibility

- The policy must be **communicated to all staff and contractors**, ensuring that every person in the building understands their role in maintaining fire safety.
- Copies should be available in induction packs, staff handbooks, and displayed on notice boards or intranet systems.
- Line managers and supervisors should reinforce policy requirements during team meetings and toolbox talks.

Review and Updating

- Fire safety policies should be **reviewed at least annually** to confirm they remain up to date with current legislation, occupancy, and operational practices.
- Policies must also be reviewed **following significant changes**, such as:
    - Alterations to the building layout or use.
    - Introduction of new processes or equipment.
    - Changes in occupancy profile (e.g., vulnerable residents, higher staff numbers).

- Reviews should be documented, with updates communicated promptly to all relevant staff.

Summary

A written fire safety policy provides a clear framework for managing fire safety on an ongoing basis. By setting out **commitment, responsibilities, maintenance, training, and evacuation arrangements**, it ensures that fire safety is understood, embedded, and consistently applied across the organisation. Regular review and effective communication make the policy a **living document**, central to keeping fire strategies effective in practice.

19.5 Training and Competence

No fire safety strategy can succeed without **trained and competent staff** who understand their responsibilities and can act decisively in an emergency. Even the most advanced fire protection systems rely on human intervention to initiate evacuation, operate equipment, and maintain order. Training must therefore be comprehensive, role-specific, and regularly refreshed.

General Awareness Training

- All staff, regardless of role, must receive **basic fire safety induction** on joining an organisation. This should cover:
    - What to do on **discovering a fire** (raising the alarm, not attempting to fight fire unless trained).
    - What to do on **hearing the alarm**, including evacuation routes, assembly points, and roll-call procedures.
    - The importance of **keeping escape routes clear** and complying with housekeeping policies.
- Refresher sessions should be provided periodically to reinforce awareness, particularly in high-turnover workplaces.

Role-Specific Training

- Some staff have designated fire safety responsibilities and require **additional training**:
    - **Fire wardens**: Responsible for sweeping areas, closing fire doors, and reporting to the incident control point.
    - **Evacuation marshals**: Direct occupants towards safe exits, prevent re-entry, and assist vulnerable persons.

    - **High-risk areas** (e.g., laboratories, kitchens, plant rooms): Staff require tailored training in local risks, emergency shut-off procedures, and safe use of extinguishers.
- These roles are essential to translating the fire strategy into safe, orderly evacuations.

Refresher Training and Drills

- Fire safety training must be **refreshed regularly**, with most organisations adopting **annual refresher sessions** for all staff.
- Fire evacuation drills should be conducted at least once a year, with frequency increased in **high-risk premises** such as care homes, hospitals, or schools.
- Drills should simulate realistic conditions, including blocked routes, reduced staffing (e.g., night shifts), or patient/resident movement in healthcare settings.
- Records of drills and training must be maintained as evidence of compliance and continuous improvement.

Specialist Training

- Certain staff require **specialist skills** to support fire safety measures:
    - **Evacuation aids**: Training in ski sheets, evacuation chairs, sleds, or bed-moving systems.
    - **Firefighting equipment**: Training in safe and appropriate use of portable extinguishers or hose reels, emphasising first-aid firefighting only.
    - **Fire control rooms**: In large or complex premises, staff must be competent in operating fire alarm panels, voice alarm systems, smoke control interfaces, and communication systems.
- Specialist training must be **practical and scenario-based**, ensuring staff can apply their knowledge under pressure.

Summary

Training and competence are the **human backbone of fire safety management**. From general awareness for all staff to specialist training for those with defined responsibilities, competence ensures that fire strategies function in practice, not just on paper. Regular refresher training and realistic drills embed confidence and preparedness, enabling staff to act swiftly and effectively when it matters most.

19.6 Fire Drills and Evacuation Exercises

Fire drills are a critical part of fire safety management, ensuring that **evacuation strategies are tested in practice**, staff remain confident in their roles, and occupants are familiar with emergency procedures. Drills also provide valuable data on the effectiveness of escape routes, communication systems, and staff coordination, forming the basis for continuous improvement.

Frequency

- Fire drills should be conducted **at least annually** in all workplaces and non-domestic premises.
- In higher-risk environments—such as **schools, healthcare facilities, residential care homes, and sleeping accommodation**—drills should be conducted more frequently (e.g., termly in schools, quarterly in care homes).
- Drills should also be scheduled to reflect **different staffing patterns**, particularly night shifts in healthcare and residential occupancies, where reduced staff numbers may alter evacuation dynamics.

Realism

- Drills should simulate **realistic conditions** while ensuring the safety of participants.
- Scenarios may include:
    - Blocked exits to encourage the use of alternative escape routes.
    - Smoke simulation (e.g., cold smoke machines in training environments).
    - Testing of voice alarm and PA systems under crowd conditions.
- Care must be taken not to **endanger vulnerable occupants**. For example, in hospitals and care homes, drills may involve staff role-playing patient movement rather than physically relocating all residents.

Recording

- Each drill must be **documented**, recording:
    - The date and time of the drill.

  - Which staff and departments participated.
  - The time taken to evacuate or relocate occupants.
  - Any issues observed, such as congestion, communication delays, or equipment misuse.
- Records should be retained for audit and inspection by enforcing authorities (e.g., Fire and Rescue Services, Care Quality Commission).

Continuous Improvement

- The outcomes of drills should be reviewed critically, identifying strengths and weaknesses in the evacuation process.
- Lessons learned must be fed back into **fire safety training programmes**, **evacuation strategies**, and, where necessary, **building modifications** (e.g., signage, exit widening, or compartment adjustments).
- Regular improvement cycles ensure that evacuation procedures evolve to reflect changes in occupancy, staffing, or building use.

Summary

Fire drills and evacuation exercises transform written fire strategies into **practical, tested procedures**. Conducted regularly and realistically, drills confirm staff competence, test evacuation systems, and highlight areas for improvement. By recording outcomes and embedding lessons learned, organisations ensure their fire safety arrangements remain **effective, reliable, and continually improving**.

19.7 Maintenance and Testing

A fire strategy is only effective if the protective measures it relies upon are **maintained in good working order** throughout the life of a building. Active and passive systems can degrade over time due to wear, damage, or lack of servicing. Regular testing, inspection, and maintenance—supported by accurate record-keeping—are therefore essential to ensure ongoing compliance and operational reliability.

Fire Alarm Systems

- Fire detection and alarm systems must be tested and serviced in line with **BS 5839-1**.
- **Weekly tests**: A manual call point is tested each week on a rotational basis, ensuring audibility and system functionality.

- **Annual servicing**: A full system service by a competent contractor is required at least once every 12 months. Large or complex systems may require servicing every six months.

Sprinklers and Suppression Systems

- Sprinklers and fixed suppression systems must comply with **BS EN 12845 (commercial/industrial systems)** or **BS 9251 (residential and care homes)**.
- **Monthly inspections**: Visual checks of control valves, pumps, water supplies, and alarm connections.
- **Annual servicing**: A full system service by a specialist contractor, including flow tests and verification of pump performance.
- Suppression systems (e.g., gaseous or mist systems) require similar inspection and recharging after any activation.

Emergency Lighting

- Maintained in accordance with **BS 5266-1**.
- **Monthly functional test** ("flick test"): Each luminaire is briefly tested to ensure correct operation.
- **Annual full discharge test**: The system is operated on battery supply for its full rated duration (typically three hours) to confirm performance under emergency conditions.

Fire Doors

- Fire doors are critical to compartmentation and must be kept in effective condition.
- Inspections should check for:
  - Integrity of door leaf and frame.
  - Functionality of **self-closing devices**.
  - Condition of **intumescent strips and smoke seals**.
  - Damage or misuse (e.g., wedging doors open).
- Formal inspections should be carried out **at least annually**, with higher-risk premises adopting more frequent checks.

Smoke Control Systems

- Smoke ventilation systems (natural, mechanical, or pressurisation) must be maintained to **BS EN 12101** standards.
- **Regular functional testing** is required, including automatic opening vents (AOVs), fans, dampers, and control panels.
- Integration with the fire alarm must also be verified, ensuring systems respond correctly on activation.

Portable Fire Extinguishers

- Maintained in accordance with **BS 5306-3**.
- **Annual servicing**: Conducted by a competent person, checking weight, pressure, and condition.
- **Visual checks**: Staff should conduct simple visual inspections monthly, confirming extinguishers are present, accessible, and undamaged.

Fire Logbook and Records

- All inspections, tests, and maintenance activities must be **recorded in a fire logbook**, either in hard copy or electronic form.
- Records should be available for inspection by Fire and Rescue Service officers or other regulators (e.g., CQC in healthcare).
- Accurate documentation provides evidence of compliance and demonstrates that systems are managed proactively.

Summary

Maintenance and testing ensure that fire precautions remain **functional, compliant, and reliable**. Weekly alarm tests, monthly lighting and sprinkler checks, annual servicing of suppression systems and extinguishers, and regular fire door and smoke control inspections form the backbone of ongoing fire safety assurance. Without these arrangements—backed by thorough record-keeping in a fire logbook—fire strategies cannot be relied upon when most needed.

19.8 Coordination in Complex Premises

In large, multi-occupied, or complex premises, fire safety management is rarely the responsibility of a single organisation. Instead, multiple duty holders must work together to provide a **coordinated approach** that ensures the safety of all occupants. Without clear communication and agreed responsibilities, gaps or conflicts in arrangements can undermine even the best technical fire protection measures.

Multi-Tenant Buildings

- In offices, mixed-use developments, or other multi-tenant premises, each occupier acts as a **Responsible Person** under the Fire Safety Order for their own areas.
- A central **fire safety manager** or **coordinating committee** is typically required to oversee shared fire precautions, such as:
    - Common alarm and detection systems.
    - Firefighting shafts, stairwells, and evacuation routes.
    - Smoke control systems and suppression equipment in shared spaces.
- Clear agreements must define **who is responsible for maintenance, testing, and record keeping**, as well as how costs are apportioned.

Shopping Centres

- Shopping centres provide a particular challenge, as dozens or even hundreds of individual tenants occupy separate units, each responsible for their own fire precautions.
- Central management typically takes responsibility for:
    - **Alarm and detection systems** in common areas.
    - **Evacuation coordination**, including public address/voice alarm systems.
    - **Fire drills and training**, ensuring tenants' staff are familiar with centre-wide procedures.
- Tenants must ensure compliance within their units (e.g., safe storage of stock, maintained extinguishers) while cooperating with centre management to avoid conflicting arrangements.

Hospitals and Universities

- Large institutions such as **hospitals and universities** often involve multiple Responsible Persons across different departments, faculties, or services.

- For example, a hospital trust may manage clinical areas, while external contractors manage catering, retail, or plant services. Similarly, universities may lease space to private companies or research partners.
- In such cases, formal agreements—such as **memoranda of understanding (MOUs)** or **service-level agreements (SLAs)**—are essential to define responsibilities for:
    - Fire risk assessments.
    - Alarm response protocols.
    - Evacuation coordination.
    - Maintenance of shared systems.
- Coordination meetings and joint drills reinforce these arrangements, ensuring all parties are aligned.

Summary

Complex premises require **coordinated fire safety management** to prevent gaps or duplication in responsibilities. Multi-tenant buildings need central oversight, shopping centres rely on cooperation between tenants and management, and hospitals or universities often require **formal agreements** between multiple Responsible Persons. Effective coordination ensures that fire safety arrangements operate as a seamless whole, regardless of the number of stakeholders involved.

19.9 Emergency Planning

Emergency planning is a vital element of fire safety management, ensuring that **preparedness extends beyond compliance** and that staff and occupants know exactly how to respond when an incident occurs. A robust emergency plan addresses not only evacuation and life safety but also communication, coordination, and recovery.

Fire Action Notices

- Fire action notices provide **simple, visible instructions** for building occupants.
- They must be prominently displayed in workplaces, accommodation, and high-traffic areas such as lobbies, stairwells, and lift lobbies.
- Notices typically include:
    - What to do on discovering a fire.

    - What to do on hearing the alarm.
    - The location of the nearest exit and assembly point.
    - Prohibitions (e.g., "Do not use lifts in the event of fire").
- Clear signage supports awareness, particularly for visitors and temporary staff unfamiliar with the building.

Personal Emergency Evacuation Plans (PEEPs)

- Organisations must prepare **PEEPs** for disabled or vulnerable occupants who may require assistance during evacuation.
- PEEPs should be **individualised**, detailing:
    - The person's mobility, sensory, or medical needs.
    - Equipment required (e.g., evacuation chairs, communication devices).
    - Staff assigned to provide assistance.
    - Routes and refuges to be used.
- Generic Emergency Evacuation Plans (GEEPs) may also be required in premises open to the public, where the exact identity of disabled visitors cannot be known in advance.

Assembly Points

- Safe assembly points are essential to ensure accountability after an evacuation.
- They must be **clearly designated, signposted, and communicated** to all occupants.
- Assembly points should be located:
    - At a safe distance from the building.
    - Away from vehicle routes, delivery bays, or potential hazards.
    - Without obstructing fire service access or hydrant locations.
- Roll calls or electronic headcounts can then be conducted to confirm that all occupants are accounted for.

Business Continuity

- Fire safety planning should extend beyond life safety to include **business continuity and recovery**.

- Business continuity plans (BCPs) identify how operations will be resumed after a fire incident, including:
    - Alternative premises or remote working arrangements.
    - Protection and recovery of critical data, IT, and records.
    - Communication strategies for staff, customers, regulators, and the media.
- Integrating BCPs with fire safety management ensures that an incident does not only protect lives but also minimises long-term disruption.

Summary

Emergency planning transforms fire strategies into **practical action** during an incident. Fire action notices give clear occupant guidance, PEEPs ensure inclusion for vulnerable persons, and assembly points provide accountability. Beyond immediate evacuation, business continuity planning ensures that organisations can recover quickly, protecting both people and operations in the aftermath of a fire.

19.10 Culture and Human Factors

Even the most comprehensive fire strategy can fail if it is not supported by a strong **safety culture** within the organisation. Fire safety is not just a matter of written policies and technical systems; it depends on the **attitudes, behaviours, and engagement of people at every level**. Embedding fire safety into daily practice ensures that risks are identified early and that procedures are followed instinctively in an emergency.

Visible Management Commitment

- Leadership sets the tone for safety culture. When managers and senior staff actively demonstrate their commitment—by attending drills, discussing fire safety openly, and holding teams accountable—it encourages compliance across the organisation.
- Conversely, if management is seen to disregard fire precautions (e.g., ignoring wedged fire doors), it undermines the culture and signals that fire safety is not a priority.

Regular Communication

- Fire safety must remain a **live issue** rather than a "tick-box exercise."

- Communication should include:
  - Regular reminders in team meetings, newsletters, or staff briefings.
  - Updates following fire drills or risk assessments.
  - Visible signage and posters reinforcing key messages.
- Keeping fire safety visible ensures that awareness is maintained and complacency avoided.

Reporting Systems

- Staff at all levels should feel empowered to **report fire safety concerns**, such as blocked exits, damaged fire doors, faulty alarms, or poor housekeeping.
- An open and blame-free reporting system encourages staff to highlight issues before they develop into serious risks.
- Feedback loops are critical: concerns raised must be acted upon and outcomes communicated back, reinforcing trust in the system.

Human Factors in Fire Safety

- Human behaviour often determines the effectiveness of fire safety arrangements:
  - Occupants may delay evacuation while seeking confirmation of a fire.
  - Groups often move together, waiting for colleagues or family members.
  - Staff under stress may forget procedures if training has not been reinforced.
- Recognising these factors, fire safety management must account for real-world behaviour, not just theoretical compliance. This includes:
  - Designing escape routes with redundancy to accommodate hesitation or bottlenecks.
  - Reinforcing staff training so responses become second nature.
  - Supporting behavioural awareness through scenario-based exercises.

Summary

A strong safety culture is as important as technical fire precautions. **Visible management commitment, regular communication, and effective reporting systems** ensure that fire safety is embedded in daily operations. By acknowledging human factors and building an environment where staff take ownership of fire safety, organisations create resilience that extends beyond systems and policies.

19.11 Case Example

Office Complex with Multiple Tenants

A large city-centre office complex accommodates several tenant organisations, each occupying different floors. To ensure consistency and compliance across the whole building, a **central fire safety manager** is appointed by the landlord.

Fire Safety Management Structure

- The central fire safety manager oversees the coordination of policies, ensuring alignment between tenants and building management.
- Each tenant organisation nominates **fire wardens** who receive role-specific training, enabling them to lead evacuations within their occupied areas.
- Responsibilities are clearly documented, preventing duplication or gaps in arrangements between landlord and tenants.

Training and Drills

- Fire wardens participate in annual refresher training, while all staff receive basic awareness training during induction.
- **Fire drills are conducted twice yearly**, including one **unannounced drill** to test real-time readiness and highlight potential weaknesses in occupant response.
- Debriefs are held after each drill, and lessons learned are fed back into the fire safety management plan.

Maintenance and Record-Keeping

- The building operates an **electronic fire logbook**, used to record:
    - Weekly alarm tests.
    - Monthly suppression system inspections.
    - Emergency lighting and extinguisher checks.
    - Fire door inspections, including closer and seal integrity.

- The logbook provides real-time visibility of compliance and is accessible to both building management and enforcing authorities.

Continuous Improvement

- A **quarterly fire safety committee** brings together the landlord, tenant representatives, and the central fire safety manager.
- The committee reviews:
    - Recent fire drills and incident reports.
    - Results of inspections and maintenance checks.
    - Updates to legislation and standards, ensuring ongoing compliance.
- This coordinated approach ensures that all tenants share ownership of fire safety and that policies evolve with changing risks.

Outcome

The combination of **central oversight, trained wardens, regular drills, electronic record-keeping, and collaborative governance** provides a resilient and transparent fire safety management system. The structure fosters a strong safety culture across multiple organisations, ensuring that evacuation strategies and fire precautions remain effective in practice.

19.12 Summary

Fire safety management ensures that the **precautions designed and installed at construction stage remain effective throughout the life of a building**. Without structured policies, trained staff, and ongoing maintenance, even the most advanced passive and active fire protection systems can deteriorate or fail under real emergency conditions.

Key elements of effective fire safety management include:

- **Clear identification of the Responsible Person** to ensure accountability under the Regulatory Reform (Fire Safety) Order 2005.
- A **written fire safety policy**, outlining responsibilities, procedures, and organisational commitment.
- **Regular staff training and evacuation drills**, building competence and confidence in responding to fire incidents.

- **Systematic maintenance and record keeping**, ensuring alarms, suppression, lighting, and compartmentation remain fully operational.
- **Coordination across multiple occupancies**, supported by central management structures and formal agreements where necessary.
- A **strong safety culture**, where leadership commitment, open communication, and staff engagement reinforce compliance and vigilance.

Together, these factors create a **living fire strategy**, one that evolves with the building, its use, and its occupants.

The next chapter will examine **Testing, Maintenance, and Record-Keeping** in greater detail, focusing on the **practical arrangements** required to keep fire precautions reliable, auditable, and fit for purpose throughout a building's operational life.

Chapter 20

Testing, Maintenance, and Record-Keeping

20.1 Introduction

The long-term effectiveness of fire safety measures depends not only on their **initial design and installation** but also on their **ongoing reliability**. Passive systems such as fire doors and compartment walls can be damaged or degraded over time, while active systems such as alarms, sprinklers, and smoke control require regular testing to ensure functionality. Even the most comprehensive fire strategy will fail if these measures are not **systematically tested, maintained, and properly recorded**.

The purpose of testing and maintenance is twofold:

1. To ensure that **fire precautions remain operational** when needed, preserving life safety and property protection.
2. To provide **documented evidence of compliance**, demonstrating to enforcing authorities, insurers, and stakeholders that the organisation is meeting its legal obligations.

Record-keeping is equally critical. A well-maintained fire logbook—whether paper-based or digital—serves as the official record of inspections, tests, defects, and remedial actions. Without reliable documentation, organisations cannot prove compliance, even if systems are being serviced correctly.

This chapter sets out **best practice for inspection regimes, testing schedules, and record-keeping**, drawing on statutory requirements, British Standards, and industry codes of practice. It provides a practical framework to ensure that fire precautions remain:

- **Compliant with legislation** and relevant standards.
- **Functional in practice**, not just on paper.
- **Reliable throughout the life of the building**, adapting to changes in use, occupancy, or risk.

20.2 Legal Framework

Testing, maintenance, and record-keeping of fire safety systems are not optional best practices—they are **legal requirements** under UK legislation, reinforced by British Standards and enforced by local fire authorities. Together, these obligations ensure that fire precautions remain reliable and effective throughout the operational life of a building.

Regulatory Reform (Fire Safety) Order 2005 (RRO)

- The RRO places a duty on the **Responsible Person** to ensure that all fire precautions are:
    - Maintained in an **efficient state**.
    - Kept in **efficient working order**.
    - Preserved in **good repair**.
- This duty applies to all fire safety measures, whether passive (e.g., fire doors, compartmentation) or active (e.g., alarms, sprinklers, smoke control).
- A "suitable and sufficient" **Fire Risk Assessment (FRA)** must confirm that maintenance arrangements are in place, with schedules tailored to the risks of the premises.

British Standards

Compliance with relevant British Standards provides the recognised benchmark for maintaining fire precautions. Key references include:

- **BS 5839** – Fire detection and fire alarm systems:
    - Specifies testing intervals, weekly user checks, and annual servicing by a competent contractor.
- **BS 5266** – Emergency lighting:
    - Details monthly functional checks ("flick tests") and annual full-duration discharge tests.
- **BS EN 12845** – Sprinkler systems:
    - Provides requirements for monthly inspections, pump flow tests, and annual servicing.
- **BS 5306** – Portable fire extinguishers:
    - Sets out inspection requirements, annual servicing, and extended discharge testing every 5 or 10 years depending on type.

- **BS EN 12101** – Smoke and heat control systems:
  - Covers natural and mechanical smoke ventilation, requiring regular functional testing to verify integration with alarms.

Adherence to these standards demonstrates **compliance with the RRO** and provides assurance to insurers, auditors, and enforcing authorities.

Enforcement

- **Fire and Rescue Authorities** routinely inspect maintenance records during audits or inspections.
- Inspectors will expect to see:
  - A **fire logbook** or electronic record detailing maintenance activities, test dates, outcomes, and remedial actions.
  - Evidence that defects have been rectified promptly.
  - Records of staff training in routine user checks (e.g., alarm testing, extinguisher visual inspections).
- Failure to maintain adequate records—or failure to keep systems in efficient working order—may result in:
  - **Enforcement notices**, requiring corrective action.
  - **Prohibition notices**, restricting or preventing use of the premises until compliance is achieved.
  - **Prosecution**, with unlimited fines or imprisonment for serious breaches.

Summary

The legal framework for testing and maintenance is built on the **RRO 2005**, which requires ongoing care of all fire precautions, supported by detailed **British Standards** for individual systems. **Fire authorities enforce compliance** by auditing records and inspecting premises, making accurate and up-to-date maintenance documentation essential for both safety and legal protection.

20.3 Fire Alarm and Detection Systems

Fire detection and alarm systems are the **first line of defence** in any fire strategy, providing early warning to occupants and staff so evacuation procedures can be initiated. To remain effective, systems must be tested routinely and maintained in accordance with **BS 5839-1** for non-domestic premises and **BS 5839-6** for residential buildings.

Weekly Tests

- A **manual call point** (MCP) should be operated once per week, with the location varied on a rotational basis to ensure the whole system is tested over time.
- Audible alarms must be confirmed as functioning throughout the building, and in premises with **visual alarm devices (VADs)**, these should also be checked.
- The system should be reset correctly after each test, and staff reminded of the alarm signal to reduce confusion during a real incident.
- Any failures (e.g., silent zones, inaudible signals) must be reported immediately for rectification.

Quarterly and Annual Inspections

- In addition to weekly user checks, **competent contractors** must carry out periodic servicing:
    - **Quarterly (or six-monthly)**: Inspection of alarm panels, detectors, and manual call points to confirm functionality. Smaller or lower-risk buildings may use six-monthly intervals; larger or high-risk premises should adopt quarterly inspections.
    - **Annual servicing**: Full inspection of the system, including:
        - **Detector sensitivity testing** to ensure smoke and heat detectors activate within specification.
        - Verification of panel operation, fault monitoring, and control interfaces.
        - Testing of standby power supplies (batteries or backup generators).
        - Confirmation of integration with other systems (e.g., door release mechanisms, smoke control, or suppression).

Logbook Records

- Every test and inspection must be recorded in a **fire logbook** or digital compliance system.
- Records should include:
    - Date and time of the test or inspection.

  - Location of the MCP or detector tested.
  - Results and confirmation of audibility/visibility of alarms.
  - Any faults identified and the remedial actions taken.
- Accurate record-keeping not only demonstrates compliance but also provides essential evidence in the event of enforcement action or legal proceedings.

Summary

Fire alarm and detection systems must be **tested weekly** by building staff and **serviced quarterly/annually** by competent contractors. Proper logbook records ensure compliance with **BS 5839** and provide assurance that the system will operate as intended in an emergency.

20.4 Emergency Lighting

Emergency lighting is essential for ensuring occupants can evacuate safely in the event of power failure during a fire. It provides illumination for escape routes, signage, and critical areas, enabling occupants and staff to navigate safely, while also supporting fire and rescue service intervention. Maintenance and testing must follow **BS 5266-1**, which sets out the inspection regime and performance requirements.

Monthly Flick Test

- A **short functional test** (often called a flick test) must be carried out at least once every month.
- During this test, the mains power supply to the emergency lighting system is briefly interrupted to confirm that each luminaire operates correctly in emergency mode.
- Checks should confirm that:
  - All lamps illuminate immediately.
  - Exit signs are visible and correctly positioned.
  - No damage, obstruction, or deterioration is present.
- After the test, mains power should be restored and the system allowed to recharge.

Annual Test

- A full **three-hour discharge test** must be carried out once a year by a competent person.

The system is operated on its battery backup supply for the full rated duration (usually three hours) to confirm it can maintain illumination for the required evacuation period.

Any luminaires that fail to operate for the full duration must be **repaired or replaced immediately**, as a single failure can compromise evacuation safety.

Following the test, the system must be recharged and certified as fully operational.

Records

- All tests, inspections, and remedial actions must be documented in the **fire logbook** or a digital compliance system.
- Records should include:
    - The date and time of the test.
    - Areas and luminaires tested.
    - Any defects identified.
    - Details of corrective actions taken and the date completed.
- These records serve as evidence of compliance and ensure accountability in maintaining escape route safety.

Summary

Emergency lighting systems require **monthly flick tests** and an **annual full-duration discharge test** in accordance with **BS 5266-1**. Immediate replacement of failed luminaires and thorough record-keeping ensure that escape routes remain illuminated and safe under fire conditions.

20.5 Fire Doors and Compartmentation

Fire doors and compartmentation are critical elements of a building's passive fire protection system, designed to contain fire and smoke, protect escape routes, and enable progressive or phased evacuation. Over time, these features can deteriorate through wear, damage, or poor maintenance, so **systematic inspection and record-keeping** are essential to ensure their ongoing reliability.

Routine Inspections

- **Monthly visual checks** should be carried out by trained staff or caretakers. These inspections confirm that:
    - Fire doors close fully and securely onto their frames.
    - Door closers operate correctly without sticking or slamming.
    - Intumescent and smoke seals are present and undamaged.
    - Required signage ("Fire Door – Keep Shut" or "Fire Door – Keep Locked") is in place and legible.
- **Six-monthly inspections** must be conducted by a **competent person** with detailed knowledge of fire door assemblies. These inspections assess door integrity, glazing, hardware, and frames, identifying issues that might compromise performance under fire conditions.

Common Issues

Fire door inspections frequently highlight recurring problems, such as:

- **Worn or missing intumescent strips or cold smoke seals**, reducing the door's ability to resist smoke and fire spread.
- **Damaged glazing or frames**, which can compromise the integrity of the door assembly.
- **Propped-open doors**, which negate the function of self-closing devices and break compartmentation.
- **Inappropriate alterations**, such as untested letterboxes, security viewers, or non-fire-rated hardware.
- **Warping or misalignment**, preventing the door from achieving a full seal.

Recording

- All inspections must be documented in a **fire door inspection schedule**, which is incorporated into the building's fire logbook or digital compliance system.
- Records should include:
    - The date of inspection.
    - The location of each door checked.
    - Any defects identified.

    - Remedial actions taken and confirmation of completion.
- Tracking ensures that recurring issues are addressed promptly and that the integrity of compartmentation is maintained across the building.

Summary

Fire doors and compartmentation require **monthly visual checks** and **six-monthly inspections** to remain effective. Common issues such as damaged seals, glazing, or misuse can quickly undermine fire safety if left unchecked. By maintaining inspection schedules and thorough records, organisations ensure that passive fire protection continues to perform as intended throughout the life of the building.

20.6 Firefighting Equipment

First-aid firefighting equipment, including **portable extinguishers** and **hose reels**, provides occupants and trained staff with the means to tackle small fires at an early stage. While these systems are not a substitute for evacuation, they can prevent fire escalation if used safely and appropriately. Their reliability depends on **regular inspection, servicing, and record-keeping**, in accordance with relevant British Standards.

Portable Fire Extinguishers (BS 5306-3 & BS 5306-8)

- Extinguishers must be provided in appropriate types and numbers, matched to the building's risks (e.g., water, CO_2, foam, powder, or wet chemical).
- **Monthly visual inspection (by staff):**
    - Check that extinguishers are present in their designated locations, clearly visible, and unobstructed.
    - Confirm that pressure gauges are in the green zone (where fitted).
    - Inspect for signs of damage, corrosion, leakage, or missing pins and tamper seals.
- **Annual servicing (by a competent contractor):**
    - A full check of weight, pressure, condition, and accessibility.
    - Verification of labels, signage, and operating instructions.
    - Replacement or recharge as required.

- **Extended servicing/discharge testing:**
  - Carried out every **five years** for water, foam, and powder extinguishers, and every **ten years** for CO_2 extinguishers.
  - Involves complete discharging, pressure testing of cylinders, and refilling/recharging before returning to service.

Hose Reels (BS EN 671-3)

- Hose reels may be installed in some commercial or industrial premises to provide continuous water for firefighting.
- **Annual functional test:**
  - Conducted by a competent contractor to confirm flow rates, hose condition, nozzle operation, and valve integrity.
  - Ensures that hose reels deliver adequate water supply and are free from leaks or obstructions.
- **Monthly staff checks** should confirm accessibility, signage, and that reels are not obstructed or damaged.

Training and Safe Use

- Staff must receive **basic training** in the safe use of extinguishers and hose reels.
- Emphasis should be placed on **when not to use** equipment—for example, never attempting to fight large or rapidly spreading fires, and always prioritising evacuation.
- Extinguisher training should include practical demonstrations under controlled conditions to build confidence.

Recording

- All inspections and servicing must be logged in the **fire logbook** or digital compliance system, including:
  - Dates of staff checks.
  - Contractor service reports.
  - Details of any defective or replaced equipment.
- Accurate records provide proof of compliance with **BS 5306** and demonstrate readiness for enforcement audits.

Summary

Portable extinguishers and hose reels require **monthly staff checks**, **annual servicing**, and **extended discharge testing** in accordance with British Standards. Regular training and clear record-keeping ensure that firefighting equipment remains serviceable, safe to use, and capable of providing first-aid firefighting while supporting evacuation.

20.7 Sprinklers and Suppression Systems

Sprinklers and fixed suppression systems are among the most effective active fire protection measures, capable of **controlling or extinguishing fires automatically**, reducing smoke production, and significantly extending the available safe egress time (ASET). Their performance, however, is entirely dependent on **rigorous inspection, maintenance, and record-keeping** in line with **BS EN 12845** (commercial and industrial systems) and **BS 9251** (residential and domestic systems).

Weekly / Monthly Checks

- Carried out by trained in-house staff or facilities teams to confirm system readiness.
- Checks include:
    - **Control valves**: Must be locked open or fitted with tamper-evident seals to ensure water supply availability.
    - **Pressure gauges**: Verified for correct readings, confirming that system pressure is maintained.
    - **Water supplies**: Levels and condition of storage tanks, and status of pumps or standby supplies.

Any abnormalities must be reported immediately and investigated by a competent contractor.

Quarterly / Annual Inspections

- Performed by **specialist contractors** with competence in sprinkler systems.
- **Quarterly inspections** generally cover:
    - Functional testing of alarm valves and flow switches.
    - Confirmation that pumps start automatically and operate correctly.

    - Checks on water supply arrangements, including duplicate feeds or emergency connections.
- **Annual servicing** expands to include:
    - **Full flow tests** of pumps to confirm capacity and reliability.
    - Testing of alarm bells and transmission devices.
    - Examination of pipework, hangers, and nozzles for corrosion, obstruction, or accidental damage.
    - Verification that the system design still aligns with building use and risk profile.

Record Keeping

- A **sprinkler logbook** must be maintained, recording:
    - Results of weekly/monthly in-house checks.
    - Contractor service reports, including any defects identified.
    - Remedial works and dates of completion.
- Logbooks and test sheets must be available for inspection by enforcing authorities, insurers, and auditors.
- Electronic compliance platforms are increasingly used to improve transparency and accessibility of records.

Summary

Sprinklers and suppression systems require **weekly/monthly user checks** and **quarterly/annual contractor inspections** to remain reliable. Robust record-keeping in a **sprinkler logbook** provides assurance of compliance with **BS EN 12845** or **BS 9251**, ensuring that these life-saving systems are ready to perform at any time.

20.8 Smoke Control and Ventilation

Smoke control systems are critical to preserving **tenable conditions** for evacuation and enabling safe fire service intervention. These systems, which may be natural (automatic opening vents, roof vents, or windows) or mechanical (fans, shafts, and pressurisation), must be maintained in accordance with **BS EN 12101** and tested regularly to ensure reliability in an emergency.

Weekly / Monthly Tests

- Routine tests should be carried out by trained building staff or facilities teams, with the frequency determined by system type and complexity.
- Checks should include:
    - Operation of **automatic opening vents (AOVs)** to confirm that they open fully and reliably when activated.
    - Functional testing of **natural vents and rooflights**, ensuring motors, hinges, and actuators are free from obstruction or damage.
    - Deployment of **smoke curtains**, verifying full descent and correct retraction after reset.
    - Verification that **manual overrides** and control switches are clearly labelled and operational.

Annual Tests

- A full **operational test under simulated fire conditions** must be carried out by a **competent specialist contractor** at least once a year.
- Testing should confirm:
    - Automatic activation from the fire detection and alarm system.
    - Correct sequencing between vents, fans, dampers, and make-up air supplies.
    - Pressure differentials in **pressurisation systems** are maintained within design limits to keep stairwells and lobbies smoke-free.
    - Integration with standby power supplies, ensuring resilience in the event of mains failure.
- Where systems are part of a performance-based design (e.g., atriums, car parks), testing should demonstrate that the system continues to meet its design objectives.

Records

- Every test must be documented in the **fire logbook or digital compliance system**, including:
    - Date and time of the test.
    - Identification of which **vent, fan, damper, or curtain** was tested.

- Test results and any faults identified.
- Actions taken to rectify defects and confirmation of completion.

- Comprehensive record-keeping ensures transparency and provides evidence for enforcement audits and insurance reviews.

Summary

Smoke control and ventilation systems require **weekly/monthly functional tests** of AOVs, vents, and smoke curtains, supported by **annual full-scale operational testing** under simulated fire conditions. Accurate records of each test ensure compliance with **BS EN 12101** and confirm that these critical life safety systems remain fully functional.

20.9 Firefighting Shafts and Lifts

Firefighting shafts and lifts are essential features in medium- and high-rise buildings, hospitals, and complex facilities, providing **protected access routes** for fire and rescue services and enabling the rapid transport of personnel and equipment. To ensure they are available and reliable during an emergency, they must be inspected and maintained in accordance with **BS EN 81-72** (firefighting lifts) and relevant rising main standards.

Firefighting Lifts

- Firefighting lifts are designed to remain operational in fire conditions, providing crews with direct vertical access and facilitating rescues.
- Maintenance requirements include:
    - **Monthly functional checks** carried out by building staff, ensuring the lift responds to call points, doors close correctly, and recall functions activate when tested.
    - **Annual tests by a competent contractor**, including:
        - Verification of recall and override controls for fire service use.
        - Testing of emergency communication systems between the lift car and the fire control point.
        - Confirmation that **backup power supplies** (secondary electrical feeds or generators) operate correctly.

- Operation of the lift under full **fire mode conditions** to ensure it remains reliable under stress.

Rising Mains

- Rising mains (dry or wet risers) provide water supplies to upper floors for firefighting operations. Their reliability is critical for fire service effectiveness.
- **Dry risers**:
  - **Six-monthly visual inspection**, checking inlets, outlets, landing valves, and signage for damage, vandalism, or obstruction.
  - **Annual pressure testing**, ensuring the system can withstand required operating pressures without leaks or failures.
- **Wet risers**:
  - **Six-monthly functional tests**, confirming pumps, water supplies, and valves are operational.
  - **Annual flow and pressure tests**, verifying that the system delivers adequate performance for firefighting demands.

Record Keeping

- All inspections and tests must be documented in the **fire logbook** or electronic compliance system, including:
  - Dates of checks.
  - Results of functional or pressure tests.
  - Details of any defects, repairs, or upgrades.
- Records must be readily available for **fire authority inspections**, as rising main or firefighting lift failures represent critical life safety risks.

Summary

Firefighting shafts and lifts provide the fire service with **safe, protected access and water supplies** in emergencies. **Monthly checks, six-monthly inspections, and annual full-function tests** are required to maintain operational reliability. Thorough record-keeping ensures compliance with **BS EN 81-72** and rising main standards, protecting both building occupants and fire crews.

20.10 Record-Keeping

Effective fire safety management relies on a **robust record-keeping system**, providing evidence that fire precautions are being tested, maintained, and reviewed in accordance with legal requirements and industry standards. Accurate records not only demonstrate compliance but also support accountability, transparency, and continuous improvement.

Fire Logbook

- A fire logbook remains the **primary record of fire safety activity** within a building.
- It should include entries for:
  - **Alarm system tests** (weekly, quarterly, and annual).
  - **Emergency lighting checks** (monthly flick tests and annual discharge tests).
  - **Extinguisher inspections and servicing**.
  - **Sprinkler and suppression system testing**.
  - **Smoke control and ventilation testing**.
  - **Fire drills and evacuation exercises**.
  - **Staff training sessions and attendance records**.
- Each entry should detail the **date, location, test outcome, and remedial action taken**.

Digital Systems

- Increasingly, organisations are moving towards **digital compliance platforms**.
- These provide benefits such as:
  - **Real-time monitoring** of system status.
  - **Automatic reminders** for scheduled inspections or servicing.
  - Centralised storage of service reports, certificates, and compliance data.
  - Easier access for multiple stakeholders (e.g., Responsible Person, facilities managers, auditors).
- Digital systems reduce the risk of missing deadlines and improve audit readiness.

Retention Periods

- Records should generally be retained for a **minimum of three years**, in line with standard audit cycles and potential legal proceedings.
- Longer retention is considered **best practice**, particularly for critical systems such as sprinklers, suppression systems, and smoke control, where performance histories may be relevant for insurers, courts, or inquiry investigations.
- In high-risk premises (e.g., healthcare, high-rise residential), indefinite retention of core system records is recommended.

Accessibility

- Records must be **readily available** for inspection by fire and rescue authorities, insurers, and other regulators (e.g., CQC in healthcare settings).
- They should be stored in a **secure but accessible location**, such as:
  - A dedicated fire safety file in reception or facilities offices.
  - Electronic access through compliance platforms or cloud-based systems.
- Lack of available records is often treated by fire inspectors as evidence of non-compliance, regardless of whether systems have been tested.

Summary

Record-keeping underpins fire safety management by ensuring that all tests, inspections, and training are **documented, traceable, and auditable**. Whether using a traditional logbook or digital platform, records must be **comprehensive, retained for several years, and accessible** for inspection. Without accurate records, even well-maintained fire systems may be deemed non-compliant.

20.11 Third-Party Certification

The competence of those who install, service, and maintain fire protection systems is just as important as the systems themselves. Inadequate workmanship or poorly maintained systems can render fire precautions ineffective, even if they were correctly designed. Using **third-party certified contractors** provides assurance that fire safety systems are installed and maintained to recognised industry standards by competent professionals.

BAFE (British Approvals for Fire Equipment)

- BAFE operates a range of certification schemes covering different aspects of fire safety.
- Examples include:
  - **BAFE SP203-1** – Fire detection and alarm systems.
  - **BAFE SP101** – Portable fire extinguisher organisations.
  - **BAFE SP203-3** – Fixed gaseous fire suppression systems.
- Certification confirms that contractors have been independently assessed for technical competence, quality management, and compliance with relevant standards.

LPS 1014 and LPS 1048 (Loss Prevention Certification Board – LPCB)

- **LPS 1014** – Covers the design, installation, commissioning, and servicing of fire detection and alarm systems.
- **LPS 1048** – Covers the design, installation, commissioning, and servicing of sprinkler systems.
- LPCB certification is widely recognised by insurers and regulators as a robust guarantee of competence.
- Approved companies are subject to **regular audits**, ensuring standards are maintained over time.

Why Third-Party Certification Matters

- Provides independent assurance that contractors are **competent and compliant** with industry standards.
- Reduces liability for the Responsible Person, who can demonstrate due diligence in selecting accredited providers.
- Increases confidence that systems will function correctly in an emergency.
- Often a requirement of insurers, particularly in high-value or high-risk premises.

Fire Strategy Recommendations

- A fire strategy should always recommend that **installation, servicing, and maintenance** of fire safety systems are carried out by **third-party accredited providers**.

- This ensures not only compliance with legislation but also alignment with **best practice and insurance requirements**.
- Where non-certified contractors are used, additional due diligence and independent verification may be required to mitigate risk.

Summary

Third-party certification provides assurance that fire safety contractors are **competent, audited, and compliant** with recognised standards. Schemes such as **BAFE** and **LPCB's LPS 1014/1048** demonstrate quality and reliability, supporting both compliance and confidence. Fire strategies should always recommend the use of **third-party accredited providers** for the installation and maintenance of critical life safety systems.

20.12 Continuous Improvement

Testing and maintenance regimes must go beyond a **"tick-box exercise."** While statutory checks and servicing ensure that systems meet minimum requirements, effective fire safety management involves **analysing trends and learning from recurring issues** to strengthen resilience. Continuous improvement ensures that fire precautions remain not only compliant but also robust, reliable, and adapted to the building's evolving risks.

Using Records to Identify Patterns

Fire safety records are not simply for compliance—they are valuable tools for performance monitoring. By reviewing logs, organisations can identify recurring problems that highlight deeper issues:

- **Frequent False Alarms**
    - High rates of false activations may point to incorrect detector placement (e.g., smoke detectors near kitchens or shower rooms).
    - Inappropriate detector type may also be an issue (e.g., using smoke detectors instead of heat detectors in dust-prone or humid environments).
    - Persistent false alarms undermine staff confidence and may delay evacuation response, so system design should be reassessed.

- **Repeated Fire Door Failures**
    - Consistent inspection failures may indicate poor-quality hardware, inadequate maintenance, or occupant misuse (e.g., doors wedged open).
    - Addressing root causes may require enhanced maintenance standards, replacement with more durable door sets, or targeted occupant awareness campaigns.
- **System Outages and Failures**
    - Repeated equipment downtime may suggest insufficient resilience, reliance on a single power supply, or inadequate redundancy in system design.
    - Critical systems (e.g., alarms, sprinklers, smoke control) should be reviewed for backup capacity and monitored for signs of premature component failure.

Embedding a Cycle of Improvement

- Continuous improvement requires a **feedback loop** where testing results, incident reports, and staff feedback inform system upgrades and management practices.
- Examples include:
    - Upgrading alarm systems after a series of nuisance activations.
    - Replacing poor-performing fire doors across a building rather than repeatedly repairing individual units.
    - Introducing monitoring technology to detect faults in real time, reducing response times.
- This approach shifts fire safety from reactive compliance towards **proactive risk management**.

Benefits of Continuous Improvement

- Increases system reliability and confidence among occupants and staff.
- Reduces unnecessary disruptions caused by false alarms or system downtime.
- Ensures fire precautions evolve with building use, occupancy, and technology.

- Demonstrates to regulators and insurers that the organisation treats fire safety as a **living process**, not a static obligation.

Summary

Continuous improvement ensures that testing and maintenance are used not only for compliance but also as tools for **performance optimisation**. By analysing trends such as false alarms, repeated failures, or system outages, organisations can address underlying causes, enhance resilience, and ensure that fire safety systems remain effective throughout the building's life.

20.13 Case Example

High-Rise Office Building with Digital Compliance System

A 15-storey office building adopted a **digital fire safety logbook** to manage inspection, testing, and maintenance tasks across its life safety systems.

System Implementation

- All weekly alarm tests and monthly emergency lighting checks were recorded using a **mobile compliance app**, enabling facilities staff to log results in real time.
- The app also generated **automatic reminders** for scheduled inspections, ensuring that no test was missed.
- Contractors carrying out quarterly and annual servicing uploaded certificates and reports directly into the platform, providing a centralised record of compliance.

Management Oversight

- The system produced **quarterly summary reports** automatically, which were reviewed by senior management and the Responsible Person.
- These reports highlighted:
    - Test completion rates.
    - Outstanding defects and their resolution times.
    - Trends in system performance (e.g., recurring detector faults).
- This oversight allowed management to act proactively rather than reactively, allocating resources to address emerging risks.

Audit Outcome

- During a scheduled inspection, the local fire authority reviewed the digital records.
- Inspectors noted the **high quality and completeness of documentation**, which demonstrated both legal compliance and effective management practices.
- Records showed that defective equipment, such as a failed emergency luminaire, had been replaced **within 24 hours of detection**, illustrating a strong commitment to resilience.

Benefits

- The digital system reduced administrative burden, eliminated gaps in record-keeping, and improved transparency across multiple stakeholders.
- Senior management gained greater confidence in the reliability of fire safety systems.
- The proactive approach enhanced the building's resilience, ensuring systems were consistently operational and ready in an emergency.

Summary

This case study demonstrates how a **digital compliance platform** can transform fire safety record-keeping. By combining real-time logging, automated reporting, and rapid defect resolution, the building not only achieved compliance but also strengthened its operational resilience and culture of fire safety.

20.14 Summary

Testing, maintenance, and record-keeping form the **lifeblood of operational fire safety**. A building's fire precautions, no matter how well designed or constructed, are only as reliable as the regime in place to maintain them. Without regular inspection and competent servicing, systems may degrade silently and fail at the critical moment.

Key practices include:

- **Routine inspections** at defined intervals (weekly, monthly, annually), ensuring that all active and passive systems are verified as functional.
- **Competent servicing by certified contractors**, using recognised third-party accreditation schemes to guarantee quality and reliability.

- **Robust logbooks and digital systems**, providing accurate, auditable records of tests, defects, and remedial actions.
- **Continuous improvement**, using trend analysis to address recurring faults, reduce false alarms, and strengthen resilience.

Together, these practices ensure that fire strategies remain **living systems**, capable of protecting occupants, supporting fire service intervention, and demonstrating compliance throughout the life of the building.

The next chapter will explore **Integration of Fire Strategies into the Building Lifecycle**, examining how fire safety principles must be embedded at every stage—design, construction, occupation, and refurbishment—to ensure consistent and sustainable protection.

Chapter 21

Integration of Fire Strategies into the Building Lifecycle

21.1 Introduction

A **fire strategy is not a static report**; it is a **living document** that should evolve in step with the building it protects. Its purpose is not only to demonstrate compliance at the design stage but to provide a **long-term framework** for maintaining fire safety throughout the building's lifecycle.

Too often, strategies are produced solely to satisfy **planning or building control submission requirements**, then archived or forgotten once approvals are secured. When this happens, the fire strategy no longer reflects the **operational reality** of the building, leaving critical gaps between **design intent and day-to-day practice**. Changes in layout, occupancy, or building use can quickly make an outdated strategy ineffective.

Integrating the fire strategy into the **full building lifecycle** ensures that fire safety principles are preserved, adapted, and reinforced at every stage:

- **Design**: Fire safety objectives are established, systems are specified, and performance criteria are defined.
- **Construction**: Fire-resistant materials, compartmentation, and active systems are installed correctly, with quality assurance to verify compliance.
- **Occupation and Use**: The strategy is translated into management policies, maintenance schedules, and evacuation procedures.
- **Refurbishment and Alteration**: Updates to the building are reflected in a revised fire strategy, ensuring that new risks are addressed and that original protections are not compromised.

By embedding the fire strategy into each stage of the lifecycle, buildings achieve a **continuity of fire safety** that extends beyond compliance paperwork and becomes part of the building's ongoing safety culture.

21.2 The Building Lifecycle and Fire Safety

A fire strategy must be understood as a **dynamic framework** that develops in parallel with a building. Its effectiveness depends on recognising that each stage of the building lifecycle presents **distinct fire safety needs**, requiring tailored actions to ensure the continuity of protection. The lifecycle can broadly be divided into four stages:

1. Design Stage

- The design stage is where the **fire safety objectives, assumptions, and systems** are first defined.
- Key tasks include:
  - Identifying occupancy type, risk profile, and evacuation strategy.
  - Establishing performance-based design assumptions (e.g., evacuation times, smoke control performance).
  - Specifying fire safety systems such as detection, alarms, sprinklers, smoke ventilation, and structural fire resistance.
  - Documenting these requirements in the fire strategy to provide a benchmark for later stages.

2. Construction Stage

- At the construction stage, the focus shifts to **delivery and quality assurance**.
- Fire safety features must be installed correctly, in accordance with design intent and standards.
- Critical considerations include:
  - Proper installation of passive fire protection (compartment walls, floors, doors, and cavity barriers).
  - Verification of active systems (alarms, sprinklers, smoke control) through commissioning and integrated system testing.
  - Avoiding on-site deviations that compromise fire safety (e.g., penetrations through compartment walls left unsealed).
- The fire strategy should be **referenced during site inspections** and updated to record as-built details.

3. Occupation and Use

- Once occupied, fire safety becomes an **operational responsibility**.
- The fire strategy must be translated into:
  - Fire safety management policies, staff roles, and evacuation procedures.
  - Testing, maintenance, and record-keeping regimes for fire precautions.
  - Adaptations to reflect real-world occupant behaviour and usage patterns.
- The strategy should evolve with changes in risk, for example, increased occupant loads or new fire loads introduced by tenants.

4. Refurbishment and Change of Use

- Buildings are rarely static; refurbishment, extension, or a change in use can **invalidate original fire safety assumptions**.
- At this stage, the fire strategy must be **reviewed and updated**, taking account of:
  - Altered layouts, compartmentation, or escape routes.
  - Additional or different occupant risk profiles (e.g., conversion from offices to residential).
  - The need for upgraded fire precautions to meet current codes and standards.
- Integration with building control and fire authorities is essential to ensure ongoing compliance.

Summary

By treating the fire strategy as a **living document across the four stages of design, construction, occupation, and refurbishment**, organisations can ensure that fire safety is embedded throughout the building lifecycle. This approach prevents the common pitfall of "static strategies" and delivers a **continuous chain of protection** from concept to long-term use.

21.3 Design Stage

The **design stage** is the foundation of a fire strategy. Decisions taken here establish the objectives, assumptions, and measures that will guide the entire lifecycle of the building. A well-prepared fire strategy at this point ensures that fire safety is **embedded in the design**, rather than applied as a retrofit solution.

Establishing Objectives

- The strategy should clearly set out the **objectives** for the building, which typically include:
    - **Life safety** – ensuring occupants can escape before conditions become untenable.
    - **Property protection** – safeguarding the structure and contents against major damage.
    - **Business continuity** – enabling the organisation to recover quickly after a fire.
    - **Environmental protection** – limiting the impact of smoke, water, or fire spread on surrounding areas.

Defining Assumptions

- Fire strategies rely on a set of **baseline assumptions**, which must be realistic and evidence-based. These include:
    - **Occupant load and characteristics** (e.g., numbers, mobility, familiarity).
    - **Likely fire scenarios**, such as ignition sources, fire load, and potential growth rates.
    - **Fire service response times** and access arrangements.
    - **Availability of active systems** such as sprinklers or smoke control, and their expected reliability.

Identifying Passive and Active Measures

- At design stage, the fire strategy specifies both **passive protection measures** (compartmentation, structural fire resistance, fire doors) and **active systems** (detection, alarms, sprinklers, smoke control).
- These measures are integrated to provide **layered protection**, ensuring that if one element fails, others remain in place to safeguard life and property.

Integration with Architectural, Structural, and M&E Design

- Fire safety must be woven into the building design, not treated as an afterthought.
- Examples include:
  - Architectural layouts designed to minimise excessive travel distances.
  - Structural elements specified with required fire resistance ratings.
  - Mechanical and electrical (M&E) systems integrated with fire alarms, smoke control, and suppression.
- Coordination between disciplines ensures that systems do not conflict and that fire safety features are built in efficiently.

Supporting Applications and Consultation

- A formal fire strategy supports **planning applications** by demonstrating that fire safety has been considered at concept stage.
- It provides the technical evidence required for **building control approval** and forms the basis for consultation with the **fire and rescue authority**.
- By presenting a clear, documented approach, the fire strategy provides assurance to regulators, insurers, and stakeholders.

Best Practice

- The most effective fire strategies are prepared **early in the design process**, before layouts, systems, and materials are fixed.
- This ensures that fire safety is **embedded**, influencing design choices, rather than being retrofitted later at higher cost and with reduced effectiveness.

Summary

At the design stage, the fire strategy establishes the **objectives, assumptions, and measures** that will guide the project, embedding fire safety into architectural, structural, and services design. Early preparation ensures compliance, efficiency, and resilience, setting the stage for successful construction and operation.

21.4 Construction Stage

The **construction stage** is critical in translating the fire strategy from paper into a functioning reality. Even the most robust design will fail if fire safety measures are incorrectly installed, substituted with unsuitable products, or compromised during construction. The fire strategy must therefore act as a **reference document** throughout the build, guiding material selection, installation standards, and quality assurance.

Specifications

- Fire safety depends on the correct use of **fire-rated materials and systems**.
- Contractors must follow specifications precisely, ensuring:
    - Structural elements achieve the required **fire resistance periods**.
    - Fire doors, partitions, and glazing are sourced from tested and certified suppliers.
    - Fire-stopping products (intumescent sealants, collars, wraps) are appropriate for the penetrations being sealed.
- Substitution of materials must be strictly controlled and only permitted where evidence of equivalent performance is provided.

Installation Quality

- Even certified products will fail if **poorly installed**.
- Key areas where installation quality is critical include:
    - **Fire doors**: correct fitting of frames, hinges, closers, and intumescent seals.
    - **Dampers and ductwork**: ensuring fire and smoke dampers are correctly located, accessible for maintenance, and properly sealed.
    - **Fire stopping**: sealing around services to restore compartment integrity.
- Competent installers should provide **certification or photographic evidence** of installations, forming part of the building's golden thread of information.

Change Control

- Construction projects often involve design or layout changes. Each change must be reviewed against the fire strategy to assess its impact on fire safety.
- Examples include:
    - Altered partition layouts affecting compartmentation or escape routes.
    - Relocated plant rooms requiring new smoke or suppression provisions.
    - Value engineering substitutions of fire-rated products.
- Fire engineers should review and approve changes to ensure that **design intent and regulatory compliance are preserved**.

Site Fire Safety

- Fire safety during construction is as important as permanent measures, since partially completed buildings may be vulnerable to fire.
- Temporary precautions should include:
    - **Clearly marked escape routes**, kept clear of obstructions.
    - **Temporary alarms and detection systems** suited to the construction phase.
    - **Hot works permits** and fire watch arrangements to control ignition risks.
    - Adequate extinguishers on each floor or work area.
- The fire strategy should provide guidance on **phased fire precautions**, evolving as construction progresses.

Role of Fire Engineers

- Fire engineers play a vital role in ensuring design assumptions are met during construction.
- Their responsibilities include:
    - **Site inspections** to verify compartmentation, penetration sealing, and system installations.
    - Reviewing **certification and test evidence** for fire safety products.

    - Providing input into design changes, substitutions, or non-standard details.
- Regular inspections ensure the **integrity of fire safety features** is protected before handover.

Summary

At the construction stage, the fire strategy ensures that **specifications are met, installations are correct, changes are controlled, and temporary precautions are in place**. Active involvement from fire engineers through site inspections and approvals provides assurance that the building's fire safety objectives are achieved in practice.

21.5 Occupation and Use

Once a building is brought into service, the fire strategy transitions from being a design and compliance document to an **operational management tool**. Its purpose shifts from securing approvals to guiding day-to-day fire safety arrangements, ensuring that the original design intent is preserved and adapted to real-world use.

Baseline for Fire Risk Assessments

- The fire strategy provides the **starting point for all future Fire Risk Assessments (FRAs)**.
- It records the assumptions made at design stage, such as occupancy types, evacuation methods, and system dependencies, against which risk assessors can measure ongoing suitability.
- Any changes in building use, occupancy load, or layout can then be reviewed against this baseline.

Guidance for Training and Evacuation Planning

- The fire strategy defines the intended **evacuation strategy** (simultaneous, phased, stay-put, or progressive horizontal evacuation), which informs staff training.
- It provides details on escape routes, refuges, and firefighting access points that staff must be familiar with.
- Fire wardens, marshals, and managers can use the strategy as a reference for evacuation drills and training exercises.

Maintenance Requirements

- Both active (alarms, sprinklers, smoke control) and passive (fire doors, compartmentation) fire safety systems have defined maintenance schedules.
- The fire strategy sets out these requirements and links them to statutory obligations and British Standards.
- Facilities management teams can therefore integrate fire safety into planned preventive maintenance regimes.

Reference for Drills, Audits, and Inspections

- Fire strategies serve as a **reference document** during fire drills, internal audits, and external inspections by regulators or insurers.
- Inspectors often request to see the strategy to verify that operational arrangements reflect the design intent.
- Regular audits against the strategy ensure that fire safety remains aligned with both legislation and building-specific risks.

Support for Insurance and Business Continuity

- Insurers frequently request evidence of a fire strategy when assessing property protection and business continuity risks.
- The strategy demonstrates that fire safety has been considered systematically, providing reassurance of reduced risk exposure.
- It can also assist in developing **business continuity planning**, particularly in identifying critical systems (e.g., server rooms, manufacturing lines) that require enhanced fire protection.

Responsibilities of the Responsible Person

- Under the Fire Safety Order, the **Responsible Person** must ensure that the fire strategy remains:
    - **Available** to managers, safety officers, and contractors.
    - **Understood** by those responsible for training, maintenance, and emergency planning.
    - **Reviewed** regularly to reflect operational changes or new risks.
- Embedding the strategy into daily management processes prevents the disconnect between **design intent and operational reality**.

Summary

In the occupation and use phase, the fire strategy becomes a **living management document**, guiding fire risk assessments, training, maintenance, and audits. It provides assurance to regulators, insurers, and building users while ensuring that fire safety remains effective throughout day-to-day operations. The Responsible Person plays a central role in ensuring the strategy is both accessible and applied in practice.

21.6 Refurbishment and Change of Use

Buildings evolve; **fire strategies must evolve with them**. Refurbishment works and changes in occupancy can invalidate original assumptions about compartmentation, evacuation, and system performance. A structured review and approval process is essential to keep the **design intent aligned with operational reality**.

Refurbishment Risks

- **Compromised fire stopping**
  - New service penetrations (data cabling, pipework, ducts) often breach compartment lines.
  - Require tested fire-stopping solutions matched to substrate, service type, and movement; photographic evidence and labelling should be recorded in the building's information set ("golden thread").
- **Dampers and ducts**
  - Relocated or enlarged ducts must retain **fire/smoke damper** integrity and accessibility for maintenance; re-commission shutdown/override logic.
- **Layout changes**
  - Reconfigured floors can **increase travel distances**, create **dead ends**, or **reduce exit capacity**/distribution. Run updated calculations (ADB/BS 9999) and, where needed, evacuation modelling.
- **Ceiling/void alterations**
  - Removal of fire-resisting ceilings or bulkheads can reduce compartment performance; detector and sprinkler head spacing/obstructions must be re-checked.

- **Cladding and façade works**
    - Replacement systems may alter **external fire spread** and **cavity barrier** strategy; ensure materials, fixings, and cavity barriers meet current standards and detailing.
- **Services and plant upgrades**
    - New battery systems, generators, medical gases, or kitchen plant can change the **risk classification**, demanding suppression, ventilation, or detection enhancements.
- **Cause & effect drift**
    - Integration changes (doors, access control, lifts, smoke control) require an updated **cause-and-effect matrix** and **integrated systems testing (IST)**.

Change of Use

- **Different occupancy profiles** (e.g., offices → residential; warehouses → leisure) change:
    - **Evacuation strategy** (phased/stay-put/PHE vs simultaneous).
    - **Fire load and ignition sources** (kitchens, retail stock, stage effects).
    - **Management dependencies** (transient crowds vs sleeping risk).
- Re-evaluate: travel distances, exit widths/numbers, compartment sizes, detection/suppression category, smoke control approach, and management arrangements.
- Engage early with **Building Control** and, where appropriate, the **fire and rescue service** to confirm the regulatory route and evidence required.

Review Cycle

- **Trigger-based reviews**: update the fire strategy **whenever significant changes** occur—layout alterations, façade works, new systems, occupancy change, or repeated management issues (e.g., false alarms).
- **Periodic reviews**: formally review the strategy at least **every 3–5 years** to confirm ongoing suitability, even without major works.
- **Quality assurance**:
    - Use a **design change control** process with fire engineer sign-off for substitutions and deviations.

  - Require **as-built drawings**, product certification, and installation records; update the fire strategy and O&M manuals.
  - Conduct **commissioning and IST** after works (alarms, voice, smoke control, lifts, door holds) and refresh **staff training**/drill plans.

Site Fire Safety During Works

- Maintain temporary **means of escape**, alarms/detection suited to construction phase, hot-works permits, and housekeeping.
- Phase works to avoid compromising critical compartments or escape routes; where unavoidable, provide **temporary barriers** and **management controls**.

Best-Practice Checklist

- Early fire engineer involvement for scope definition and compliance route.
- Documented change control with impact assessment on the fire strategy.
- Verified fire-stopping and cavity barrier installations (tagged and photographed).
- Updated calculations (travel distance/exit capacity), cause-and-effect, and IST.
- Revised evacuation plans, signage, and staff briefings before reoccupation.

Summary

Refurbishment and change of use can erode the protections your original strategy relies on. By applying **formal review triggers, rigorous change control, robust commissioning, and updated management arrangements**, you maintain a continuous, evidence-based link between **design intent** and **in-use performance**.

21.7 Bridging the "Golden Thread"

The **Golden Thread of Building Safety**, introduced under the **Building Safety Act 2022**, establishes a framework where information about safety-critical systems is **created, maintained, and made accessible** throughout the entire lifecycle of a building. Its purpose is to ensure that fire and structural safety are not treated as one-off compliance hurdles, but as continuous responsibilities supported by accurate, up-to-date information.

A **fire strategy** sits at the very heart of this Golden Thread, providing the **narrative and evidence base** that connects design intent, construction delivery, operational management, and later alterations.

Role of the Fire Strategy in the Golden Thread

- **Design Stage**
    - The fire strategy is first created, setting out objectives, assumptions, and performance requirements.
    - Provides the foundation for fire safety design, regulatory submissions, and consultation with fire authorities.
- **Construction and Handover**
    - The strategy must be updated with **as-built information**, reflecting any approved changes or substitutions.
    - Installation records, product certification, and commissioning results should be appended to ensure that what was designed has actually been delivered.
- **Occupation and Operation**
    - The fire strategy becomes a **management tool**, guiding risk assessments, maintenance, and evacuation procedures.
    - Updated logbooks, inspection results, and training records should be linked back to the strategy, showing how design measures are being maintained.
- **Refurbishment and Change of Use**
    - Whenever alterations occur, the strategy must be **reviewed and revised** to ensure assumptions remain valid.
    - This ensures that compartmentation, evacuation capacity, and fire safety systems continue to meet both regulatory and operational requirements.

Benefits of Integration

- **Traceability**: Maintains a clear record of fire safety decisions, designs, and actions over time.
- **Accountability**: Identifies who was responsible for decisions at each stage, supporting both compliance and liability management.

- **Regulatory Compliance**: Meets the expectations of the Building Safety Regulator and aligns with the statutory duty to preserve building safety information for Higher-Risk Buildings (HRBs).
- **Transparency**: Provides accessible information to building owners, managers, contractors, regulators, and—where relevant—residents.

Summary

Embedding the fire strategy within the **Golden Thread** ensures that fire safety is not static but a **continuous, documented process**. By creating the strategy at design stage, updating it through construction, maintaining it during occupation, and revising it during refurbishment, organisations achieve **traceability, accountability, and compliance**—strengthening confidence in both life safety and regulatory oversight.

21.8 Case Example

University Science Building – Golden Thread in Practice

A new university science building was designed with a robust fire safety strategy that specified:

- **Automatic sprinkler protection** throughout.
- **Mechanical smoke control systems** in atrium and laboratory areas.
- **120-minute compartmentation** between teaching spaces, laboratories, and plant areas.

Construction Phase

- During construction, the fire engineer carried out **regular inspections**, focusing particularly on:
    - Fire-stopping around service risers and penetrations.
    - Correct installation of fire doors and dampers.
    - Verification of compartment lines against design drawings.
- Deviations and defects were documented, corrected on site, and recorded in the project's **fire safety file** as part of the Golden Thread.

Handover Phase

- At project completion, the **as-built fire strategy** was submitted to the university estates department.

- The document became the basis for the institution's **fire safety policy**, guiding fire risk assessments, staff training, and maintenance regimes.
- The fire strategy was uploaded into the university's **digital building information system**, making it accessible to facilities staff and auditors.

Occupation and Use

- The building was managed in line with the documented strategy:
  - Regular sprinkler and smoke control tests were logged.
  - Fire drills were carried out each semester.
  - Compartmentation checks were included in maintenance schedules.
- This ensured that the original design intent remained aligned with day-to-day operation.

Refurbishment – Five Years Later

- A section of the building was converted into **specialist laboratories**, introducing higher fire loads, hazardous chemicals, and pressurised gas systems.
- The fire strategy was formally **reviewed and updated**, accounting for:
  - Enhanced fire detection in chemical storage areas.
  - Increased reliance on active suppression to manage additional risks.
  - Confirmation that 120-minute compartmentation remained suitable to contain higher fire severity.
- Revised evacuation procedures and staff training were implemented, reflecting the change in risk profile.

Outcome

- The building's fire strategy was not left static but treated as a **living document**, updated at each stage of the lifecycle.
- This ensured regulatory compliance, continuity of the Golden Thread, and maintained a high level of safety for both students and staff despite significant changes in building use.

21.9 Summary

Integration of fire strategies into the building lifecycle ensures that **fire safety is continuous, not temporary**. A strategy is most effective when treated as a **living document** that adapts alongside the building, rather than a static report filed away after design approval.

Key practices include:

- **Preparing fire strategies early** in the design process so that fire safety is embedded, not retrofitted.
- **Verifying compliance during construction** through inspections, certification, and change control.
- **Using the strategy as a management tool** during occupation, guiding maintenance, training, and emergency planning.
- **Reviewing and updating strategies** during refurbishment or change of use to reflect altered layouts, risks, and occupancies.
- **Embedding the strategy within the Golden Thread** of building safety, ensuring traceability, accountability, and regulatory compliance throughout the building's life.

By following these principles, fire strategies become a **dynamic framework** that preserves life safety, supports resilience, and provides assurance to regulators, insurers, and building users alike.

The next chapter will present **Case Studies**, demonstrating how fire strategies have been applied in practice across a variety of building types, from high-rise residential blocks to complex healthcare facilities.

Chapter 22
Case Studies

22.1 Introduction

Case studies provide practical insight into how fire strategies are applied in **real-world buildings**. While the preceding chapters have outlined principles, standards, and methodologies, case studies illustrate how these translate into workable solutions under different conditions of scale, use, and complexity.

This chapter presents a series of case studies across a range of building types, including:

- **High-Rise Residential** – where compartmentation and stay-put strategies are critical.
- **Commercial Offices** – balancing open-plan layouts with evacuation modelling and smoke control.
- **Healthcare and Care Homes** – requiring progressive horizontal evacuation and resilient detection/suppression.
- **Assembly and Recreation Venues** – where crowd safety, communication, and rapid evacuation dominate.
- **Transport Hubs** – with large atriums, transient populations, and performance-based smoke control.

Each case study will explore:

1. **Building context** – purpose, scale, and occupancy profile.
2. **Fire risks and challenges** – identifying the unique hazards posed by the building type.
3. **Strategy features** – key passive and active measures, evacuation arrangements, and management systems.
4. **Lessons learned** – good practice, innovation, and areas where risks were successfully mitigated (or, in some instances, where incidents highlighted deficiencies).

Through these examples, the importance of treating fire strategies as **dynamic, integrated, and practical documents** will be reinforced, showing how theory and standards are applied in practice to protect life, property, and business continuity.

22.2 Case Study 1 – High-Rise Residential Tower

Context

A 32-storey residential tower with a **mixed-use retail podium** accommodates approximately **600 residents**. Standing at **100 metres in height**, the building falls into the category of a higher-risk residential building (HRRB) under UK legislation, requiring enhanced fire safety measures.

Fire Strategy Features

- **Stay-Put Policy**:
    - The building adopts a defend-in-place approach, with residents instructed to remain in their flats unless directly affected.
    - This is supported by **120-minute fire-resisting compartment floors and walls**, ensuring fire and smoke remain contained within individual units.
- **Sprinklers**:
    - A full **residential sprinkler system** (BS 9251) provides automatic suppression in apartments and common areas.
    - Sprinklers are designed not only for life safety but also to protect against rapid fire spread in densely populated dwellings.
- **Firefighting Access**:
    - The tower includes **two firefighting shafts**, each with a firefighting lift, protected stair, and rising main.
    - **Dry risers** serve levels up to 60 m, while **wet risers** operate above, ensuring reliable firefighting water supply at all heights.
    - Riser placement was determined in consultation with the fire service, balancing access convenience with hydraulic performance.

- **Smoke Control**:
    - **Natural smoke shafts** are installed in each residential corridor, designed to vent smoke in the event of a fire.
    - **Computational Fluid Dynamics (CFD) modelling** was used to verify that smoke clearance times provided adequate evacuation conditions and firefighter access.
- **Façade Safety**:

External walls are constructed with **non-combustible cladding** to **Euroclass A2-s1,d0**, in compliance with post-Grenfell regulations prohibiting combustible materials on high-rise residential buildings.

Lessons Learned

- **Compartmentation Integrity**
    - Robust compartmentation remains the cornerstone of residential high-rise safety, enabling a stay-put policy to function effectively.
    - Regular inspections and a clear repair strategy are essential to ensure penetrations, service routes, and door performance are not compromised over time.
- **Early Fire Service Engagement**
    - Early consultation with the fire service on **riser placement, firefighting lift locations, and shaft design** ensured that operational access was practical and efficient.
    - This reduced potential delays during an incident and enhanced confidence in the building's design.
- **Resident Confidence and Communication**
    - A clear **public communication strategy** was vital in gaining resident trust in the stay-put policy.
    - Leaflets, resident meetings, and evacuation drills helped to explain the rationale behind the design and reassure occupants of their safety.

22.3 Case Study 2 – Shopping Centre with Atrium

Context

A regional shopping mall of **80,000 m²** across five levels features a large **central glazed atrium** as its architectural focal point. At peak times, the centre accommodates up to **15,000 visitors**, many of whom are transient and unfamiliar with the layout. The combination of high occupancy, open volumes, and multiple tenants presents significant fire safety challenges.

Fire Strategy Features

- **Extended Travel Distances**
  - Large open retail floors often exceeded the prescriptive travel distances set out in Approved Document B.
  - Using **BS 7974 performance-based analysis**, extended travel distances were justified through evacuation modelling and tenability assessments.
- **Sprinkler Protection**
  - All retail units, back-of-house areas, and mall circulation spaces were fitted with sprinklers.
  - This reduced heat release rates and smoke production, enabling smoke control systems to operate effectively.
- **Mechanical Smoke Extraction**
  - A high-capacity **mechanical smoke extract system** was installed, with fans capable of moving **120 m³/s**.
  - Deployable **smoke curtains** at atrium edges created reservoirs, directing smoke to extract points.
  - This arrangement maintained a smoke-free layer of at least **2.5 m** above floor level for safe evacuation.
- **Phased Evacuation with Voice Alarm**
  - A **phased evacuation strategy** was adopted to prevent overcrowding on escape routes.
  - A **digital voice alarm system** provided clear, multilingual instructions, overriding background music and tenant audio systems.

- **Central Fire Control Room**
    - A dedicated **fire control room** housed panels for detection, suppression, and smoke control systems.
    - From here, operators and attending fire crews could monitor events in real time and implement overrides where necessary.

Lessons Learned

- **CFD Modelling as a Justification Tool**
    - **Computational Fluid Dynamics (CFD) modelling** was critical in demonstrating that smoke control systems would maintain tenable conditions in the atrium under credible fire scenarios.
    - This modelling provided evidence to satisfy regulators and build confidence in performance-based solutions.
- **Management Across Multiple Tenants**
    - Coordinating fire safety in a shopping centre requires **clear agreements** between centre management and individual retailers.
    - Tenants were made responsible for in-unit precautions, while central management oversaw alarms, evacuation, and smoke control.
- **Voice Alarm Effectiveness**
    - The use of **voice alarms** significantly improved compliance compared to traditional bells or sirens.
    - In a noisy, distraction-heavy environment such as a shopping mall, clear verbal instructions were essential to reduce hesitation and direct occupants effectively.

22.4 Case Study 3 – Healthcare Facility

Context

An NHS acute hospital houses **20 wards**, alongside **intensive care units (ICUs)** and **operating theatres**. The building supports approximately **500 patients**—many of whom are non-ambulant or require medical support—and **1,200 staff**. The complexity of patient dependency requires a fire strategy that prioritises **progressive evacuation, staff-led response, and system resilience**.

Fire Strategy Features

- **Compartmentation and Sub-Compartmentation**
    - The hospital is divided into **60-minute fire compartments**, further subdivided into **30-minute sub-compartments**.
    - This arrangement enables **Progressive Horizontal Evacuation (PHE)**, allowing staff to move patients into an adjacent fire-resisting area on the same level, without immediate need for vertical evacuation.
- **Sprinkler Protection**
    - **Sprinklers** are installed throughout patient and staff areas, reducing the likelihood of uncontrolled fire growth and supporting defend-in-place measures in high-dependency units.
- **Double Knock Detection**
    - The hospital uses a **double knock system**, where two independent detectors must activate before a full fire alarm is triggered.
    - This reduces the chance of false alarms disrupting sensitive clinical activities such as surgeries or intensive care treatments.
- **Staff Paging System**
    - A **silent paging system** provides early alerts to staff, enabling investigation and response before a general alarm is sounded.
    - This avoids unnecessary patient panic or disruption while ensuring staff remain aware of developing risks.
- **Fire Control Room**
    - A dedicated **fire control room** provides live monitoring of detection zones and compartmentation status.
    - Large-format displays allow staff and fire service personnel to identify affected wards and evacuation routes in real time.

Lessons Learned

- **Value of Sub-Compartmentation**
    - The layered structure of 60-minute compartments with 30-minute sub-compartments proved critical for enabling staff to carry out **phased, manageable evacuation**.

    - This reduced reliance on vertical movement, which is particularly challenging for bedbound patients.
- **Silent Staff Alerting**
    - The use of paging systems significantly reduced unnecessary patient panic and disruption, particularly in high-dependency areas.
    - Staff could act quickly without triggering large-scale evacuations for false or minor alarms.
- **Importance of Staff Training**
    - Structural measures and suppression systems were only effective when paired with **staff competence**.
    - Quarterly drills, including scenario-based training for ICUs and operating theatres, were found to be as critical as the physical measures themselves in ensuring readiness.

22.5 Case Study 4 – Warehouse and Logistics Hub

Context

A modern **50,000 m² automated warehouse** was constructed to support a national logistics network. The facility incorporates **high-rack storage up to 18 metres** and is predominantly stocked with **packaging materials and plastics**, resulting in a high fire load. Approximately **200 staff** are employed on site, with most based in adjacent office and welfare areas and only limited personnel regularly present in the storage hall itself.

Fire Strategy Features

- **Suppression Systems**
    - The building is protected by **Early Suppression Fast Response (ESFR) sprinklers** at roof level, specifically designed for rapid suppression of high-challenge fires in large-volume warehouses.
    - **In-rack sprinklers** supplement this system, targeting concealed ignition points deep within storage racks.
    - The combined approach provides faster suppression and reduces the likelihood of vertical fire spread through multiple rack levels.

- **Smoke Ventilation**
  - **Automatic roof vents** provide smoke clearance, activated in conjunction with sprinkler operation.
  - Venting design was supported by modelling to ensure smoke movement did not compromise escape routes or fire service access.
- **Compartmentation**
  - Due to the scale and height of the facility, traditional **compartmentation was impractical**.
  - The fire strategy therefore relied on active suppression, smoke ventilation, and well-rehearsed evacuation procedures.
- **Evacuation Arrangements**
  - A **simultaneous evacuation strategy** was implemented, with all staff required to leave the premises immediately upon alarm activation.
  - **Quarterly evacuation drills** ensured staff were familiar with routes and could navigate the vast floorplate confidently.

Lessons Learned

- **Critical Role of Suppression**
  - In very large, high-rack warehouses, **sprinklers—particularly ESFR and in-rack systems—are essential** to control fire growth. Passive measures alone cannot provide effective protection at this scale.
- **Housekeeping and Waste Control**
  - Regular **housekeeping audits** reduced hazards associated with combustible packaging waste and shrink-wrap materials.
  - Effective management of these minor risks was vital to prevent escalation into major incidents.
- **Staff Training and Evacuation**
  - Despite the building's scale, **structured training and regular drills** ensured rapid and confident evacuation.
  - Clear signage, well-maintained escape routes, and designated assembly areas contributed significantly to egress efficiency.

22.6 Case Study 5 – Underground Transport Station

Context

A major **underground metro interchange** was designed to accommodate high passenger volumes and complex train operations. The station includes **two concourses and three platforms**, serving as a critical interchange node within the network. At peak times, the facility manages up to **20,000 passengers**, many of whom are transient and unfamiliar with the layout.

Fire Strategy Features

- **Mechanical Smoke Extraction**
  - The underground environment demanded a **high-capacity mechanical smoke control system**, with fans rated at **180 m^3/s**.
  - Backup generators and **system redundancy** ensured continuous operation, even in the event of mains power failure.
  - Extraction design was performance-based, with scenarios modelled using CFD to maintain tenable conditions during phased evacuation.
- **Cross-Passages**
  - **Cross-passages every 150 metres** were incorporated into tunnel design, providing safe evacuation routes between parallel tunnels.
  - These also offered refuge for passengers and additional access routes for fire and rescue services.
- **Evacuation Systems**
  - Evacuation relied on **illuminated floor-level wayfinding** to guide passengers in low-visibility smoke conditions.
  - A **digital voice alarm system** was integrated, capable of providing clear, multi-lingual instructions to large crowds in real time.
- **Firefighting Access**
  - **Firefighting shafts and lifts** were provided at all primary access points to deliver safe, dedicated routes for emergency services.

   - These shafts integrated with the smoke control system to ensure pressurisation and protect firefighters entering the tunnels.

Lessons Learned

- **Performance-Based Smoke Control**
   - The **scale and complexity** of underground stations make prescriptive solutions insufficient.
   - A performance-based approach, supported by CFD analysis, was essential to demonstrate safe evacuation and firefighting conditions.
- **Passenger Behaviour Insights**
   - Research into crowd behaviour showed that many passengers attempted to **evacuate via their entry routes**, even when safer alternatives existed.
   - This highlighted the need for **intuitive signage, illuminated wayfinding, and active staff guidance** to redistribute flows effectively.
- **Collaboration with Emergency Services**
   - Early and continuous consultation with the **fire and rescue service** during design was critical.
   - Joint reviews of smoke control scenarios, shaft locations, and access routes not only streamlined regulatory approval but also improved emergency preparedness.

22.7 Case Study 6 – Historic Theatre

Context

A **Grade II listed theatre** with seating capacity for **1,000 people** underwent a major refurbishment to enhance its fire safety provisions while preserving historic architectural features. The building primarily accommodates **visitors unfamiliar with the layout**, increasing reliance on clear communication and staff guidance during emergencies.

Fire Strategy Features

- **Sprinkler Protection**
   - A **full sprinkler system** was installed throughout the theatre.

  - To protect heritage aesthetics, **concealed sprinkler heads** were used, ensuring fire suppression capability without detracting from the ornate plasterwork and decorative finishes.
- **Aspirating Smoke Detection (ASD)**
  - Due to the **high ceilings, concealed voids, and decorative linings**, traditional point detectors were unsuitable.
  - An **ASD system** provided highly sensitive, early-warning detection by sampling air through discreet pipework, allowing faster response times.
- **Voice Alarm System**
  - A **voice alarm system** was integrated with the theatre's existing public address (PA) infrastructure.
  - This provided clear instructions in the event of fire, ensuring unfamiliar visitors were directed quickly and calmly to exits.
- **Compartmentation Enhancements**
  - During refurbishment, **fire-resisting linings** were added discreetly behind decorative finishes to strengthen compartmentation.
  - These measures preserved the original character of the theatre while ensuring modern standards of fire separation.

Lessons Learned

- **Balancing Heritage and Safety**
  - Protecting historic fabric while upgrading fire precautions required **bespoke, sensitive solutions**. Collaboration between fire engineers, architects, and conservation specialists was essential.
- **ASD as a Critical Tool**
  - Aspirating smoke detection proved invaluable in providing **early warning** in a challenging environment where conventional detectors would have been visually intrusive and technically less effective.
- **Consultation with Heritage Bodies**
  - Early and transparent consultation with **heritage and planning authorities** ensured that proposals were acceptable, balancing safety improvements with conservation obligations.

22.8 Case Study 7 – Hotel and Conference Centre

Context

A **20-storey hotel and conference centre** combines **500 guest bedrooms** with **extensive meeting and event facilities**, including a ballroom, multiple breakout rooms, and a 1,500-seat conference hall. The mixed-use nature of the building presents combined risks from **sleeping accommodation** and **large transient crowds**, each requiring distinct evacuation approaches.

Fire Strategy Features

- **Phased Evacuation**
    - A **phased evacuation strategy** was adopted to avoid overwhelming stair capacity.
    - Guest bedrooms were provided with a **stay-put provision** supported by 60-minute compartmentation, while the conference areas followed a **simultaneous evacuation** model.
- **Detection and Alarm**
    - Bedrooms were fitted with **point smoke detectors** connected to an addressable fire alarm system.
    - Conference and ballroom spaces used a **voice alarm system** capable of overriding background music and AV systems to deliver clear evacuation messages.
- **Suppression**
    - **Sprinklers** were installed throughout, designed to BS 9251 for residential areas and BS EN 12845 for large event spaces.
    - Kitchens and catering areas included **wet chemical suppression** for deep fat fryers.
- **Means of Escape**
    - Multiple stair cores were provided, sized to accommodate phased evacuation loads.
    - Exits from the ballroom and conference halls were **widely distributed**, ensuring rapid dispersal of large crowds.

- **Fire Safety Management**
    - Staff were trained for **dual roles**: hotel staff supported progressive evacuation of bedroom floors, while event staff directed crowds from the conference facilities.
    - Regular drills tested both evacuation models, including cross-over scenarios where hotel guests and delegates were present simultaneously.

Lessons Learned

- **Tailored Strategies for Mixed-Use Buildings**
    - A single evacuation approach was impractical. Distinct strategies for **sleeping risk** and **large assembly areas** were required, but they had to be carefully coordinated to avoid conflict.
- **Importance of Voice Alarms**
    - The **voice alarm system** proved essential in directing conference delegates, particularly international visitors unfamiliar with the layout.
    - Pre-recorded multi-lingual messages increased compliance and reduced hesitation.
- **Staff Coordination is Critical**
    - Staff training was found to be as important as physical systems.
    - Clear division of roles between hotel staff and event staff prevented confusion and ensured both guests and delegates were supported appropriately during drills.

22.9 Case Study 8 – University Campus Building

Context

A newly built **university campus building** combines **teaching spaces, laboratories, and student accommodation** in a single complex. The building includes:

- A four-storey teaching block with lecture theatres and seminar rooms.
- Two floors of laboratories with fume cupboards and chemical storage.
- An adjoining 12-storey student residence with 400 bedrooms. The combination of **educational, laboratory, and residential uses**

created a challenging risk profile requiring careful coordination of fire safety measures.

Fire Strategy Features

- **Compartmentation and Zoning**
    - Teaching areas, laboratories, and residential accommodation were separated by **120-minute compartment walls and floors**, preventing fire spread between different uses.
    - Sub-compartmentation was introduced in the laboratories, enabling hazardous areas to be isolated quickly.
- **Suppression Systems**
    - **Sprinklers** were installed throughout the student residence in line with BS 9251.
    - Laboratories were fitted with a combination of **sprinklers and gaseous suppression systems** in high-value equipment rooms, protecting sensitive research areas.
- **Detection and Alarm**
    - **Aspirating smoke detection (ASD)** was installed in laboratories, providing early warning despite airflows from fume cupboards and ventilation systems.
    - The student accommodation was fitted with **point smoke and heat detectors**, linked to a central panel with cause-and-effect programming to avoid unnecessary whole-building evacuations.
    - **Voice alarm systems** were provided in teaching blocks to support safe and orderly evacuation of lecture theatres.
- **Evacuation Strategy**
    - **Simultaneous evacuation** was adopted for lecture theatres and teaching areas.
    - Laboratories used a **staff-led evacuation**, supported by specialist training and procedures for shutting down experiments safely before leaving.
    - The residential block operated on a **stay-put policy**, supported by robust compartmentation, with phased evacuation only if needed.

- **Management and Training**
  - University security staff were trained to act as fire wardens across all uses.
  - Laboratory users received additional training in emergency shutdown procedures and chemical fire hazards.
  - Students were given fire safety inductions on arrival, supported by drills each term.

Lessons Learned

- **Mixed-Use Complexity**
  - Combining teaching, laboratory, and residential spaces required **tailored fire strategies** for each risk type, integrated into a single overarching plan.
- **Value of Robust Compartmentation**
  - Strong compartmentation between uses ensured that risks in laboratories did not compromise student accommodation or teaching areas.
- **Training and Communication**
  - Staff and students needed clear, **differentiated evacuation procedures**. Regular drills proved essential to ensure both groups understood when to evacuate and when to remain in place.

22.10 Case Study 9 – Airport Terminal

Context

A newly constructed **international airport terminal** was designed to handle **25 million passengers per year**. The facility features a **central atrium** with retail, check-in halls, and circulation areas, as well as secure-side departure lounges and airside gates. Peak occupancy can exceed **30,000 passengers**, the majority of whom are transient, unfamiliar with the layout, and often carrying luggage that can hinder evacuation.

Fire Strategy Features

- **Atrium Smoke Control**
    - A **mechanical smoke extraction system** was installed, with fans rated at **250 m³/s** to manage the large atrium volume.
    - Deployable **smoke curtains** created reservoirs, directing smoke to extract points while maintaining a smoke-free layer of 2.5 m.
    - **CFD modelling** validated the system's performance across multiple fire scenarios.
- **Evacuation Strategy**
    - A **phased evacuation** was adopted to avoid overloading circulation routes and security choke points.
    - Departure lounges and retail areas closest to the fire zone were evacuated first, followed by adjacent areas.
    - **Voice alarm systems** with multilingual messaging and visual signage supported wayfinding for international passengers.
- **Detection and Alarm**
    - **Aspirating smoke detection (ASD)** was used in high atrium spaces to ensure early warning despite high ceilings.
    - Heat detection supplemented alarms in baggage-handling areas where dust and mechanical activity made smoke detection less reliable.
- **Suppression Systems**
    - **Sprinklers** were installed throughout retail, back-of-house, and baggage-handling areas.
    - High-value IT and control rooms were protected by **clean agent gaseous suppression systems**, avoiding water damage.
- **Fire Control Centre**
    - A central **fire control room** allowed real-time monitoring and control of alarms, smoke extraction, and suppression systems.
    - Integrated CCTV and communication links supported coordination with airport fire crews and local fire services.

Lessons Learned

- **Performance-Based Design is Essential**
  - The scale of the atrium and high transient occupancy required performance-based analysis using CFD and evacuation modelling to demonstrate compliance and maintain tenable conditions.
- **Passenger Behaviour and Wayfinding**
  - Studies showed that many passengers attempted to exit via familiar **entry routes**, even if alternatives were closer.
  - Multilingual **voice alarms, illuminated signage, and staff guidance** were critical in directing crowds efficiently.
- **Coordination with Emergency Services**
  - Close collaboration with **airport fire crews and local fire and rescue services** during design ensured that operational access and firefighting provisions were practical.
  - Regular joint exercises reinforced confidence in the fire strategy.

22.11 Case Study 10 – Mixed-Use High-Rise

Context

A **45-storey mixed-use tower** combines three distinct occupancies:

- **Residential apartments** (floors 20–45, ~400 residents).
- **Commercial offices** (floors 5–19, ~2,000 staff).
- **Retail units and restaurants** (podium levels 1–4): The combination of **sleeping accommodation, workplaces, and public spaces** created a complex fire risk profile, requiring an integrated yet flexible fire strategy.

Fire Strategy Features

- **Compartmentation**
  - Residential, office, and retail zones were separated by **120-minute compartment floors and walls.**
  - Vertical risers and service shafts were fire-stopped at each level to prevent cross-contamination between uses.

- **Suppression Systems**
  - **Residential sprinklers** (BS 9251) were installed in apartments.
  - **Commercial sprinkler systems** (BS EN 12845) protected offices and retail spaces.
  - Kitchens in restaurants were equipped with **wet chemical suppression** for deep fat fryers.
- **Smoke Control**
  - **Pressurised stairwells** provided protected escape routes across all occupancies.
  - **Mechanical smoke extract systems** were installed in office floors and the retail atrium, validated through CFD modelling.
- **Evacuation Strategy**
  - **Stay-put policy** was applied in the residential tower, supported by robust compartmentation.
  - **Phased evacuation** was used in office floors, prioritising the fire floor and adjacent levels.
  - **Simultaneous evacuation** was adopted in the retail podium, with support from a voice alarm system.
  - All three strategies were coordinated in a unified **cause-and-effect matrix** to avoid conflict during incidents.
- **Firefighting Access**
  - Two dedicated **firefighting shafts** with lifts and wet risers served the entire building.
  - A central **fire control centre** provided zone plans, system monitoring, and override functions for all occupancies.

Lessons Learned

- **Tailored Strategies for Each Occupancy**
  - A "one-size-fits-all" approach was impossible. Distinct strategies were applied to residential, office, and retail zones, yet carefully integrated to avoid operational conflicts.
- **Centralised Fire Control**
  - The fire control centre was essential in managing complex multi-occupancy responses, giving the fire service a single point of control.

- **Stakeholder Engagement**
    - Early engagement with residents, office tenants, retailers, and the fire service ensured all parties understood their roles, increasing trust in the building's safety systems.

22.12 Summary

The case studies presented in this chapter illustrate how **fire strategies are applied in practice** across a wide range of building types, from high-rise residential towers and healthcare facilities to warehouses, transport hubs, and heritage buildings. Each example demonstrates the importance of tailoring fire safety solutions to the unique risks, occupancies, and operational realities of the building in question.

Key themes that emerged include:

- **Compartmentation as a cornerstone** – whether in residential towers, hospitals, or mixed-use developments, robust compartmentation underpins strategies such as stay-put or progressive horizontal evacuation.
- **Suppression as a critical safeguard** – sprinklers, ESFR systems, and specialist suppression play a vital role in controlling fire growth, especially in large-volume or high-risk occupancies.
- **Performance-based design** – complex spaces such as atriums, underground stations, and airports require CFD modelling and evacuation simulations to demonstrate safety beyond prescriptive codes.
- **Human behaviour and communication** – strategies must account for how people respond in emergencies, with voice alarms, signage, and staff training proving as important as structural measures.
- **Integration across uses** – mixed-use buildings and multi-tenant facilities demand coordinated strategies to avoid conflicting responses during an incident.
- **Management and training** – even the best-engineered solutions rely on competent staff, rehearsed evacuation procedures, and strong fire safety culture.

By examining these real-world applications, it becomes clear that fire strategies are **not static documents**, but **living frameworks** that must adapt to building use, occupancy, and change over time. Case studies bridge the gap between theory and practice, reinforcing that effective fire safety depends on both **sound engineering principles** and **practical implementation on the ground**.

The next chapter will look ahead to **Conclusion and Future Perspectives**, exploring how advances in technology, modelling, and materials are shaping the future of fire strategy design and management.

Chapter 23

Conclusion and Future Perspectives

23.1 Introduction

This book has explored the **principles, strategies, and practical applications of fire engineering** across a wide variety of building types and contexts. From residential towers and healthcare facilities to transport hubs and heritage assets, we have seen how fire strategies combine **passive design, active systems, evacuation planning, and management arrangements** into coherent frameworks that protect life, property, and business continuity.

The closing chapter serves two purposes:

1. **Conclusion** – drawing together the core lessons that underpin modern fire engineering practice.
2. **Future Perspectives** – looking ahead to emerging challenges and opportunities that will shape fire safety in the coming decades.

Fire engineering has always evolved in response to societal, technological, and regulatory change. As buildings become taller, more complex, and more sustainable, and as digital technologies transform design and management, the role of the fire engineer becomes ever more critical.

This chapter will therefore reflect on:

- The enduring principles that remain the foundation of fire safety.
- The importance of integrating fire strategies across the **building lifecycle**.
- The need for strong **management, competence, and accountability** to ensure strategies remain effective.
- Emerging opportunities in areas such as **digital modelling, smart building systems, new materials, and resilience planning**.

By combining these perspectives, we can see how fire engineering will continue to adapt, ensuring that the **buildings of the future remain safe, functional, and resilient**.

23.2 Key Themes from the Book

Throughout this book, a number of **core themes** have consistently emerged. Together, they form the backbone of modern fire engineering practice and provide a foundation for how fire strategies should be developed, implemented, and maintained.

1. Structure and Contents of Fire Strategies

- A well-structured fire strategy provides a **clear framework** that is transparent, auditable, and practical for end users.
- Beyond being a regulatory requirement, it is a **working document** that supports decision-making at every stage of a building's life, from design approval to day-to-day management.

2. Core Technical Principles

- Despite advances in technology, the **fundamental pillars of fire safety** remain unchanged:
  - **Means of escape** that allow occupants to reach safety.
  - **Compartmentation** to contain fire and smoke.
  - **Active systems** such as detection, alarms, sprinklers, and smoke control to manage fire growth.
- Equally important is the **realistic understanding of human behaviour**. Evacuation planning must reflect how people actually move, respond, and make decisions in emergencies.

3. Building-Specific Challenges

- Different building types bring unique risks that demand tailored strategies:
  - **High-rise towers** require robust compartmentation, stay-put policies, and dedicated firefighting facilities.
  - **Large floor areas** such as malls or warehouses depend on effective smoke control and suppression to maintain tenability.
  - **Healthcare and sleeping occupancies** rely on progressive evacuation models and trained staff to protect vulnerable occupants.

  - **Mixed-use developments and underground spaces** cannot be addressed with prescriptive codes alone, requiring **performance-based analysis** and bespoke solutions.

4. Lifecycle Integration

- Fire strategies must not be treated as a static report delivered at the design stage.
- They must be **reviewed, updated, and integrated** into building management systems across the entire lifecycle.
- From refurbishment works to changes of use, strategies should evolve to remain accurate, ensuring that fire safety measures continue to perform as intended.

5. Management and Maintenance

- Even the best-designed fire strategy will fail without **strong management arrangements.**
- Regular **testing, maintenance, and record-keeping** are essential to ensure systems remain functional and reliable over decades.
- A positive **safety culture**—supported by competent staff, effective training, and clear accountability—is as critical as technical measures in achieving long-term resilience.

23.3 Lessons from Tragedies

The evolution of fire engineering has often been shaped by **tragedies**, where catastrophic incidents have exposed weaknesses in design, management, or regulation. Events such as the **Grenfell Tower fire in London (2017)**, and major international fires in **high-rise residential buildings, industrial complexes, and healthcare facilities**, continue to serve as stark reminders of what can happen when fire safety is compromised.

These incidents underline several critical lessons:

- **Prescriptive Compliance Alone is Not Enough**
  - Meeting the minimum requirements of codes and standards does not guarantee safety.
  - Prescriptive rules may fail to address complex or unique building risks, particularly in mixed-use, high-rise, or large-volume spaces.

    - Performance-based approaches, supported by modelling and risk assessment, are necessary to demonstrate real-world safety.
- **Integration and Maintenance are Critical**
    - Fire strategies that are poorly integrated into the **design, construction, and operational stages** can leave dangerous gaps between intent and practice.
    - Even robust systems will fail if they are not **maintained, tested, and managed** consistently.
    - In many tragedies, failures were not due to the absence of fire safety systems, but to **their degradation, disconnection, or neglect** over time.
- **Transparency and Accountability**
    - Tragedies reveal the consequences of unclear responsibilities, fragmented oversight, and poor communication.
    - The **Golden Thread of building safety**, mandated by the Building Safety Act 2022, reinforces the need for traceable decisions, accessible records, and accountable duty holders.
    - Fire safety cannot be effective in a culture of secrecy or complacency—it requires openness, competence, and continuous oversight.

Summary

Each of these lessons points towards a single truth: **fire safety is never a one-time achievement but a continuous responsibility**. Only through integration, maintenance, accountability, and a culture of proactive risk management can future tragedies be prevented.

23.4 The Future of Fire Strategies

The next generation of fire strategies will evolve in response to **regulatory reform, technological innovation, sustainability goals, and societal expectations**. While the fundamentals of fire safety remain rooted in compartmentation, escape, and suppression, the way these are designed, managed, and monitored will change significantly in the decades ahead.

The Building Safety Act 2022 and the "Golden Thread"

- The introduction of the **Golden Thread** represents a cultural and procedural shift in how fire safety is documented and managed.
- Fire strategies will no longer be static reports but must form part of a **permanent, accessible, and continually updated record** for higher-risk buildings.
- Ongoing accountability will rest with **building owners and accountable persons**, ensuring that safety responsibilities are traceable and enforceable throughout the building lifecycle.

Digital and Smart Buildings

- The integration of fire safety systems with **Building Management Systems (BMS)** will provide unprecedented oversight.
- Real-time monitoring of alarms, sprinklers, smoke control, and evacuation systems will allow rapid fault detection and intervention.
- The emergence of **digital twins**—virtual replicas of buildings—will enable live testing and simulation of fire scenarios, improving preparedness and resilience.

Artificial Intelligence (AI) and Modelling

- **AI-enhanced evacuation and smoke modelling** will advance performance-based design, allowing engineers to simulate human behaviour and fire dynamics with greater accuracy.
- Predictive modelling will help anticipate how design changes, occupancy patterns, or system failures affect safety outcomes.
- **Sensor-driven predictive maintenance**, supported by AI, will flag potential failures in fire systems before they occur, ensuring continuous reliability.

Sustainability and Fire Safety

- The drive towards **net-zero carbon buildings** and the increased use of **timber and other innovative materials** introduces new fire risks that must be accounted for in strategies.
- Airtight building envelopes, designed for energy efficiency, can complicate **smoke control and ventilation**, requiring new design approaches.

- Fire strategies will need to balance **environmental sustainability with fire resilience**, ensuring that one is not achieved at the expense of the other.

Resilience and Business Continuity

- Beyond life safety, fire strategies will increasingly address **property protection and business continuity**, recognising the economic and social consequences of fire.
- Organisations will expect strategies to support rapid recovery, with systems designed not just to protect occupants but also to **preserve operations, assets, and community function** in the aftermath of fire incidents.

23.5 Final Reflections

Fire strategies are more than compliance documents. They are both **technical artefacts**—grounded in engineering science, codes, and modelling—and **living management tools** that must adapt to the realities of construction, occupation, and change.

They require:

- **Competence and rigour** in their design, ensuring that every assumption and calculation is robust.
- **Clarity in structure and communication**, so that strategies are not only technically sound but also usable by those who must implement them.
- **Commitment from owners, managers, and regulators** to keep them accurate, relevant, and enforced across the building lifecycle.

Ultimately, the goal of every fire strategy is simple yet profound:

To ensure that people can live, work, and gather in buildings that are safe, resilient, and fit for the future.

This responsibility—shared by engineers, designers, regulators, and building operators—will only grow in importance as our built environment becomes more complex, interconnected, and ambitious. Fire strategies must therefore continue to evolve, embracing innovation while never losing sight of their enduring purpose: **the protection of life above all else**.

23.6 Epilogue – The Value of Fire Strategies

This book has explored the **science, practice, and application of fire strategies** across the full spectrum of modern buildings—from high-rise towers and hospitals to heritage theatres, transport hubs, and warehouses. At every stage, one theme has remained constant: **fire safety is not a single measure, but an integrated system of people, processes, and technologies working together.**

We have seen how:

- **Core principles**—compartmentation, means of escape, and suppression—remain the foundation of fire safety worldwide.
- **Building-specific strategies** must adapt to unique risks, whether in sleeping accommodation, healthcare, assembly buildings, or industrial complexes.
- **Human behaviour** is as critical as engineering; clear communication, training, and management often determine whether strategies succeed in practice.
- **Lifecycle integration** ensures that strategies do not end at design but evolve through construction, occupation, and refurbishment.
- **Management, maintenance, and accountability** are essential in keeping fire precautions reliable over decades.
- **Lessons from tragedies** remind us that prescriptive compliance alone is never enough; fire safety demands competence, transparency, and vigilance.

Looking forward, the future of fire strategies will be shaped by:

- **The Building Safety Act 2022 and the Golden Thread**, embedding accountability and traceability.
- **Digital technologies and AI**, enabling real-time monitoring, predictive maintenance, and smarter modelling of fire and evacuation scenarios.
- **Sustainability**, as innovative materials and low-carbon construction methods require fresh approaches to fire resilience.
- **Resilience and business continuity**, expanding the role of fire strategies beyond life safety to include protection of assets, operations, and communities.

Above all, this book reinforces a simple truth:

Every fire strategy has one ultimate purpose—to allow people to live, work, and gather in buildings that are safe, resilient, and fit for the future.

Fire strategies are not just technical reports. They are commitments—commitments by engineers, architects, regulators, and owners to protect life, preserve property, and create confidence in the places we inhabit.

As the built environment continues to evolve, so too must fire engineering. By combining enduring principles with emerging technologies, we can ensure that tomorrow's buildings are not only innovative and sustainable, but also **fundamentally safe.**

www.ingramcontent.com/pod-product-compliance
Lightning Source LLC
LaVergne TN
LVHW100246210926
848480LV00003B/834
9781807230067